Religion Explained?

Scientific Studies of Religion: Inquiry and Explanation

Series editors: Donald Wiebe, Luther H. Martin and William W. McCorkle

Scientific Studies of Religion: Inquiry and Explanation publishes cutting-edge research in the new and growing field of scientific studies in religion. Its aim is to publish empirical, experimental, historical and ethnographic research on religious thought, behaviour, and institutional structures.

The series works with a broad notion of 'scientific' that will include innovative work on understanding religion(s), both past and present. With an emphasis on the cognitive science of religion, the series includes complementary approaches to the study of religion, such as psychology and computer modelling of religious data. Titles seek to provide explanatory accounts for the religious behaviors under review, both past and present.

The Attraction of Religion, edited by D. Jason Slone and James A. Van Slyke
Contemporary Evolutionary Theories of Culture and the Study of Religion, Radek Kundt
Death Anxiety and Religious Belief, Jonathan Jong and Jamin Halberstadt
New Patterns for Comparative Religion, William E. Paden
Religion in Science Fiction, Steven Hrotic
The Mind of Mithraists, Luther H. Martin

Religion Explained?

The Cognitive Science of Religion after Twenty-Five Years

Edited by
Luther H. Martin and Donald Wiebe

Bloomsbury Academic
An imprint of Bloomsbury Publishing Plc

B L O O M S B U R Y
LONDON • OXFORD • NEW YORK • NEW DELHI • SYDNEY

Bloomsbury Academic
An imprint of Bloomsbury Publishing Plc

50 Bedford Square
London
WC1B 3DP
UK

1385 Broadway
New York
NY 10018
USA

www.bloomsbury.com

BLOOMSBURY and the Diana logo are trademarks of Bloomsbury Publishing Plc

First published 2017

British Library Cataloguing-in-Publication Data
A catalogue record for this book is available from the British Library.

ISBN: HB: 978-1-3500-3246-0
ePDF: 978-1-3500-3247-7
ePub: 978-1-3500-3248-4

Library of Congress Cataloging-in-Publication Data
Names: Martin, Luther H., 1937– editor.
Title: Religion explained? : the cognitive science of religion after twenty-five years / edited by Luther H. Martin and Donald Wiebe.
Description: New York : Bloomsbury Academic, 2017. |
Series: Scientific studies of religion: inquiry and explanation |
Includes bibliographical references and index.
Identifiers: LCCN 2017005668 | ISBN 9781350032460 (hb) |
ISBN 9781350032484 (epub)
Subjects: LCSH: Religion. | Cognitive science.
Classification: LCC BL51 .R34793 2017 | DDC 200.72–dc23
LC record available at https://lccn.loc.gov/2017005668

Cover image © Shutterstock

Series: Scientific Studies of Religion: Inquiry and Explanation

Typeset by Newgen Knowledge Works Pvt. Ltd., Chennai, India.
Printed and bound in Great Britain

Contents

Illustrations vii

Contributors viii

Introduction: Religion Explained? The Cognitive Science of Religion after Twenty-Five Years *Luther H. Martin and Donald Wiebe* 1

Part 1 Retrospectives 5

1 The Cognitive Science of Religion and the Growth of Knowledge *E. Thomas Lawson* 7

2 Twenty-Five Years In: Landmark Empirical Findings in the Cognitive Science of Religion *Robert N. McCauley* 17

3 Twenty-Five Years of CSR: A Personal Retrospective *Harvey Whitehouse* 43

4 The Beautiful Butterfly: On the History of and Prospects for the Cognitive Science of Religion *Uffe Schjødt and Armin W. Geertz* 57

Part 2 State of the Art 69

5 Religion Explained? Some Variants of Cognitive Theory *Stewart Elliott Guthrie* 71

6 Cognitive Attractors in the Evolution and Diffusion of Religious Representations *Pascal Boyer and Nicolas Baumard* 81

7 The Long Way from Cognitive Science to History: To Shorten the Distance and Fill in the Blanks *Panayotis Pachis and Olympia Panagiotidou* 89

8 The Indispensability of Cognitive Science for a Genuine History of Religion *Anders Klostergaard Petersen* 97

Part 3 CSR 2.0 105

9 Exiting the *Motel of the Mysteries*? How Historiographical Floccinaucinihilipilification Is Affecting CSR 2.0 *Leonardo Ambasciano* 107

10 Minimal Counterintuitiveness Revisited, Again: The Role of Emotional Valence in Memory for Conceptual Incongruity *Michaela Porubanova and John H. Shaver* 123

11 Constraints on Theory-Building in the Cognitive Science of Religion with Reflections on the Influence of Physics Envy and the Principle of Sufficient Reason *Benson Saler and Charles A. Ziegler* 133

12 The Effects of Relative Stable Feedback Loops: Cognitive Science and Historical Explanations *Jesper Sørensen* 143

Part 4 Looking Forward 153

13 The Road Not Taken: Possible Paths for the Cognitive Science of Religion *Richard Sosis* 155

14 Looking Back to Look Forward: From Shannon and Turing to Lawson and McCauley to …? *Justin E. Lane* 169

15 A Neo-Victorian Cognitive Science of Religion *Steven Hrotic* 181

Conclusion: On Keeping Cognitive Science of Religion Cognitive and Cultural *Justin L. Barrett* 193

Notes 203

References 211

Index 251

Illustrations

9.1 "Harriet proudly strode around the site wearing the Sacred Collar and matching Headband. She also wore the magnificent *plasticus* ear ornaments and the exquisite silver chain and pendant" (Macaulay 1979: 36) 108

9.2 The "vast funerary complex" as recovered by future archeologists, actually a motel whose rooms are misinterpreted as being individual tombs. Seen from the same biased perspective, the structure comprises a "sanctuary" (the huge hall), a (sacrificial) kitchen, a "ceremonial pool" (more prosaically, a swimming pool), and a "specially marked funerary game areas intended to occupy the spirits of the dead during eternal life" (a miniature golf course; Macaulay 1979: 39) 114

9.3 Planimetry and stratigraphic units of the archeological site of Göbekli Tepe 115

9.4 The rise and fall of the *Arctometatarsalia* clade in two cladograms. The top image illustrates the result of the original analysis by Holtz (1994) in which he proposed the monophyletic ascendance of the clade. The bottom image, instead, represents a more recent re-analysis of the original dataset by Holtz that, with the benefit of hindsight, indicates that the arctometatarsalian foot evolved independently four times (black rectangles) from the original non-arctometatarsalian condition 118

10.1 Combined recall for Incongruous and Congruous concepts that vary in emotional valence 128

10.2 Performance in delayed recognition (average percent recognized) for Incongruous and Congruous items that vary in emotional valence 128

Contributors

Leonardo Ambasciano earned his PhD in historical studies at the University of Turin (Italy) in 2014 with a cognitive and evolutionary analysis of the ancient Roman female cult of Bona Dea. In 2016 he served as visiting lecturer at Masaryk University, Brno (Czech Republic). He is the European editor of the *Journal of Cognitive Historiography*.

Justin L. Barrett is chief project developer for Fuller Theological Seminary's office for Science, Theology, and Religion and a professor of psychology. He is also a research associate of the University of Oxford's School of Anthropology, where he formerly served as a senior researcher and course convener. His scholarly work primarily concerns cognitive scientific approaches to the study of cultural expression.

Nicolas Baumard is a CNRS research scientist at the Department of Cognitive Science of the Ecole Normale Supérieure in Paris. He is interested in using evolutionary and psychological approaches in the social sciences, including economics, philosophy, anthropology, and history. More specifically, he draws on life history theory to explain behavioral variability across cultures, time, and social classes.

Pascal Boyer studied philosophy and anthropology at the University of Paris and Cambridge, where he did his graduate work with Professor Jack Goody on memory and oral literature. He has done anthropological fieldwork in Cameroon on the transmission of the Fang oral epics and on Fang traditional religion. Since then he has worked mostly on the experimental study of cognitive capacities underlying cultural transmission, particularly in the domain of religious beliefs and behaviors. He is the author of *Religion Explained*, *The Naturalness of Religious Ideas*, and *Tradition as Truth and Communication*. After teaching in Cambridge, San Diego, Lyon, and Santa Barbara, he moved to his present position as Henry Luce Professor of Individual and Collective Memory at Washington University, St. Louis.

Armin W. Geertz is professor at the Department for the Study of Religion, and Jens Chr. Skou Senior Fellow at Aarhus Institute of Advanced Studies, Aarhus University, Denmark. He is cofounder of the Religion, Cognition and Culture Research Unit at Aarhus and author, editor, or coeditor of several books, and the author or coauthor of a large number of articles and book chapters. He is coeditor of the *Journal for the Cognitive Science of Religion* and of the *Advances in the Cognitive Science of Religion* series.

Stewart Elliott Guthrie (PhD Yale 1976) is professor emeritus of anthropology at Fordham University. His 1980 cognitive theory of religion holds that religion is anthropomorphism, which stems from an evolved strategy that privileges personalistic interpretations of phenomena. This strategy is adaptive because such interpretations, when mistaken, are relatively harmless and when right are invaluable. Guthrie's theory has been influential in the cognitive science of religion and, according to some scholars, foundational.

Steven Hrotic is a cognitive anthropologist with special interests in religion and aesthetics. After studying at Queen's University Belfast, he held research positions at opposite ends of the disciplinary spectrum, working with biologists at the University of Utrecht, then with philosophers at the University of North Texas. His first book, *Religion in Science Fiction*, investigated recent conceptions of what religion is and should be and the appeal of genre fiction qua shared culture.

Anders Klostergaard Petersen is professor for the study of religion at Aarhus University. His areas of expertise are Late Second Temple Judaism, including early Christ-religion; Graeco-Roman religion and philosophy; and general questions in the philosophy of science pertaining to the humanities and social sciences. He has approximately 485 publications and is, at present, particularly concerned with matters pertaining to biocultural evolution in the study of religion.

Justin E. Lane is currently finishing his DPhil at the University of Oxford and works in Boston, MA as the Center for Mind and Culture's post-doctoral research fellow in modeling and simulation. His research focuses on the use of computational methods (specifically multi-agent AI) to study the stability of social systems from a cognitive perspective. Specifically, he's interested in how social and environmental ecologies contribute to religious and ethnic violence and the stability and instability of religions, particularly new religious movements.

E. Thomas Lawson is professor emeritus of comparative religion at Western Michigan University, honorary professor of cognition and culture at Queen's University Belfast, executive editor of the *Journal of Cognition and Culture*, coauthor with Robert N. McCauley of *Rethinking Religion: Connecting Cognition and Culture* and *Bringing Ritual to Mind: Psychological Foundations of Cultural Forms* as well as the author *of Religions of Africa: Traditions in Transformation.*

Luther H. Martin is professor emeritus of religion at the University of Vermont. He is author of *The Mind of Mithraists* (2015) and of numerous articles in this field of his historical specialization. He has also published widely in the field of theory and method in the study of religion, now collected in *Deep History, Secular Theory* (2014). He is a former president of the International Association for the Cognitive Science of Religion and has been recognized as an honorary life member of the International Association for the History of Religions.

Robert N. McCauley is William Rand Kenan Jr. University Professor of Philosophy in the Center for Mind, Brain, and Culture at Emory University. His books include *Why Religion Is Natural and Science Is Not* (2011), *Philosophical Foundations of the Cognitive Science of Religion* (2017), and with E. Thomas Lawson, *Rethinking Religion* (1990), *Bringing Ritual to Mind* (2002). He has been president of both the Society for Philosophy and Psychology and the International Association for the Cognitive Science of Religion.

Panayotis Pachis is professor at Aristotle University of Thessaloniki, Greece. He teaches Study of Religion (Method and Theory for the Scientific Study of Religion) and Religions of the Ancient World (Cults and Religions of Classical and Graeco-Roman Age). He has published numerous articles and books in these areas.

Olympia Panagiotidou is a post-doctoral researcher at Aristotle University in Thessaloniki. She holds a joint PhD and MA in cognitive science and the study of religion from Aristotle University at Thessaloniki and Aarhus University in Denmark and a BA in the history and archaeology from Aristotle University. She is interested in the application of modern cognitive theories to the Greco-Roman cults, especially Mithraism and the Asclepius cult.

Michaela Porubanova is an assistant professor of Cognitive Psychology at Farmingdale State College in New York where she directs the Emotion and Visual Cognition Laboratory. She is interested in elucidating the role of emotion in guiding visual attention and preferential processing of emotional information by attention. Additionally, she has been working on understanding the role of cognitive factors in cultural transmission and propagation of ideas.

Benson Saler is professor emeritus of anthropology at Brandeis University. He has carried out ethnographic field research in Guatemala, Columbia, and Venezuela. He is especially interested in epistemological issues relating to the study of religions. His most recent book is *Understanding Religion: Selected Essays* (2009).

Uffe Schjødt is director of the Religion, Cognition and Culture research unit at the Department of the Study of Religion. His research includes experimental work in the neuroscience of prayer as well as the examination of cognitive resource depletion and sensory deprivation in religious interactions. He is currently working on the application of the predictive processing framework in the study of religious practice and experience.

John H. Shaver is a lecturer in religion at the University of Otago. He holds a PhD in Anthropology from the University of Connecticut. He works on understanding intracultural variation in ritual behavior, the relationships between religion and fertility, and the evolution of syncretic religions. He has conducted research in the Czech Republic, Fiji, Mauritius, New Zealand, and the United States.

Jesper Sørensen is associate professor in comparative religion at Aarhus University, head of the research program at the Department of the Study of Religion, and member of the research unit Religion, Cognition and Culture as well as Interacting Minds. He is the author and editor of several books and articles in the cognitive science of religion focusing in particular on ritual behavior, magic, method, and theory and on the immunology of cultural representations.

Richard Sosis is James Barnett Professor of Humanistic Anthropology at the University of Connecticut. His work has focused on the evolution of religion and cooperation, with particular interests in ritual, magic, religious cognition, and the dynamics of religious systems. To explore these issues, he has conducted fieldwork with remote cooperative fishers in the Federated States of Micronesia and with various communities throughout Israel. He is cofounder and coeditor of the journal *Religion, Brain, and Behavior*, which publishes research on the biological study of religion.

Harvey Whitehouse is Chair of Social Anthropology and director of the Institute of Cognitive and Evolutionary Anthropology at the University of Oxford, where he studies the evolution of social complexity. One of the founders of the cognitive science of religion, he is well known for his theory of "modes of religiosity," which explains how the frequency and emotional intensity of collective rituals influence the scale and structure of religious organizations.

Donald Wiebe is professor of philosophy at Trinity College, University of Toronto. He has served as dean of the Faculty of Divinity and the Faculty of Arts at Trinity College and as associate director of the former Centre for the Study of Religion at the University of Toronto. He is the author of *The Irony of Theology and the Nature of Religious Thought* and *The Politics of Religious Studies*, among other books and numerous articles and reviews.

Charles A. Ziegler is a lecturer in social anthropology at Brandeis University. He is coauthor of *UFO Crash at Roswell: The Genesis of a Modern Myth* (1997).

Introduction: Religion Explained? The Cognitive Science of Religion after Twenty-Five Years

Luther H. Martin and Donald Wiebe

The theme of this volume is that of the Fifth Biennial Meeting of the International Association for the Cognitive Science of Religion (IACSR), June 20–22, 2014 in Brno, Czech Republic.[1] Our topic inverts the title of Pascal Boyer's seminal book, *Religion Explained* (2001), into an interrogative. To what extent has "religion" actually been explained by the cognitive sciences of religion (CSR) since its inception in 1990?

We selected 1990 as the beginning of a cognitive science approach to the study of religion with the publication of E. Thomas Lawson's and Robert N. McCauley's *Rethinking Religion: Connecting Cognition and Culture*. Although such a field of study could be traced as far back as the publication of David Hume's philosophical reflections on *The Natural History of Religions* (1757) or, far more recently, to Stewart Guthrie's prescient anthropological proposal in 1980 for "A Cognitive Theory of Religion," it was the publication of Lawson and McCauley's book that finally began to engage scholars of *religion* with cognitive theory and it is their respective reflections with which we open this volume. The essays in this volume, then, offer selected assessments of CSR since the publication of Lawson and McCauley's *Rethinking Religion* in 1990.

The editors' own commitment to a *naturalistic* study of religion extends deep into our respective academic histories (see Martin and Wiebe 2016). This commitment was the basis for our founding—with Tom Lawson—of the North American Association for the Study of Religion (NAASR) in 1985 as an alternative to the increasingly religious inclusivity embraced by the American Academy of Religion (Martin and Wiebe 2004). Consequently, when Lawson with McCauley published their *Rethinking Religion* five years later, we were among the first to take seriously our colleague's proposal for this new scientific approach to the naturalistic study of religion. Martin, a historian of religion like Lawson, became especially interested in the implications of cognitive sciences for a historiography of religions, while Wiebe, a philosopher like McCauley, became especially interested in questions of theory and method.

Martin didn't publically venture into the new field of cognitive studies until 2002 when he presented a paper to the conference "Minds and Gods: The Cognitive Science of Religion,"[2] entitled "Towards a Cognitive Historiography of Religion." A scholar of Hellenistic religions, Martin subsequently elected to focus his own cognitive historiographical studies on the Roman cult of Mithras. Consequently, in 2004, he published a paper in a *Festschift* for Tom Lawson, in which he analyzed Mithraic rituals in terms of Lawson and McCauley's ritual competence theory (Martin 2004b). Researches in cognitive historiography, as exemplified by the study of various aspects of the Roman cult of Mithras, subsequently characterized much of Martin's academic career; in 2015, he published a collection of his historical and cognitive studies in the Roman cult of Mithras.

The cognitive science of religions is an interdisciplinary pursuit. Consequently, Martin participated in three conferences sponsored by the "British Academy Networks Project on 'Modes of Religiosity': The Ethnographic Evidence," all co-organized by Harvey Whitehouse. The first of these conferences, in 2001, was co-organized with anthropologist James Laidlaw on "a comparative anthropology of religion" (Whitehouse and Laidlaw 2004), and a third, in 2003, was co-organized with McCauley on "psychological and cognitive foundations of religiosity" (Whitehouse and McCauley 2005a). Martin was co-organizer with Whitehouse of the second of these conferences, held at his home University of Vermont in 2002, focusing on "archaeology, history, and cognition" (Whitehouse and Martin 2004a). While a Distinguished International Fellow at the Institute of Cognition and Culture, Queen's University Belfast, Martin continued his interest in cognitive historiography by organizing, in 2007, an international conference focusing on "evolution, cognition, and history." The proceedings of this conference were published as *Past Minds: Studies in Cognitive Historiography* (2011); and in 2013, Martin became a founding editor of the *Journal of Cognitive Historiography.*

Although Wiebe had been interested in the study of religion as a scientific enterprise since 1975 with the publication of his essay "Explanation and the Scientific Study of Religion," he was not immediately attracted to the new developments in the cognitive science of religion. He made that clear in the title of a presentation to the 1995 meeting of the Society for the Scientific Study of Religion (SSSR) entitled "Connecting the Cognitive and the Cultural: How Plausible is the Enterprise?"—but he was not ready to publish his preliminary views expressed there. Like Martin, he was a participant in the 2002 "Modes of Religiosity" conference in Vermont, where he presented a paper on "Critical Reflections on the Modes of Religiosity Argument," and at the 2005 meeting of the SSSR, he presented a slightly less skeptical paper on "Cognitive Science and the Study of Religion" (which, like his 1995 SSSR paper, also remains unpublished). In 2006, now less uncertain about the value of the cognitive sciences for the study of religion, Wiebe raised questions about whether the relationship of the cognitive sciences to the neurosciences was essential to the development of a scientific study of religion; this paper, presented to the Laboratory on Theories of Religion at the University of Aarhus, is another paper that remains in the proverbial desk drawer. In 2007, Wiebe accepted Martin's invitation to present a paper to the "Past Minds" conference in Belfast. In this paper, "Beneath the Surface of History?" (2011), Wiebe offers a clearer idea of the possibilities that the cognitive sciences might offer for the development of a genuinely scientific study of religion.

Both Martin and Wiebe, then, have maintained, pretty much from its beginnings (with one of us more skeptical but still hopeful), that CSR is the best available paradigm for a scientific study of religion—although not necessarily the only one, with other examples being evolutionary theory (which actually frames CSR), behavioral economics (that incorporates CS), Big Number Analysis, cliodynamics, and contemporary network theory, among others. Consequently, we both became founding members of the International Association for the Cognitive Science of Religion (2005), for which Martin is past president (2010–2012). We also afforded a significant proportion of the academic program of the Twentieth Quinquennial Congress of the International Association for the History of Religions (IAHR) in 2010 (with Wiebe as Director of the Congress and Martin as Chair of the Program Committee) to papers on and related to the cognitive science of religion.

However, as this new field strove to establish its presence in the academy, we were not sanguine in accepting all of its directions—a critical questioning that is, or should be, characteristic of any science. For example, we have questioned the ahistorical "presentism" and generalizing claims of much experimental research in CSR, a confidence in neo-Darwinian evolutionary theory as a curious analogy for explaining historical and cultural change, logical contradictions between—or even within—differing CSR hypotheses. Nevertheless, we affirm our judgment that CSR currently offers the most fully developed paradigm for a scientific study of religion.

The insights of CSR have increasingly been acknowledged by many scholars of religion and incorporated into their study. In addition to recognition of its significance by NAASR and by the IAHR, this approach to the study of religion has now even formally established its presence at the notably conservative American Academy of Religion. While there is only a single chair devoted specifically to CSR at the undergraduate level in North America (occupied by Claire White at California State University, Northridge), CSR remains a graduate-level topic that is actively pursued at only a few international research centers—Aarhus University, Boston University, Masaryk University, Oxford University, Queen's University, Belfast, and the University of British Columbia—and by a diaspora of individual scholars. These researchers and university-based research centers are increasingly attracting large research grants, which have in turn increased the academic respectability of the study of religion in the context of the modern research university. Significantly, however, a large amount of this funding comes from the self-interested religiously oriented John Templeton Foundation rather than from scientific funding agencies, which may, in the long run, undermine that reputation.[3]

There are now a large number of scholars, both senior and junior, with a history of research in CSR who might have been invited to contribute to this volume. Initially, the editors invited presenters at the 2014 Brno conference who had directly addressed its theme to contribute their presentations. Although some of the participants in the Brno conference regrettably declined our invitation, Geertz, Lane, McCauley, Porubanova-Shaver, Sosis, and Whitehouse graciously accepted. Subsequently, we invited contributions from those unable to attend the conference but who are considered to have played a founding role in the development of CSR (e.g., Boyer, Guthrie, and Lawson), a few who have made significant contributions to the field (e.g., Pachis, Petersen, Saler-Ziegler, and Sørensen), and some junior scholars in the field, since we thought it important to include the perspective of those just beginning their research careers in and teaching of CSR (e.g., Ambasciano, Hrotic). We would especially like to thank Justin Barrett—who is widely considered the founder of experimental studies in

CSR—for agreeing to write an evaluative conclusion for this volume. Constraints on volume size precluded the inclusion of many other researchers in CSR whose contributions would have further enriched this volume.

We would like to thank the contributors to this volume for taking time from their ever busy schedules to reflect on this still relatively new scientific paradigm for the study of religion. We hope that it marks a formal consolidation and recognition of this new field of study and a promise for its future integration into the research and teaching of a universal human phenomenon.

And, we would like to thank Dr. Steven Hrotic for his invaluable editorial contribution to the preparation of this manuscript and Ms. Kalyani of Newgen Knowledge Works, Chennai, India for her brilliant copy-editing for this book.

L. H. M and D. W.
December 2016

Part One

Retrospectives

1

The Cognitive Science of Religion and the Growth of Knowledge

E. Thomas Lawson

The mark of good science is its capacity to make our knowledge of the world grow. Philosophers of science have paid a great deal of attention to the issue of how our scientific knowledge of the world increases. For example, Imre Lakatos (1978), building on Karl Popper's *The Logic of Discovery* (1934), argued that there were two kinds of research programs that were either progressive or degenerative. Progressive research programs strengthen and increase explanatory power, leading to new forms of inquiry while maintaining a core set of principles, whereas degenerating programs decrease in explanatory power and ultimately produce less and less knowledge. Progressive research programs, therefore, are inferentially rich, represent growth in our knowledge of the world, and are a standard of excellence worthy of striving for. If that is the case, then it is worthwhile to examine new research programs in order to see whether they qualify as progressive. The cognitive science of religion provides a fine opportunity for such an examination. In its present state, it certainly seems to have made significant strides in the direction of inferentially rich knowledge. It is worth asking whether this putative accomplishment is apparent or real.

While this particular progressive research program started in the humanities, the scholars who began to think cognitively had for some time been paying attention to social scientific explanations of religious thought and behavior but were dissatisfied with both the methods and the results of the various studies on the table. Much of social science had adopted an interpretive stance, which claimed that religious thought and behavior should be understood, not explained, while those who had explanatory interests insisted, under the influence of Durkheim, that only social facts could explain social facts. Fortunately, a revolution was occurring within the social sciences—the cognitive revolution—and it was this new way of explaining human thought and behavior in general that attracted a new generation of scholars. These cognitive thinkers, knowledgeable about the linguistic revolution that had shaken up the humanities, began to pay attention to new modes of psychological investigation in terms of the notion of information processing. What made the linguistic revolution interesting was its focus on the fact that languages were eminently sociocultural phenomena. If it could be shown that such collections of variation were, nevertheless, governed by a set of universal principles, then there was finally an exciting new way of looking at other cultural phenomena such as religions, which were also extensively varied, both historically and culturally.

So the cognitive science of religion, initially inspired by the achievements of the linguistic revolution, as well as by the reduction of behavioral psychology to cognitive psychology, rather quickly built theoretical relationships with the new social sciences such as cognitive psychology and eventually began a conversation with the newly developing cognitive anthropology and the emerging evolutionary psychology, as well as with the natural sciences, particularly the neurosciences and evolutionary biology. These connections led to questions and modes of inquiry at a number of levels of analysis, namely the social, psychological, and biological. The idea that only social facts could explain social facts was gleefully abandoned, as was the behaviorist notion that all explanatory variables were external to the organism. Autonomy and blank slates were, hopefully, put to rest.

Because of the conversation with cognitive scientists, whether they were philosophers, psychologists, anthropologists, or biologists, the initial focus of a new cognitive science of religion research program lay in finding ways to connect cognition and culture by approaching a subject matter "religion" that desperately needed an explanatory perspective rather than the purely interpretive approach that had characterized studies of religion at that point. Hence what started out as a conversation between a few scholars in the humanities grasping rather tentatively the reins of science ended up with a veritable crowd of chariots charging around the scientific arena. It is not always clear that these risk takers are now all participating in the same arena, but the excitement of the game is palpable.

How, then, has our knowledge of religion grown? First, simply by seeking better explanations than those provided by the autonomous social sciences, we were forced to connect sociocultural and psychological facts when looking at certain puzzling forms of thought and behavior. We were no longer satisfied to explain ritual behavior, for example, by appealing to socialization. It did not make sense to us that very widespread, cross-cultural phenomena just happened to be transmitted from generation to generation willy-nilly without paying attention to something that made people susceptible to certain ideas and the behavior associated with them. Second, as we began to develop new and more complex theories, it became apparent that the assumptions about "religion" as a coherent and systematic phenomenon was called into question because what we were learning from the cognitive, and eventually the evolutionary, sciences was that the mind that produced, entertained, and acquired religious concepts consisted of many different kinds of processing mechanisms that were sensitive and responsive to domain-specific information but not necessarily in any unified way.

This led us to pay attention to such cognitive mechanisms as animacy detection, agent detection, eye-gaze detection, face recognition, threat detection, contagion sensitivity, and many other psychological processes. For example, agency detection was a different cognitive process than either contagion sensitivity or fear-producing mechanisms, all of which, once identified, became relevant for the explanation of various forms of human thought and behavior in general and religious thought and behavior in particular.

It eventually became apparent that by focusing on cognitive mechanisms at the psychological level, we had better also take processes such as evolution on the biological level into consideration, because there had to be a story about how those mechanisms got there in the first place and what their significance and function were for survival and reproduction. What interdisciplinary scholars now saw was that we had to develop a cognitive *and* evolutionary science of religion if we were to continue to make our knowledge grow—a significant feature of an inferentially rich research program. And so a conversation among scholars in the humanities, specifically philosophers, historians, anthropologists, psychologists, and biologists, began to take place. All of those involved in these early stages were interested in why people all over the world were so willing to entertain very odd thoughts and engage in sometimes very bizarre behavior while in other respects their judgments seemed quite veridical. The subject matter of "religion" was clearly on the table, but seen now as something genuinely puzzling, not just for those "primitives" hiding in the jungle but also in the churchly houses of the new world.

Let us focus on the issue of whether the subject matter "religion" is a coherent system. Why, for example, do some people find it important and relevant to perform a set of acts that they think involves agents with some very counterintuitive qualities when they initiate someone into a special age group? And how does this concept of such special agents come about? It is no secret that there is a very serious and extremely interesting argument going on about whether certain religious ideas or concepts (that either directly or indirectly inform certain types of behavior) are either instances of adaptations or at least linked pleiotropically to types of behavior that are adaptive (Sosis 2009). As Richard Sosis notes, those who came late to the cognitive science of religion encountered a well-established and highly influential by-product view of religious thought and behavior. An adaptationist approach by D. S. Wilson in his *Darwin's Cathedral* (2002), as Sosis points out, attacked the by-product view not only of the cognitive scientists of religion, but the social-scientific views of Stark, Bainbridge, and Fink (e.g., Stark and Bainbridge 1996; Stark and Fink 1992). Not only did Wilson argue that religion is an adaptive phenomenon but also he employed the notion of group selection to fortify his case. Many who had taken evolutionary theory seriously thought that group selection explanations had never been particularly convincing and had, in fact, long since been abandoned by evolutionary scientists (West et al. 2007). Wilson not only insisted on the importance of group evolution but also failed to take the fact of cognitive mechanisms into account at all. In a sense he made a direct jump from the biological to the cultural, bypassing the psychological mechanisms that promised a bridge for a safer crossing for those committed to an interdisciplinary perspective.

Evolutionary biologists such as West, Griffin, and Gardner (2007) question the usefulness of group selection approaches to cooperation (which Wilson thinks is the main function of religion) and show how kin selection explains more formally and efficiently the various examples that Wilson raises and does so without the confusions that group selection approaches typically create.

The same cannot be said for psychologists such as Jesse Bering, who has performed psychological experiments that focused primarily on afterlife beliefs (2006). Bering does not argue that "religion" is adaptive as a group selectionist phenomenon, but only that a special set of beliefs about the afterlife may very well be. His experiments provide important clues as to why this might be the case. Sosis, however, intends to go the whole way and is now working on religion as a complex adaptive system. He sees "religion" as a coherent phenomenon and focuses on the notion of cooperation, placing an emphasis on the notion of prosociality, a theme developed initially by Durkheim. Whether this will persuade the scientists who have developed the by-product approach, which has not only identified many different kinds of mechanisms but also tied these to evolutionary processes on the kin selection model (Boyer 2001), remains to be seen. There is clearly more than one route to evolution without having to adopt either a group selectionist perspective as does Wilson or to see religion as a complex adaptive system as proposed by Sosis. These questions are, of course both theoretical and empirical, and much work remains to be done.

The central issue is whether we need to focus on a bundle of features as the unit of analysis, as Sosis (2009) suggests, or whether it makes more sense from an evolutionary perspective to focus on the garden-variety mechanisms as adaptive (given a notion of basic capacities in need of development by environmental triggers that calibrate them). And it will be important to recognize the temporal differences involved when the different capacities achieve their mature status and efficient functioning via calibration. For example, the false-belief test has a temporal aspect that comes to place around the age of four (but see Spelke and others for an earlier date). My view is that it is better, from a methodological perspective, to regard these mechanisms as separate modules, each with its own history. But pluralism is a good thing and ensures that important questions are not ignored. One of the marks of a progressive research program is that alternative avenues of investigation open up and lead not only to new answers but also to new questions.

Of particular importance in the inclusive cognitive and evolutionary research program (i.e., putting aside the differences between the adaptationists and the spandrelists as well as the debate between kin and group selectionists) is the question of what kinds of ideas held a transmission advantage. As long as you were interested in evolution and thought that the capacities that natural selection was kind enough to endow us with were sufficient to account for the persistence of some representations at the expense of others, this question made a lot of sense. The groundbreaking work here was that of Dan Sperber (1985a), who developed a sophisticated account of cultural transmission in terms of the notion of an epidemiology of representations. Sperber provided powerful arguments for a precise account of cultural transmission, going against the notion of memes originally suggested by Richard Dawkins (2006) and presently championed by Dan Dennett (2006). Some of the experimental work confirms that minimally counterintuitive ideas have a transmission advantage because of their attention-grabbing abilities and our capacity to pay attention to novelty. They are counterintuitive enough to focus our attention but not so counterintuitive that they simply become noise. Despite the applicability of these ideas for explaining how the counterintuitiveness of religious ideas could be successfully transmitted and prove capable of explaining why religious practices persist, no matter what the cultural context, there were certain nagging problems that simply kept demanding our attention. What about cost?

Actually, the initial question that captured my attention was the one so nicely put by Dan Sperber in 1975 when he asked why we find so frequently in human societies the kind of symbolic activity (of which religious symbolism was an instance) where the means put into play seem to be disproportionate to the explicit or implicit end, whether this be knowledge, communication, or production. Surely, where the costs of a cultural form seem clearly to outweigh its benefits, one is required to explain the *persistence* of such an expensive form of human thought and behavior. Sperber provided an important account by distinguishing between dictionary, encyclopedic, and symbolic representations and later developed a very sophisticated notion of relevance (Sperber and Wilson 1986). We make our judgments on the basis of their relevance.

The issue of the costliness of certain forms of behavior has been developed in very interesting ways by Sosis. Some rituals can involve great expenditures of resources, cause great pain, and even put participants in danger. Do the benefits outweigh the costs? Sosis thinks that they do if they are sufficient to diminish the threat of free riders. I presume that the conclusion is that the prosocial benefits overcome the costs in the energy consumption ledger. This of course raises a larger question: Are we trying to show that religion is really a good thing or a bad thing? Durkheim, a long time ago, was criticized for providing a justification for religion rather than an explanation of it. Adaptationists sometimes seem to be suggesting that religion is a good thing because of its prosocial benefits. Dawkins (2006), on the other hand, thinks that religion is a bad thing because its social costs are enormous.

Actually, by-product theoreticians have not participated in this debate, first of all because they see religion as an aggregate phenomenon and argue that the unit of analysis is individual objects of selection, and second because, by focusing on the various individual mechanisms, they are able to point to the adaptive effects of these mechanisms without arguing that the *by-products* are adaptations. For example, it can be argued that agency detection is an important cognitive skill with great evolutionary advantages, without having to show that agents with some counterintuitive qualities are advantageous for survival and reproduction. (See, however, Michael Blume [2009].)

Of course there are still scholars heavily influenced by Freudian thought, who continue to raise the issue of motivation. Are explanatory accounts that focus on cognitive mechanisms sufficient to explain motivation? I think that the answer is actually quite simple. The processes of evolution designed us to be susceptible to certain representations such as ritualized forms. We want to mark transitions from one state of affairs to another because they provide satisfaction. We want to think that there are hidden forces at work that are not available to the senses but might have dangerous consequences. While scientific theorizing and the sophisticated experimental work that accompanies it is dependent on our reflective capacities, our desire for knowledge is also driven by our intuitive capacities.

This leads me to an important issue that has played a significant role in the development of the cognitive and evolutionary science of religion—the distinction between intuitive and reflective knowledge. Evolution has designed us with a set of capacities capable of extensive calibration by input external to the organism, a set of expectations that lead, for example, to the acquisition of language, as well as various forms of folk knowledge (folk physics, folk biology, and folk psychology), that for the most part equip us to deal with the world around us. We can call this the capacity for generativity.

The issue of generativity raises the question of what capacity or capacities it took to produce the ideas that acted as candidates subject to the forces of selection. Was blind variation sufficient to account for the production of these ideas? Even if it was, we still need an account of the mechanisms that produced the variations. Perhaps we can agree that evolutionary theory has gone some way toward explaining how our cognitive capacities have acquired the form that they have. In the simplest terms they consist of a set of biases or capacities strongly dependent on development and calibration by environmental variables that are valuable aids in increasing the probability of survival and reproductive success. While the cognitive capacity for the acquisition of language, originally developed by Noam Chomsky, is already on a firm footing, there are many other capacities that deserve our attention and have been the subject of investigation in the cognitive and evolutionary science of religion. Take, for instance, agency detection (alluded to earlier). The capacity to distinguish agents from everything else in the world is a notable evolutionary achievement that denotes significant advantages. In fact, this powerful mechanism detects not only agents but also kinds of agents. The infant in the cradle, sometimes referred to as the scientist in the crib (Gopnik et al. 2000), is a remarkable creature, and clever experimental studies show how early in human development the capacities in question make their presence known.

Now we spandrelists have made much of the so-called hyperactivity of our agency detection system: HADD, as it has come to be known (initially suggested by McCauley at a conference on the island of Seili in Finland and developed by Stewart Guthrie [1993] and Justin Barrett [2004a] in very fruitful ways). The basic idea is that if we have an agent detector mechanism running full tilt, then there are good reasons for it to become overactive in certain situations. Actually, I doubt whether we need to think of such a system as being hyperactive at all. Any detection mechanism worth its salt simply needs to be active enough to be sufficient for the job, that is, sufficient to have the capacity to distinguish agents from everything else in the world and attempt to compute their intentions. It simply needs to deliver slightly more false positives than negatives over the long haul to deliver a selective advantage. Consequently, some stimuli can trigger responses that suggest the presence of agents without any agents being there. The reason that they are taken seriously is because of the mechanism of decoupling, that is, we have no problem regarding real people we know are alive but not physically present. Their ontological status is not dependent on their immediate presence to the perceiver in question.

In any case, the contingent association between agency detection and danger, or if you will, threat to survival and reproduction, is of particular importance. Having the capacity to discriminate between friendly and unfriendly agents on the basis of available cues would clearly provide an organism with significant advantages. Joseph LeDoux (1998) has provided the classic study on the fearful response to imminent danger. Of much more interest, however, would be the capacity to be particularly responsive to *potential* danger. Running away, when you are being charged by a lion, while useful and potentially life enhancing, is far more energy intensive and costly than being clever enough to recognize the potential presence of a predator by learning to read its tracks and rapidly leaving the scene of the potential encounter with the predator. Learning to pay attention to eye gaze, or ear motion, or the various shapes that faces take for that matter, might be even more fitness enhancing than responding to rapid movement toward you. For example, as I can testify from personal experience, elephants rhythmically flap their ears when they contemplate charging you. So if they wiggle their ears vigorously, it is time to read the signs of potential danger and get out of the way before they charge. Waiting for the charge to occur before you remove yourself from the scene can be fatal. Or to take another example, being particularly sensitive to certain disgusting smells is certainly a useful aid to avoiding dangerous substances (Rozin et al. 1993).

Now I think that if we were endowed with the capacity to be even moderately responsive to potential danger, this would have implications for our understanding of certain forms of behavior that might be considered religious in nature. Initial work in this area has been done by Szechtman and Woody (2004) and Boyer and Liénard (2006). Both pairs of scientists were interested in ritualized behavior on a continuum from the behavior of ordinary people in special situations such as pregnancy where intrusive thoughts, for example, emerge naturally, to such examples as patients with obsessive-compulsive disorder. Work on ritualization has been very productive and has provided important clues about cultural rituals in general and religious rituals in particular. Developing this idea would not eliminate the spandrelistic character of the behavior, but it might show that culturally developed forms of such behavior would reflect or incorporate these tendencies, both thematically in the words spoken and behaviorally in the deeds done. What I have in mind here is not the compulsive focus on certain features in the environment requiring our attention, but the repetitive elements that seem to play such an important role in religious rituals. Mary Douglas, in *Purity and Danger* (1966), called our attention to the widespread phenomenon of symbolically polluting substances. What we need to do, however, is to broaden the concept of danger to include substances, individuals, and actions. Ethnographic work shows that in religious contexts all three are potentially dangerous and need to be ritually addressed. So I am arguing that there seems to be a close connection between such "symbolic" behavior and the kind of behavior in response to potential danger that we seem to be capable of recognizing and responding to. Sometimes we run from an actual predator, sometimes we run from a potential predator. And sometimes the latter action takes the shape of a dance around a fire as we sing a-wim-o-weh.

Then, of course, there is the issue of religious experience. There is no doubt that some people have special experiences that they deem religious. Certain scholars have suggested that an explanation of the origin and persistence of religion can be developed purely on the basis of experiences such as mysticism. Some neuroscientists have attempted to develop this idea by searching for structures in the brain that are dedicated to the production of such extraordinary experiences. These ideas emerge from the work of William James (1902) in his *Varieties of Religious Experience* that certainly established the tradition for this form of inquiry. While I am willing to acknowledge that individuals who are susceptible to such experiences may make some contribution toward our accounts for the continued success of religious systems no matter what the cultural context is, no matter what level of hostility may be present, as a kind of role model, I think the fact remains that, from a sociological and psychological point of view, the vast number of religious participants have never had such experiences but find participation in religious ritual worthwhile anyway. Religious practices can do their job even if they have no meaning to the participants. As any anthropologist can tell you, a predictable answer to the question, "why do you do this in the way that you do?" is "because it is the custom." Of course some religious traditions make meaning of something that is of great importance. An elaborate hermeneutic tradition can develop with elaborate theologies. However, theologies, an example of reflective thought, are not necessary for religious ideas to work.

A notion of theological correctness has been developed in significant ways by Justin Barrett and Frank Keil (1996) and developed by Jason Slone (2007). The notion here is that even though people, when they reflect on their own ideas and behavior, develop certain norms, have a set of concepts that they regard as a kind of orthodoxy, or are right thinking about counterintuitive agents, for example, reveal, under experimental conditions, that their quick and dirty judgments are governed by their intuitions and not their reflections. In fact, the notion of cognitive biases is intimately connected with our notions about intuitive and reflective capacities generally and it plays a role not only in the cognitive science of religion but also in such disciplines as behavioral economics.

While the original insights of the cognitive science of religion were largely theoretical with some attention paid to empirical issues, primarily historical and ethnographic reports, it was clear that experimental studies would have to be done in order to support this new way of looking at religious thought and behavior. One of the earliest was by Barrett and Lawson (2001), which was designed to confirm or disconfirm hypotheses about peoples' intuitive knowledge about ritual behavior. This experiment made three predictions: (1) people with little or no knowledge of any given ritual system would have intuitions about the potential effectiveness of a ritual given minimal information about the structure of ritual; (2) the representation of superhuman agency in the action structure would be considered the most important factor contributing to effectiveness; and (3) an appropriate intentional agent initiating the action would be considered relatively more important than any specific action performed. The results of the study supported the Lawson and McCauley predictions (1990) and suggested that our expectations regarding ordinary social actions apply to religious rituals. This provided grist for the by-product approach to the study of religion. Since that early study there have been a host of experimental studies testing various predictions that have emerged from the development of the cognitive science of religion.

In conclusion we may now ask the question again: Is the cognitive and evolutionary science of religion a progressive research program that has made a contribution to the growth of our scientific knowledge of a biological, psychological, and cultural phenomenon such as religious thought and behavior? From the evidence presented here, I think that the answer is a definite yes. The research program has found new and interesting ways to connect cognition and culture. Moreover, it has also connected cognition with evolutionary explanations. It has developed both notions of religious thought and behavior as a by-product and also entertained the idea of certain cognitive features as being adaptive. The caveat here is that some avenues are not as progressive as others. The by-product approach, particularly when focusing on various different capacities, has painted a far broader and more interesting picture of religion than one simply focused on prosociality. The cognitive and evolutionary science of religion has laid the ground for a serious examination of the notions of religion either as an aggregate phenomenon or as a complex adaptive system. My view is that the former is the more useful and productive approach.

The cognitive science of religion has also shown that good ethnography recognizes the differences between peoples' intuitive and reflective beliefs and a good psychology focuses on our evolutionarily endowed capacities that are subject to calibration by various environmental and cultural variables. And it has explained how much of our knowledge of the world is determined through the intricate processes of cultural transmission. We now have in place the scientific tools for further growth in our knowledge.

2

Twenty-Five Years In: Landmark Empirical Findings in the Cognitive Science of Religion*

Robert N. McCauley

Religious studies' collective advocacy on behalf of diversity and inclusion stands in poignant contrast to its persisting *exclusionary* ethos (within most quarters of the field) concerning questions of method. A legacy of prohibitions in religious studies about who can study religions and about how they must proceed when doing so has tended to curb innovation. Born of protectionism (Lawson and McCauley 1990) or special pleading (McCauley 2013) or outright religious impulses (Martin and Wiebe 2012), such prohibitions have skewed the field in favor of the idiosyncratic over the recurrent, of the idiographic over the systematic, and of the interpretive over the explanatory. My long-standing interest in the promise of the cognitive sciences for studying religion has been, in part, to redress those imbalances. Redressing imbalances, however, does not involve dismissing the idiosyncratic, the idiographic, or the interpretive, but only suggests, first, that they are not the whole story and, second, that greater attention to the recurrent, the systematic, and the explanatory will enrich—not eliminate—our understandings and our inquiries.

My aim in this chapter (and in Lawson and McCauley 1990, 1993; McCauley and Lawson 1996; McCauley 2000, 2013) is to substantiate the second proposal for paying greater attention to the recurrent, the systematic, and the explanatory in our inquiries. I offer two lines of philosophical argument in its support—one broad and one more narrow, which I briefly summarize in the next two sections. The remaining sections of this chapter advance a further set of considerations in support of the cognitive science of religion (CSR hereafter) based on its many *scientific* successes.

Explanatory pluralism

The first broad line of philosophical argument in defense of my contention that systematic explanations about recurrent patterns in religious systems will enhance our

* This chapter originally appeared as chapter 6 in R. N. McCauley's (with E. Thomas Lawson) *Philosophical Foundations of the Cognitive Science of Religion: A Head Start* (London: Bloomsbury, 2017). It is reprinted here with minor revisions by permission of Bloomsbury Press.

understanding of religious phenomena (developed in Lawson and McCauley 1990; McCauley and Lawson 1996; McCauley 2000 and 2013) follows from general positions in the philosophy of science about the character and consequences of cross-scientific relations and their implications for explanations in science (McCauley 1986, 1996, 2007). CSR and the cognitive sciences generally are but instances of cross-scientific investigations to which these general positions' verdicts apply.

Both CSR and the cognitive sciences, more generally, exemplify the *explanatory pluralism* that prevails in cross-scientific contexts throughout the sciences (McCauley 2013). For new purposes pertaining to their own inquiries, scientists frequently enlist the conceptual, theoretical, analytical, methodological, experimental, and evidential resources developed within other sciences that are pursued at what are, sometimes, quite distant analytical levels. The various cognitive sciences span multiple analytical levels, generating and integrating insights from the biological, psychological, and social sciences—*from* cognitive neuroscience and comparative psychology at the biological level, *to* cognitive and cultural anthropology at the sociocultural level, and everything in between. They have assembled an extensive collection of investigative techniques and developed pictures both of human behavior and of the structure and operations of the human mind that are far more penetrating and insightful than those available heretofore. Their impact has been transformational in linguistics and economics, and they hold comparable potential for the study of politics, for the study of society and culture generally, and for the study of religion.

The resulting integrative accounts of religious phenomena in CSR certainly *extend* our understanding of the variety of factors that may be influencing religious thought and action in diverse locales. Those accounts also increase the range of resources available for situating phenomena conceptually and theoretically. Explanatory pluralism highlights the many means by which scientific investigators exploit the varied resources of the sciences and scientists' opportunistic approach to evidence, in particular.

Explicitly aiming in cross-scientific settings to *supplant* prevailing theories and the approaches that inspire them (e.g., Bickle 2003) ignores both the normative and historical considerations that animate explanatory pluralism. In particular, it contravenes scientists' bountiful opportunism regarding evidence. The replacement of theories and approaches that these positions envision would only reduce the number and variety of assets available in cross-scientific inquiries. More importantly, perhaps, the history of science supplies few, if any, precedents for such aspirations, especially once the pertinent inquiries enjoy some measure of intellectual and institutional stability (McCauley 2007 and 2013).

This is *not* to say that the reduction or the elimination of theories in science never occurs. They sometimes do, but the consequences of the first, that is, reduction, and the contexts in which the second, that is, elimination, occur carry no deleterious implications for the relationship between CSR and conventional religious studies. With regard to the first, the smooth mappings between cognitive theories and interpretive proposals implicated in a scientific reduction serve to *vindicate* the interpretive account in the specific context it addresses. With regard to the second, the replacement, indeed, the outright elimination of theories and their accompanying ontologies, is the outcome of intense competition *within* some sciences over time between clearly incompatible theories (McCauley 1986, 1996, 2007). Historically, such intense competition between what Thomas Kuhn (1970) called "incommensurable" alternatives tends to be comparatively short-lived (often but a decade or two) in the experimental sciences (Thagard 1992). Concentrated experimental research reveals the new competitor's strengths and liabilities relative to those of the prevailing theory, and either the relevant scientific community eschews the upstart or the field undergoes a scientific revolution that eliminates that previously prevailing theory and at least a few of its ontological commitments.

Crucially, the sorts of *cross-scientific contexts* in which any incompatibilities between cognitive theories and interpretive proposals would arise at some point in time are not, historically, situations that occasion scientific revolutions. Substantial conceptual and theoretical incompatibilities in cross-scientific contexts may generate selection pressures between analytical levels, but the history of science indicates that such inter-level pressures rarely, if ever, suffice to bring about such stark outcomes. No scientific consideration requires such a draconian approach to resolving theoretical incompatibilities *across* analytical levels in science. Instead of these selection pressures pushing in the direction of revolutionary upheaval at some analytical level, they may just as well ignite efforts to forge cross-scientific connections, especially when the research programs in question are pursued within scientific disciplines that have long-standing intellectual and institutional legacies.

The consistent emphasis in CSR on *explicitly* formulating theories in detail is an unqualified virtue. It helps to clarify the points where, whatever their provenance, theoretical proposals may make contact as well as whether points of contact are likely to result in conflicts or connections.

Whether through providing speculative interpretive proposals, or counterinstances that challenge cognitive hypotheses, or recommendations for refining such hypotheses, or focused scrutiny on relevant phenomena, or through simply presenting basic findings to be made sense of, standard works in religious studies and the history of religions can engage in myriad collaborative enterprises with cognitive scientists of religion. Historians' findings about both persisting and extinct religions, in particular, is a test-track on which, sooner or later, cognitive theories must run (Whitehouse and Martin 2004a; Pyysiäinen and Uro 2007; Martin and Sørensen 2011; Czachesz and Uro 2013; Martin 2015).

The sorts of findings in CSR as well as those in the cognitive sciences generally offer scholars of religion plentiful resources for inspiring and refining interpretive proposals. The cognitive sciences have collectively uncovered grounds for contextualizing, qualifying, supplementing, and, in some circumstances, even superseding many of the stock assumptions of commonsense psychology that inform interpretive approaches to human mental life, discourse, and action. Presumptions that seeing is believing, that the conscious mind matters most, that the mind's operations can go on in comparative isolation from the body, that memory is retrieval of fixed snapshots of past events, that possessing false memories follows from either some functional impairment or intentional perversity, that we mean what we say, that people have privileged knowledge about the workings of their own minds, and so on must all be hedged in one way or another most of the time. That alone is probably not news to humanists. What the cognitive sciences deliver, though, are increasingly detailed accounts of why, when, and how, and how much they should be hedged—detailed accounts that have withstood exacting empirical scrutiny and experimental tests.[1]

The cognitive sciences pose no barriers to humanists' interpretive projects. Familiarity with those sciences' accomplishments should, however, abet the sophistication of interpretive proposals (Lawson and McCauley 1990).[2]

Avoiding the quandaries of interpretive exclusivism

The second line of philosophical argument defending CSR's potential to enrich inquiries addressing religious phenomena (developed in Lawson and McCauley 1993 and McCauley 2000) concerns its abilities to supply constructive responses to moral, epistemic, and metaphysical quandaries that other prominent approaches in the study of religion face. These include crises of conscience, riddles of identity, epistemic overconfidence, and metaphysical muddles.

E. Thomas Lawson and I (1993) pointed to the crisis of conscience in anthropology and to the inadequacies of interpretive methods for addressing it. In the subsequent two decades, scholars' moral sensitivities have appropriately expanded beyond questions of colonialism to include persisting forms of oppression of women, people of color, and people of various racial, ethnic, and religious backgrounds as well as the poor, the disabled, and people with any of a variety of sexual and gender identities (who do not benefit from heterosexual privilege as a result), *wherever* they may reside. These forms of oppression no less merit our moral concern, but they are no better confronted by antiscientific ideology than was colonialism in the mid- and late-twentieth century.

Antiscientific enthusiasts' claims, first, that with scientific knowledge comes power to oppress and, second, that the interests of oppressors and scientists often coincide, are both surely true. But stopping there neglects the further truth that no human undertaking and, certainly, no human undertaking on the scale that modern science pursues, comes close to the level of self-policing that science achieves. Of course, that self-policing pertains most directly to epistemic rather than moral matters.[3] But, crucially, to the extent that scientific rationality and morality both turn on ideals about honesty and truth (however the latter should be characterized), they are of a piece. Its self-policing helps ensure that science is unsurpassed as a tool for obtaining knowledge about the world of our experience and that, barring its obliteration (a fairly high price to pay for oppressors who putatively rely on science as a means to power), its verdicts are neither wholly nor finally subordinate to the powerful. Antiscientific enthusiasts' suggestions—that the best response to the use of science to oppress is to abandon science and its ideals—are counterproductive. For example, the cognitive sciences can aid us, not least, in gaining a deeper understanding of human moral psychology (e.g., Graham et al. 2013).

Over the last few decades scholars of religion have recognized that, to the extent that the field of religious studies has itself collaborated with cultural anthropology and pursued parallel projects, it has been something of an unindicted coconspirator in these crises of conscience. Lawson and I argue, though, that religious studies have an additional problem concerning its identity as an intellectual project. Long wary of being categorized with either theology or the social sciences, religious studies programs have mostly sought a haven within the humanities. Traditionally, the principal defense for that position has maintained that religious studies has either a unique object of study, a unique method for studying that object, or both. Those assumptions, however, have undergone withering criticism from multiple quarters—and not only from cognitive scientists of religion. Arguments about the exclusivity of religious materials and the methods for their study have faced at least two major objections.

The first objection concerns the very notion that distinctively *religious* materials even exist. Cognitive scientists of religion and many recent contributors to religious studies question the viability of "religion" as an analytical category and its (metaphysical) status as an object of study. Although their reasons for skepticism that "religion" designates a unified body of phenomena differ, they are consistent and complementary grounds for that negative conclusion.

The by-product theory is the earliest and remains the most prominent theoretical orientation in the cognitive science of religion (Boyer 1994b, 2001; Guthrie 1980, 1993; Lawson and McCauley 1990; McCauley and Lawson 2002; Whitehouse 1992, 1995). The by-product theory holds that religions, like various other cultural arrangements from folklore to militaries, engage a host of ordinary cognitive systems (theory of mind, contamination avoidance, kinship recognition, linguistic competence, etc.) that are in place on the basis of considerations that have nothing to do with one another and, crucially for current purposes, considerations that have nothing to do with religion. Those cognitive capacities' exercises in religious contexts are *by-products* of their normal functioning. Whether they concern anthropomorphism (Guthrie 1993), action representation (Lawson and McCauley 1990), episodic and autobiographical memory (Whitehouse 2004), or all of these and more (Boyer 2001), these cognitive capacities exist in human minds because they enable people to deal with the species' perennial problems. It follows, at least from a cognitive perspective, that neither religion nor religiosity is some stable, uniform sensibility or pattern of behavior.

Some dissidents in religious studies raise a different set of considerations for questioning the analytical purchasing power of talk about religion. They argue that "religion" is a concept born of the scholarly enterprise of the modern western world. The dissidents (e.g., Fitzgerald 2000) note that the concept is tarnished by virtue of its association with the crises of conscience. "Religion" is an analytical term deployed by scholars in the West in an area of study that bolsters the projects of colonialists and capitalists, let alone the projects of proselytizers.

Their complaints, however, are not only moral. The sorts of features that receive scholars' attention favor the arrangements of the religions of the book, primarily, and of the world religions, secondarily, without attention to the understandings, practices, contexts, and lives of the members of the myriad small-scale societies around the world and throughout human history. Dissidents point out that the concept of religion does not seem to exist in many cultures; nor, they note, is there a similar word in many languages. It is precisely the plethora of *apparently* relevant stories, beliefs, actions, practices, social arrangements, institutions, and more as well as the variety of ways that they are regarded from one setting to the next that foil scholars' interminable attempts even to define "religion" (Saler 2008).

From this point, the argument is straightforward. Dissidents regard the lack of consensus among scholars and their persistent failure to define the field and its primary object of study as grounds for the analytical vacuity of the concept "religion." Without any defensible, coherent grounds for identifying what should count as religion, the traditional proposal that "religion" and its cognates pick out exceptional phenomena, whose study requires exceptional methods, seems forlorn.

Casting the discussions at the level of *particular religions*, in order to circumvent the problems associated with religion construed as some general domain of human thought and endeavor, introduces its own metaphysical complexities (McCauley 2000). What kinds of things are religions? Where, exactly, do they exist? How are their boundaries determined? What is the basis of their continuities over time? What, if anything, whether beliefs, practices, heritages, and so forth, is essential? Who counts as a Muslim or a Christian or a Buddhist and, more importantly, who gets to decide? The latter two questions have acquired considerable poignancy in a time when religiously motivated assaults routinely seize headlines throughout the world, yielding never-ending arguments about who should count as a *true* X, where "X" designates the name of some particular religious affiliation or other. Such talk about who is a "true" X (a true Muslim, a true Christian, etc.) or what makes for "true" X (true Islam, true Christianity, etc.) is transparently normative and that normativity is transparently theological. That leads straightforwardly enough to the second objection.

The second objection to the traditional claims about the unique character of religious materials and the necessity of extraordinary methods for their study holds that the positive case made in their behalf faces a dilemma. Either the arguments commit the fallacy of petitio principii (i.e., they assume what they set out to defend) or they depend on what are, finally, fundamentally theological conceptions of religious matters, with repeated references to "the holy," "the sacred," "the transcendent," and so on. As a way around this dilemma, many scholars recruited broader (but, in most regards, parallel) arguments deployed across the humanities typically about the singular character either of (human) subjectivity (e.g., Nagel 1974) or of the meaningful (e.g., Geertz 1973), or of both. These considerations were alleged, on the positive side, to require phenomenological and hermeneutic methods and, on the negative side, to constitute a barrier to scientific approaches. Without reviewing the academic culture wars of the past fifty years, let it suffice here to make four observations.

First, hermeneuticists' (and their postmodernist offsprings') preoccupations with *the text* and with its interpretation as the dominant metaphor for conceptualizing all meaningful materials—so that, for example, religion in all of its facets is construed as textual—leaves scholars not with matchless methods but with woefully deficient ones. That is because, whatever religion is, it is both much more and much less than texts.[4] This seems transparent with regard both to the ancestry of religion in the prehistory of our species and to what appears to be religious goings-on among nonliterate groups. But that is not the end of it. The myriad activities, items, and settings, let alone the mental states, experiences, and utterances that do not even remotely resemble texts but that play such vital roles in people's lives, in nonliterate, *and in literate* societies, operate far beyond the borders of the hermeneuticists' textual spotlight.

An emphasis on the textual also inspires a particular view of the past. History, on this view, is textually based. History is concerned with the production of texts that gets to its interpretations of past events and agents primarily through reflections about texts—whether previous texts about those events and agents or texts that those agents produced themselves. With such an approach, the cultural traditions, the salient events, and the past lives of the nonliterate risk *invisibility*. In a scholarly sphere in which both human origins and countless human groups are, in effect, invisible, the religions of large-scale, literate societies and their textually documented traditions inevitably hog the attention. The imbalanced distribution of scholarly attention goes largely unnoticed, because, quite literally, the non-textual is mostly nowhere to be seen.

Second, it is worth noting that even if the positive conclusions about the special status of the subjective and the meaningful are sound, they render *religious* subjectivity and *religious* meaning nothing but subcategories of far more sweeping considerations concerning humanistic pursuits overall. They provide no case (that does not circle back to covert theology) for picking the religious varieties out for special disciplinary treatment. The humanists who champion the subjective and the meaningful have far more ambitious aims than merely insulating religion. They intend to safeguard the researches of *all* of the humanities.

Third, the negative conclusions about the inabilities of the sciences to address such matters is an overreaction to a form of scientific exclusivism (scientism) and a conception of the sciences (logical empiricism) that the cognitive sciences and their practitioners neither endorse nor exemplify. The rejection of unsatisfactory accounts of science and of its reach does not require the rejection of science generally, or of the cognitive sciences, or of CSR.

Fourth, as I noted at the outset, such an exclusionary ethos is ill-advised, because the cognitive sciences and CSR offer invaluable methods and findings for enhancing our understandings both of aspects of subjectivity and of the making, having, and using of meaning in human life in general and in religious contexts in particular. In the absence of compelling arguments for interpretive exclusivism or impenetrable subjectivity and in the face of the myriad successes of the cognitive sciences and of CSR, assertions about science's putative inability to further illuminate these matters begin to look like reactionary protectionism of a field, instead of profound insights about either human life or the limits of scientific inquiry.

In the sections that remain, I will situate and summarize some of those successes of CSR—that is, new theoretical proposals; durable, replicated findings; and promising results in that field. The best evidence for the usefulness of these cognitive scientists' methods is the resilience and the fruitfulness of the ensuing research. A fair sample of the numerous methods that CSR has employed will emerge in the course of reporting on this research. Its explanatory pluralism will be transparent.

The development of CSR as a scientific enterprise with new experimental findings generating theories from the bottom-up

CSR arose from a range of theoretical proposals (Boyer 1994b, 2001; Guthrie 1980, 1993; Lawson and McCauley 1990; McCauley and Lawson 2002; Whitehouse 1992, 1995) that share at least three assumptions:

1. The employment of the theories, methods, and findings of the cognitive sciences to study religious thought and behavior would yield valuable new insights.
2. The mind has no department of religion, that is, that the mind has no systems, structures, or processes specifically dedicated to managing religious materials.
3. Many forms of religious cognition are by-products of the operations of cognitive systems that are in place for reasons having nothing to do either with one another or with how they figure in religious matters.

Consistent with the second and third assumptions, these various theoretical works address a wide variety of religious matters and a comparable array of cognitive functions and systems.

CSR began very much as a top-down endeavor. To their credit, theoreticians both welcomed attempts to test their hypotheses and explored relevant empirical evidence themselves (Boyer and Walker 2000; McCauley and Lawson 2002; Whitehouse and Laidlaw 2004; Whitehouse and Martin 2004a; Whitehouse and McCauley 2005a). They also worked to bring empirical evidence to bear on the *comparisons* of theories (e.g., McCauley and Lawson 2002). In doing so, they demonstrated that the research enterprise in which those theories figured qualified as *empirical science*. Still, most of this work involved appeals to ethnographies, case studies, and historical illustrations, which, because of their particularity, disclose only thin slices of the vast landscape of phenomena that pertain to the assessment of any particular theory and, because of origins independent of the theories in question, fail to illuminate many of the parts of that landscape that are of greatest interest. The original theoreticians saw that in addition to simply marshaling available empirical evidence, CSR would clearly benefit from taking the additional step of becoming an *experimental science* (Barrett and Lawson 2001; Boyer and Ramble 2001; Atkinson and Whitehouse 2010).

Experiments not only *test* theories, they *produce* (new) empirical evidence. Testing theories experimentally helps guard against the biases that may influence theoretical partisans' selections among already available, less systematic, empirical evidence that they cite in support of their theories. Experimentation enables scientists to target precisely those parts of that landscape that are of greatest interest. The process of managing unexpected experimental results theoretically may not prevent confirmation bias in scientists, but it does tend to make it more conspicuous when it arises (McCauley 2011).

Speculative theoretical ventures like CSR eventually require the kind of systematic support from the bottom-up that elaborate programs of experimentation furnish. Scientific theories should not only be able to fly. They should also be able to land. When theoreticians assemble existing evidence, they are, in effect, carefully selecting spots, surveyed from above, to bring their theoretical aircraft in for a landing. By noting this, I do not intend to be dismissive. Touching down on the hard surfaces of the facts like that *is* an accomplishment; however, experimentation forces theories to the ground at points of interest to those who are watching from below and negotiating conditions there. The more points on the map (i.e., the more facts designated by experimentation) where a theory can land safely, the more worthy it is.

Their common assumptions notwithstanding, these early theories in CSR have never been fully integrated. Still, as research has proceeded, these theoretical proposals have been expounded in ways that have revealed plentiful points of contact and coincidences of views. What, finally, is of far greater usefulness, though, are their conflicts and disagreements both with one another and with other theories that have been proposed subsequently (e.g., Bering 2006), for they serve as invitations for experimentalists to explore competing theories' implications in unusual settings that allow for the control of the theoretically interesting variables.

As noted, CSR began as a high-level, theoretical project. Over the subsequent twenty-five years many of the associated programs of research have secured their credentials not only as empirical science but also, in many cases, as experimental science as well.[5]

Once experimentation commences in any science, it exhibits a dynamic of its own. Initial findings from experiments aimed at testing big theories inevitably spawn dozens of finer-grained hypotheses. Those hypotheses address what the experiments' findings hint about potentially relevant variables. Reliably, some of those phenomena prove to be, simultaneously, so complex and so theoretically suggestive that they become objects of sustained investigation on their own. Within CSR, various topics are receiving such ongoing experimental investigation. These include such effects as theological incorrectness, promiscuous teleology, and characteristic patterns of reasoning about dead agents' minds as well as the mnemonic effects of minimally counterintuitive representations and the consequences of ritual for building social cohesion and increasing cooperation within groups. Beyond the sheer number of experimental papers that have appeared, perhaps the best evidence over the past fifteen years of CSR's status as a maturing experimental science is the emergence of such topics within its purview that have inspired this sort of prolonged and focused experimental scrutiny that takes on a life of its own. For more than a decade these topics have in each case attracted the attention of several groups of researchers from around the world, who have examined their various facets in considerable detail.

Further evidence of abundant fertile experimental research in CSR is the development of new theories built from the bottom-up. Possessing a collection of robust experimental results that appear to be disparate is what typically leads to constructing theories in such a fashion. Theoretically-minded experimentalists propose a nonobvious principle underlying the apparent incongruities among the findings as a means of integrating them theoretically (McCauley and Lawson 1998).

The parade case from recent work in CSR is the cognitive resource depletion theory (Schjødt et al. 2013). That theory proposes a unified treatment of the cognitive mechanisms undergirding religious ritual. It groups three apparently disparate phenomena pertaining to ritual:

1. The persistent occurrence for participants in rituals of goal-demotion and causal opacity (Boyer and Liénard 2006; Liénard and Boyer 2006; Nielbo and Sørensen 2011).
2. The negative effect on memory of the requirement, especially in high-arousal rituals, that participants suppress their emotional responses (Morinis 1985; Xygalatas et al. 2013a and b).
3. The deference participants show to charismatic ritual authorities about the performance and understanding of the rituals in question (Schjødt et al. 2011).

The theorists argue that each of these patterns involve the depletion of the cognitive resources that participants can bring to performing, remembering, and interpreting rituals. They maintain that, whether by swamping or by starving the relevant cognitive systems' processors during ritual performances, the resulting deprivations of processing resources create opportunities for religious authorities to proffer accounts, either before or after the rituals' performances, about what is transpiring. The theory proposes, in effect, that if people are daunted by mastering causally opaque details in their ritual actions or by controlling themselves in the face of profoundly stimulating sensory pageantry and community engagement or by the expertise of a charismatic ritual officiant, they have little time or energy for encoding ritual details or pondering ritual meanings. The cognitive resource depletion theory readily squares with the supposition that religious ritual systems have evolved to exploit these cognitive vulnerabilities. Such arrangements are likely to reduce the variability in these rites, which creates a space for cultural authorities to prescribe and regularize their performances and interpretations.

Three effects

Basic empirical outcomes in the sciences sometimes, all by themselves, elicit extended "normal" scientific treatment (Kuhn 1970). Conditional reasoning, flashbulb memory, the false belief task, and change blindness are examples in mainstream cognitive science from the past few decades. Such continued research on these and other phenomena within the cognitive and psychological sciences is part of the reason why Robert Cummins (2000) has stressed that the psychological sciences' principal accomplishments are the discovery of *effects*—the von Restorff effect (1933), the Stroop effect (1935), the phoneme restoration effect (Warren 1970), the SNARC effect (Dehaene and Mehler 1992), and so on—rather than the formulation of laws.

CSR offers illustrations of Cummins' (2000) observation that the sciences of the mind specialize in the discovery and articulation of *effects*, as opposed to laws. Effects are patterns in human performance that are pervasive but about which people are often inattentive, if not unconscious. Effects supply insights about how human minds work. For example, the Spacing effect (e.g., Madigan 1969) is the finding that distributed practice with materials increases the probability of their long-term retention in memory more than the same amount of massed practice does. If occasions for rehearsal are spaced out over time, memory performance is likely to exceed that from employing some small number of massed practice sessions of comparable duration.

Effects also inspire extensive programs of experimental research.

Theological incorrectness

Perhaps the best known effect arising from experimental research in CSR is the proclivity for theological incorrectness. Recondite theological formulations routinely feature counterintuitive representations. Their counterintuitiveness, however, is profuse, not minimal. The theologically correct Christian God, for example, is all-good, all-seeing, all-knowing, all-powerful, and all-present. The theologically informed religiosity of educated participants in large-scale, literate societies regularly employs representations that are nearly, if not equally, as *radically* counterintuitive as the representations scientific theories employ (McCauley 2011).

Carefully formulated, theologically correct texts commonly issue from years of contestation and debate. Scholars may overplay their prominence, at least with regard to their influence on participants' religious understandings and inferential predilections. Justin Barrett and Frank Keil (1996; and Barrett 1998) furnish evidence that in online tasks, such as processing and recalling narratives, religious people overwhelmingly utilize conceptions driven by the implicit assumptions associated with various unconscious, task-specific systems that appear to underlie so much of popular religious cognition. They designed short narratives about interactions between people and God to be consistent with the theologically correct doctrines, which their experimental participants affirmed when they were directly queried about their beliefs. The participants offered up conventional, theologically correct, non-anthropomorphic conceptions of God that constitute the orthodox beliefs of their doctrinal religious systems. Instead of deploying those theologically complex, ecclesiastically approved and policed concepts, which they endorsed when questioned directly, Barrett and Keil's experimental participants frequently revert to spontaneous, *theologically incorrect* conceptions in their recollections of these narratives. Participants reason about God on-the-fly similar to how they reason about Superman. Of course, Superman is an extraordinary character too. But being stronger than a locomotive, moving faster than a speeding bullet, leaping tall buildings in a single bound, and having X-ray vision still falls a good deal short of omnipotence, omnipresence, and omniscience. Barrett and Keil obtained the same findings when the task is merely to paraphrase, rather than recall, the narratives—even when participants have full access to the texts, they are paraphrasing throughout the task. They obtained such findings with Christians and Jews in America and with Hindus in India. Jason Slone (2004) outlined multiple lines of evidence for similar patterns in additional religions. Religious participants explicitly affirm theologically correct propositions; they often memorize theologically approved doctrines and learn a good deal of theology themselves (G. Peterson 2013), but it does not follow that any of this substantially influences how they think and reason *on-line* about religious matters in ordinary settings.

Emma Cohen in her ethnography (2007) and in her experimental work with Barrett (2008a,b) marshals evidence indicating that theological incorrectness occurs even in unpretentious settings in which scholarly sophistication and ecclesiastical hierarchy are meager. They show that theologically incorrect ideas readily intrude in the thought of followers of a small Brazilian spirit-possession cult. In order to handle a variety of theological complications, such as a possessing spirit showing strikingly different personality traits when possessing different people at different times, the cult leader teaches that spirit possession involves the *fusion* of the possessing spirit with the mind of the host. Spirit fusion, however, neither squares with folk psychology nor delivers much inferential potential. It is a substantially counterintuitive notion that does not comport very well with theory of mind. Cohen found that participants in the spirit-possession cult (and she and Barrett found that experimental participants in other cultures) virtually unanimously opted for the intuitive view (regularly portrayed in Hollywood movies) that the possessing spirit *displaces* the host's spirit, instead. Interestingly, people seem less troubled by the complications that accompany this view, such as what the host's spirit is up to and where it resides when it has been displaced.

The automatic intrusion of these maturationally natural intuitive mental systems guarantees the repeated eruption of theological incorrectness, no matter how humble the religious system. These are instances of a general pattern in which recurring intuitive assumptions connected with basement level cognitive systems intrude in thought and can trump painstakingly acquired reflective knowledge, whether theological or scientific (McCauley 2011).

Promiscuous teleology

A second seminal finding in CSR, which has also sustained an ongoing program of research, concerns an effect that Deborah Kelemen (1999a) has dubbed "promiscuous teleology." Kelemen first carried out experimental studies (1999a,b) supporting the position that children find function, purpose, and design throughout the natural world. She documented preschool-age children's inclination to over-attribute functions to things as a result of their new facility with theory of mind and growing experience with purposeful agents pursuing goal-directed actions. Unlike educated American adults, most children this age are willing to attribute functions to entire organisms (e.g., tigers) as well as to natural objects (e.g., icebergs), their parts (e.g., a mountain protuberance), and their properties (e.g., the pointiness of rocks).

In subsequent research Kelemen and her colleagues have produced grounds for suspecting that the penchant for promiscuous teleology may extend beyond childhood. In experiments with Romani adults, Krista Casler and Kelemen (2008) provide evidence *against* the assumption that the discontinuities between children's teleological promiscuousness and adults' apparent abstemiousness are the inevitable outcomes of development. Like the children Kelemen has studied, uneducated Romani adults differ significantly from educated Romani adults and from educated American adults in their willingness to approve teleological explanations for natural objects.

Education matters, but does it suffice to extinguish promiscuous teleology? Kelemen and Evelyn Rosset (2009) obtained experimental evidence indicating that, at least under some conditions, it does not. They had educated participants, who had, on average, completed 2.5 college-level science classes, assess the worthiness of proposed explanations for a range of natural phenomena. Participants who were forced to do the task fast (they had 3.2 seconds to read and respond to each item) proved significantly more likely to endorse incorrect teleological explanations than those not under such time pressures. They did so even though the speeded conditions had *no effect* on those participants' accuracy with regard to control items. They also found that similar percentages of participants assented to some unwarranted teleological explanations (e.g., "the earth has an ozone layer to protect it from UV light," 140) in *all* conditions, time pressured *or not*. An additional study revealed that educated adults with some experience of college-level science appear to think that "natural phenomena exist to benefit each other ..." and are "intrinsically directed towards survival ... and maintaining the Earth's natural equilibrium" (141). Crucially, they did *not* restrict such judgments to biological phenomena. Kelemen et al. (2013) found a similar proclivity for teleological explanation of nonbiological natural phenomena in experienced, PhD-level, physical scientists with appointments at major American research universities when they too had to make time-pressured assessments of explanations.

Assembling evidence from a wide array of developmental research in addition to that for children's teleological promiscuousness, Kelemen (2004: 295) has proposed that they are "intuitive theists," that is, they are naturally inclined to regard natural objects as "nonhuman artifacts" that reflect "nonhuman design." She argues that, by school age, children possess the requisite mental capacities for thinking about intangible agents,[6] their mental states and design intentions, and the (possible) role of the latter in determining objects' purposes. She notes that the view squares with Margaret Evans' (2000, 2001) findings that, up to the age of 10, children prefer "creationist" explanations of natural objects, regardless of their upbringing or of their parents' views about religion. It also comports with Barrett's arguments that young children's difficulties with the possibility of others having false beliefs, ironically, indicates that they are better equipped to understand infallible minds than they are the minds of humans (Barrett, Richert and Driesenga 2001; Barrett 2012). Presumably, Kelemen's subsequent research on promiscuous teleology in adults (previously outlined) adds to the plausibility of her suggestion (2004: 299) that adults, at least in their own less cautious ruminations, are also inclined to presume design intentionally imposed on things throughout their natural surroundings.

Dead agents' minds

Jesse Bering and David Bjorklund (2004) produced a body of findings suggesting a third, hitherto unnamed, effect concerning human reasoning about dead agents' minds that points to intuitive presumptions about minds outliving bodies. They first demonstrated what might be dubbed the "Dead Agents' Minds Effect" in preschoolers and kindergarteners. Although these young children held discontinuity views about biological functions concerning a dead mouse (for example, they did not think that the mouse would ever need to go to the bathroom again), large majorities spurned such discontinuities with regard to the mouse's psychobiological or cognitive states.

Substantial majorities of these young children thought that the mouse was still hungry (as he was when he died) and that he still wanted to go home (as he was attempting when he met his end).

It is not until late elementary school age that clear majorities of Bering and Bjorklund's participants affirmed discontinuity views about the psychobiological and cognitive states of dead agents' minds. This group was, however, the only group in this study that showed a significant difference in their responses to these two question types. Significantly more late elementary-school-age children certified discontinuities about psychobiological states than did so with regard to the cognitive states of dead agents' minds.

Bering and Bjorklund's work parallels Kelemen's proposal about intuitive theism in at least two important respects. First, they supplied evidence that children's inclinations toward continuity views about dead agents' minds, although probably enhanced by religious indoctrination, did not depend on it (Bering et al. 2005). Second, what initially looked like a pattern among youngsters proved, on further experimental investigation, to be manifest in adults as well.

In a further experiment, Bering and Bjorklund examined participants' views about postmortem organismic and mental states at a much finer grain than in their earlier studies, and they did so not only with kindergarteners and late elementary-school-age children but also with college-age adults. In this experiment Bering and Bjorklund posed multiple questions about biological states (e.g., "Do you think that Baby Mouse will ever need to *drink water* again?"), psychobiological states (e.g., "Do you think that Baby Mouse is still *hungry*?"), perceptual states (e.g., "Do you think that Baby Mouse can *see* where he is now?"), desires (e.g., "Do you think that Baby Mouse still *wants* to go home?"), emotional states (e.g., "Do you think that Baby Mouse is still *sad* because he can't find his way home?"), and epistemic states (e.g., "Do you think that Baby Mouse *knows* that he's not alive?"). This experiment basically replicated the findings of Bering and Bjorklund's earlier experiments with the kindergarteners and the late elementary-school-age children.

The pivotal findings of the experiment, though, concerned the adults' responses. First, like the late elementary-school-age children, the adults were significantly more likely than the kindergarteners to support discontinuity views with regard to the biological, psychobiological, perceptual, and emotional states as well as with regard to desires. This was *not* true, though, with regard to the mouse's epistemic states. Second, again like the late elementary-school-age children, the adults were significantly more likely than the kindergarteners to be *consistent* discontinuity theorists, that is, to give discontinuity responses to *every* question of a particular type.

Their findings with their adult participants provided evidence for two conclusions. First, discontinuity views about the organismic and mental states of dead agents would appear to be *learned*, as is, presumably, the view that death involves the *extinction* of the mind, from which such discontinuity views follow. Bering and Bjorklund found a significant effect for age group with regard to discontinuity responses. Late elementary-school-age children gave more discontinuity responses than kindergarteners and adults gave more still, and both differed significantly from the kindergarteners on this front.

Second, although large numbers of adults explicitly avowed extinctionist views about dead agents' minds, many did not seem to subscribe to that view when making judgments about the possibilities pertaining to the more purely psychological states of dead agents' minds. Only half of the adults were consistent discontinuity theorists with regard to epistemic states, in particular. Subsequent research has shown that what is, in effect, religious priming can amplify such effects (Harris and Gimenéz 2005; Astuti and Harris 2008). K. Mitch Hodge (2011) argues that both Bering and Bjorklund's findings as well as these priming effects depend, more fundamentally, on humans' abilities to carry out off-line social reasoning about absent agents.

Minimally counterintuitive religious representations

Pascal Boyer's (1994b, 2001) account of the cognitive bases of religious representations has inspired several studies exploring his contention that religions' minimally counterintuitive representations enjoy a mnemonic advantage over unproblematic, intuitive representations (no matter how strange or unusual) and over substantially counterintuitive representations.[7]

The diversity of religious representations can seem overwhelming. Boyer argues, however, that they are significantly constrained. Humans' unconscious inferences about "intuitive ontology" (2001) figure centrally in Boyer's explanation. Intuitive ontologies constitute foundational theories about the kinds of things in the world.

Boyer maintains that religious ontologies follow a standard pattern. Religious concepts violate expectations associated with some member of a small set of intuitive ontological categories, *while preserving* all of that category's further default inferences. That set consists of animal, person, tool, natural object, and plant. Violations of physical, biological, or psychological properties yield concepts with *counterintuitive* properties, exemplified by walking on water, being immortal, and *knowing* other peoples' thoughts, respectively. Those violations are of two sorts. *Breaches* occur when something transgresses a principle of folk physics, folk biology, or folk psychology that ordinarily applies. A person who passes through walls violates intuitive physics. A person who is the offspring of a lion breaches our folk biological expectations. *Transfers* occur when properties are assigned to items that do not possess them. Talk about a mountain that is alive transfers a collection of biological properties to a natural object. Claims about a snake that talks transfer a collection of sophisticated psychological capacities to an organism without them. Usually these representations involve only one violation in each instance; thus, they are *minimally* counterintuitive (MCI).[8]

Boyer hypothesizes that MCI concepts enjoy an advantage from the standpoint of competition for humans' attentions, as they approximate a cognitive optimum. First, *all* counterintuitive concepts are *attention grabbing*. Counterintuitiveness is not the only way to get noticed, but it suffices. Second, MCI concepts retain substantial *inferential potential*. An MCI concept's single violation leaves its abundant inferential power basically intact. Moses may have parted the Red Sea, but we can still infer that he would have made a splash had he jumped in, that his heart was beating throughout the episode, and that he would have expected that the subsequent inundation of the

Egyptians would interrupt their pursuit. These are but three unsurprising inferences, which follow from this story that contains the concept "person who parted the red sea." Boyer accentuates the instantaneousness and alacrity with which humans carry out such inferences and the wealth of inferences available.

The *memorability* of MCI concepts is a third consideration contributing to their selective advantage. MCI concepts not only fascinate, they tend to stick, which is necessary for their transmission. Various experimentalists have tested this hypothesis about MCI concepts' mnemonic advantages. Early studies obtained the predicted effects (Boyer and Ramble 2001; Barrett and Nyhof 2001). In assorted cultural and religious settings on four continents, MCI concepts were remembered significantly better than

- normal, intuitive concepts (e.g., a person who delivers thoughtful sermons and sleeps at night);
- highly unusual but not counterintuitive concepts (e.g., a chocolate table); and
- substantially counterintuitive concepts that involve many violations of intuitive assumptions (e.g., a statue that hears and answers prayers, weeps and bleeds, and flies around at night).[9]

Researchers have examined what role other variables might play in facilitating the recollection of MCI concepts. Those variables include imagery (Slone et al. 2007), causal integration (Harmon-Vukić and Slone 2009), background knowledge and narrative context (Gonce et al. 2006; Upal et al. 2007), and the amount of cognitive processing the concept demands (Harmon-Vukić et al. 2012). In each case, the mnemonic advantage accruing to MCI concepts generally stands. Moreover, in these and other studies (e.g., Norenzayan et al. 2006) that advantage increases as retention intervals increase. This was especially true with retention intervals measured in months, which would seem to be the time frames most relevant to matters of cultural transmission (Barrett and Nyhof 2001). Recent experiments indicate that the heightened memorability of MCI concepts holds for children as young as seven (Banarjee et al. 2013).

Occasionally religious representations incorporate more than one violation of humans' ontological intuitions. Moses, for example, has a conversation with a burning bush that is not consumed. Thus, I have suggested that "MCI" might be better construed as *modestly* counterintuitive (McCauley 2011). Konika Banarjee and her colleagues (2013), in fact, have provided experimental evidence that suggests that the mnemonic advantage for counterintuitive concepts extends to two violations. They found that seven- to nine-year-old children showed significantly better recall both directly and after one week for concepts involving either one *or two*, but not three, violations of intuitive ontology, relative to intuitive concepts.

New directions

Like cognitive science more generally, CSR has expanded in a variety of new directions in the twenty-first century. The field has attracted greater numbers of researchers, and those researchers have simultaneously advanced new theoretical proposals and introduced many new ways to test them. They have enlisted methods from across the social, cognitive, and brain sciences. The following three subsections will briefly discuss empirical research that simultaneously exhibits: (a) three of these new methods (economic games, brain imaging, and physiological measures in the field) and (b) three of the most conspicuous new directions for research in CSR (evolutionary theorizing, cognitive neuroscience, and religious experience).

These three new directions for research are by no means unique to CSR. They echo research trends across the cognitive sciences.

Evolutionary theorizing

The many controversies they have provoked notwithstanding, the emergence of (1) sociobiology (Wilson 1975a), (2) theories of cultural evolution (Boyd and Richerson 1985; Richerson and Boyd 2006; Henrich 2016), and (3) evolutionary psychology (Barkow et al. 1992; Buss 2005) has reintroduced reflection on the evolutionary foundations of cognition and mental life that had, for various reasons, been largely moribund for more than seventy years. Because it connected so directly with an existing experimental paradigm of long-standing interest, namely the Wason selection task (1966 and 1968), Leda Cosmides' (1989) discoveries about the crucial influence of social exchange on conditional inference in that task and the voluminous literature that it subsequently spawned thrust evolutionary considerations into cognitive science. The productive research programs associated with the alliance between evolutionary psychology (Barrett 2012) and cultural group selection and cultural evolution (Henrich 2016) ensure they will not be going away.

It is probably not a coincidence that all of the first generation contributors in CSR were by-product theorists. Among that group, it was Boyer (1994b, 2001) who developed what was the most elaborate evolutionary account of religious cognition. Evolutionary psychologists' commitments to the domain specificity of numerous cognitive systems, especially, have informed Boyer's proposals about religious cognition from the outset. Subsequent researchers (such as Bulbulia 2004 and Bering 2006) have wedded their views of religious cognition more directly to natural selection, arguing on a variety of grounds that humans have religious cognitive proclivities because those proclivities are individually adaptive. They maintain that religious sensibilities have aided individuals in passing on their genes.

Those hypotheses typically move in either or both of two directions. The first stresses the beneficial impact of religious participation on human health and well-being (e.g., Bulbulia 2006). The second concerns the ways in which religious beliefs, especially those about the gods' concerns with human conduct, encourage behaviors that are likely to make individuals trusted members of their social groups (e.g., Bering 2006). Whether their behaviors concern exhibitions of fidelity to the group and the group's gods, trustworthiness in moral matters, or both, the general proposal is that persons with such dispositions will, on average, have greater success obtaining resources and mates. Their compatriots will be more likely to enter into productive relationships—in all of the relevant senses—with such individuals, since their penchant for religious belief and deportment makes them good people with whom to cooperate. These circumstances should enable them to leave more copies of their genes, all else being equal, in the next generation.

Embracing explanatory pluralism inevitably produces pressures for broad interpretations of cognitive science, the fairly traditional conceptions of some theories in CSR notwithstanding (e.g., Lawson and McCauley 1990). With regard to any particular explanatory question of how narrowly or how widely cognition and the scientific enterprises that study it should be construed should turn primarily on the productivity of theoretical proposals, the empirical findings those proposals motivate, and how those theories and findings bear on the range of questions inquirers wish to explore. There is no such thing as a *complete explanation* in science. Thus, principled arguments for or against narrower or broader conceptions of cognitive science are probably misplaced. Instead of casting cognitive explanations exclusively in terms of internal rules and representations, a "4E" cognitive science stresses that cognition is typically *embodied*, *enacted*, *embedded*, and *extended* (Menary 2010). It surely is. For many purposes, including some that have arisen within CSR, however, 4E cognitive science is at least 1E too few.[10] The evidence that many forms of human cognition are *evolved* equals or exceeds that for any of the more celebrated Es (Buss 2005).

CSR theorists, who construe at least some religious belief and behavior as adaptations at the individual level, have enlisted methodological, theoretical, and evidential resources from evolutionary research in the biological sciences and, especially, in biological anthropology (e.g., Bulbulia and Sosis 2011). A particularly influential study for subsequent cognitive theorists has been a collaboration between Richard Sosis, a behavioral ecologist, and Bradley J. Ruffle, an economist, in exploring the role of religious ritual in forging intragroup cooperation to the benefit, presumably, of each of the cooperators (2003). The study exhibits how tools and methods from the social and cognitive sciences, namely economic games, can be deployed to test such evolutionary hypotheses.

Sosis and Ruffle take advantage of existing arrangements in Israel to compare cooperation among members within religious as opposed to secular kibbutzim. They test the hypothesis that ritual participation builds cooperation. Current circumstances already lend indirect evidence to hypotheses about the beneficial effects of common religious affiliation. On virtually every relevant front (profitability, retention of members, longevity, etc.), religious kibbutzim on average fare better than do secular kibbutzim in contemporary Israel. Consequently, in order to have greater control on as many theoretically extraneous social and economic variables as possible, Sosis and Ruffle carried out the nonreligious half of their study in some of the most successful secular kibbutzim.

Sosis and Ruffle compared kibbutz members' performance in an economic game as a measure of their cooperativeness. They used real money in a common-pool resource dilemma game in which two members of the same kibbutz play together anonymously. Both players know that the initial pool is 100 Israeli shekels (equivalent to about $25) and that each of them will propose to withdraw some amount. The rules are simple. If the sum of the two players' proposed withdrawals exceeds 100 shekels, then neither player receives anything. If the sum of the two equals 100 shekels, then each player receives exactly the amount that he or she proposed to withdraw. If the sum of the two withdrawals is less than 100 shekels, then each player receives not only the amount that he or she proposed to withdraw but in addition three-fourth times whatever remains from the 100 shekels after both withdrawals have been made. Sosis and Ruffle presume that common-pool resource dilemma games are reasonably good models for the use of common resources, such as water or electricity, on a kibbutz. Crucially, both players stand to benefit more, if they can trust one another to make small withdrawals from the original 100 shekels.

Sosis and Ruffle's study revealed significant differences between the performances in this game among members of religious as opposed to secular kibbutzim. The members of religious kibbutzim withdrew significantly smaller amounts from the initial 100 shekel pool than did members of the secular kibbutzim. The study supplied further evidence, though, that this result may well have been the effect of participation in collective public rituals. In fact, the difference between the two groups was a function of the proposed withdrawals of the *male* members of the religious kibbutzim and, in particular, of the proposed withdrawals of the male members who participated in *collective, public* ritual praying three times each day. Female players from religious kibbutzim did not propose withdrawals that differed significantly from those of players from secular kibbutzim. Although females in religious kibbutzim carry out rituals, those activities are mostly done domestically in private.

In some brief introductory comments, Sosis and Ruffle frame their findings in terms of costly signaling theory (e.g., Irons 2001). This theory proposes that participating in rituals communicates to other members an individual's commitment to the group. Basically, the more costly the ritual is, for example, costly initiation rites that include adopting group markers (Whitehouse [1996] 2014), the more convincing the signal is to the group. Participation in such rituals are hard-to-fake, high cost signals to the community that the participants are reliable group members, who will not defect. After all, by participating in such rituals, participants have paid a non-trivial cost in time, energy, and material resources. Both participating in the ritual and adopting group markers, such as scarification, characteristic clothing, or food taboos, requires that group members surrender various opportunities to pursue their own interests.

Since evolutionary thinking examines changes in large-scale systems over the long-term (McCauley 2009) rather than proximate cognitive mechanisms, much of that research is cognition blind. The influence of Sosis and Ruffle's findings on subsequent research in CSR (e.g., Whitehouse and Lanman 2014), notwithstanding, they do not, in fact, discuss cognition. Joseph Henrich (2009) stresses that costly signaling theory leaves questions about proximate mechanisms unaddressed. He notes that costly signaling theory does not tackle either the group dynamics or the historical processes from which such patterns arise. It also offers no account of the underlying psychological processes involved. It explains neither why costly requirements will increase commitments to beliefs, nor why costly signals seem less costly to insiders, nor when or why the production of costly signals reaches a ceiling.[11]

Henrich argues for the pivotal psychological role of credibility-enhancing displays (CREDs) in explaining the import of costly signals. In addition to the *content* biases in cognition, which Boyer deploys so effectively to explain the character of religious representations, Henrich (2016) argues that human beings also possess evolved *context* biases in cognition. Specifically, as *cultural learners*, Henrich proposes that human beings have an evolved disposition to attend to prestigious individuals (Henrich and Gil-White 2001). Because prestigious people have either expertise in some area of human endeavor or sound judgment or both, prestigious people are good people to model. Focusing on CREDs, Henrich suggests, constitutes a kind of cultural immune system in that CREDs signal to cultural learners the models' reliability with regard to their avowed commitments to the group, to the cause, to the beliefs, and so forth. Psychologically, the fact that everyone understands that cultural learners attend to models' CREDs (or lack thereof) decreases the possibilities that those models are self-interestedly manipulating cultural learners.

Certainly, costly signals are a variety of CREDs, but by no means do they exhaust the category. If prestigious models recommend some unfamiliar food and then, in fact, do such things as eat it themselves and feed it to their kin, not only have they exhibited a CRED but also they and their kin have benefited from the nourishment. Not all CREDs are costly. Preliminary experimentation with both adults and children indicates the value of CREDs for the cultural transmission of both practices and beliefs (Willard et al., n.d.)

CREDs explain the prominence and influence that leaders, who have made costly sacrifices, possess. Religious leaders, who forego wealth and sex or, in some cases, even their lives, demonstrate their good faith, so to speak, and increase the probabilities of the transmission of their religions. As Henrich comments (2016: 330), "CREDs can turn pain into pleasure and make martyrs into the most powerful of cultural transmitters." An evolved psychology of prestige underlies cultural learners' willingness to follow religious leaders who consistently produce CREDs.

Cognitive neuroscience

In their landmark article on cognitive science in the twentieth century, William Bechtel, Adele Abrahamsen, and George Graham (1998) stress that different disciplines among the several cognitive sciences enjoyed particular prominence for intervals across the time period in question. For example, the advent of the digital computer and the advances in computational theory after the Second World War endowed work in computer science and artificial intelligence during the first two decades of cognitive science with a certain pride of place. The prominence of the neurosciences in twenty-first-century cognitive science has also resulted from new technologies, though these have to do with brain imaging.

The ability to view structure and activity in human brains noninvasively has not only provided far more direct access to the central mechanisms of human cognition. It has also occasioned the development of inter-level theorizing and research integrating insights and findings from across the social, psychological, and brain sciences. These new imaging technologies have also furnished substantial, new bodies of evidence bearing on those hypotheses. Researchers from across the cognitive sciences have brought a variety of familiar tasks from experimental work in psychology and economics into the scanner to ascertain the impact of various stimuli on cognition and decision making. No work in CSR better illustrates such developments than Uffe

Schjødt and his colleagues' study (2011) of the influence of perceived charisma on the cognitive processing of believers.

Employing functional magnetic resonance imaging (fMRI), the Schjødt team explored the effect of different speakers' perceived (religious) qualifications on listeners' responses to those speakers' spoken prayers. The researchers examined participants' responses to the spoken intercessory prayers of three individuals, whom they were told differed in their religious statuses. One was described as a non-Christian; the second was described as a Christian, and the third was described as a Christian "known for his healing powers" (Schjødt et al. 2011: 120). (Assignments of these religious qualifications to the speakers were counterbalanced between participants.) Half of the participants were self-described Christian believers, while the other half were nonbelievers who were comparatively inexperienced with prayer and related religious matters. As a control, the listeners also heard nonreligious speech with the same structure as prayer. Participants also responded to two questionnaires. The first, which was administered before the scan, assessed the character and level of their religiosity and experience with religious matters. The second, which was administered after the scan, inquired about their experiences of the three speakers in the experiment and of God's presence while they were listening to the three.

Responses to the first questionnaire provided ample evidence for the religiosity of the religious participants, who held traditional beliefs with self-described conviction and who had considerable experience with standard religious forms and practices. By contrast, the secular participants did not believe in God, and they did not pray.

Responses to the second questionnaire indicated that the Christian participants rated the charisma of the reputed Christian speakers known for their healing powers significantly higher than that for the alleged non-Christian speakers, whereas the secular participants showed no significant differences between their ratings of the various speakers. The two groups showed even greater disparity with regard to their feelings of God's presence during the various prayers that they heard during the experiment.

To ascertain whether the researchers' hypothesis that participants' views of the various speakers' religious qualifications would have an impact on their neural activity, they compared activity levels across a host of the participants' brain areas, as measured by the blood oxygen level dependent (BOLD) imaging of their brains in fMRI scans. Their study uncovered a number of striking patterns.

Only the Christian participants' neural activity showed significant differences in their responses to the speakers and only in the contrast between the supposed non-Christian speakers and the speakers who were putatively Christians known for having healing powers. To get a sense of the comparative size of the effects of these different speakers on the Christian participants' neural activity, Schjødt and his colleagues compared activations in the five areas of participants' brains that showed the most extreme differences in response to the two speakers with their measures of baseline activity for those areas, which they had also obtained. Crucially, in all five areas, levels of neural activity were *less* than baseline for the Christian speakers known for their healing powers and *more* than baseline for the non-Christian speakers. These patterns also held for the Christian participants' responses to the post-scan questionnaires. Deactivation also correlated inversely with participants' reports about their feelings of

God's presence, and the Christian participants' post-scan ratings of God's presence and of the speakers' charisma were strongly positively correlated.

The brain areas (medial prefrontal cortex, the temporoparietal junction, the temporopolar area, and the precuneus), which exhibited what the researchers described as "massive deactivation" in response to the reputed Christian speakers with healing abilities, concern social cognition and executive function. These are areas that are centrally involved in humans' negotiations of their complex social worlds and their experiences of decision making in that domain and others. The researchers note that their study involved a "passive paradigm" in which participants simply listened to the speakers praying without knowing that they would be asked to assess them afterward. The Schjødt team (2011: 126) proposes that participants "trust in passive paradigms down regulate executive and social cognitive processing, because [they] suspend or 'hand over' their critical faculty to the trusted person." Their Danish Christian participants rated the alleged Christian speakers with healing powers significantly more charismatic than the alleged non-Christian speakers. Schjødt and colleagues (2011: 127) suggest that such down regulation of neural activity in these brain regions may well be an earmark of followers' susceptibilities to charismatic authority.

Religious experience

Submission to charismatic authority may not leap to mind as a paradigmatic illustration of religious experience, but it is often a salient dimension of what happens to many people in the course of their religious lives. Theorists, besides Max Weber, accord considerable prominence to the influence of charismatic leaders in their general accounts of religion (e.g., Stark and Bainbridge 1996). Still, such features of people's religious experience seem pedestrian by comparison with the wondrous goings-on routinely reported by venerated religious figures, saints, and mystics.

The first theories in CSR, given their focus on the *transmission* of religious *ideas*, tended (at least compared to most other approaches in the study of religion) to downplay the importance of religious experience. The general contention (Sperber 1996) was that no matter how exhilarating or inspiring participants' experiences might prove, their transmission is always subject to *cognitive* constraints on religious representations' recognition, ability to attract attention, memorability, motivational impact, and communicability. Without packaging exhilaration and inspiration in a form that is readily transmittable, the relevant religious experiences are a good deal less likely to make any decisive differences in a religious system's fate.

Still, these theorists did not ignore religious experience altogether. Boyer (2001) noted that powerful emotions are frequently elicited automatically when engaging many of the domain-specific cognitive capacities that religions target—from contamination avoidance, to kin detection, to fear of snakes, and more. Even if their principal concerns were mnemonic matters, both Whitehouse ([1996] 2014) and McCauley and Lawson (2002) were particularly interested in the emotional experiences that various religious rituals elicit in ritual patients. Again, though, such aspects of religious experience seem inconsequential when compared with the confrontations with the Cosmos, with the Holy, with the mysterium tremendum that many religious people supposedly go through.

For a variety of reasons, CSR has, more recently, turned its attentions to religious experience (e.g., Taves 2009). Beyond the traditional, widespread interest in the topic, ample evidence exists for package-able religious technologies (from rituals, to disciplines, to mind-altering drugs) that kindle some of those attention-grabbing experiences. Two considerations, however, are paramount. First, as the advocates of both 4E (and 5E!) cognitive science maintain, new tools and approaches (not just those of cognitive neuroscience) offer resources for understanding the experiential dimensions of our cognitive processing. Second, not only religious experience but also its ability to intrigue are never going away. That the topic would resurface prominently in CSR was inevitable.

Dimitris Xygalatas, Ivana Konvalinka, and their colleagues' studies (Xygalatas et al. 2011; Konvalinka et al. 2011) of extreme rituals exemplify the sort of exciting new findings about the associated experiences that the tools of the cognitive sciences can produce. The Xygalatas-Konvalinka team studied a fire-walking ceremony that concludes the annual festival of San Juan in the small Spanish village of San Pedro Manrique. The fire walking occurs at midnight on the summer solstice in an arena that was specially constructed for this ceremony and accommodates 3,000 (approximately six times the population of the village). Earlier in the evening the twenty-eight fire walkers processed through the village to the venue, accompanied by the townspeople. Over a half hour, one by one, the twenty-eight, usually carrying a friend or loved one on their backs, walked across a seven meter bed of hot coals, which reached temperatures of 677°C at the surface.

The Xygalatas-Konvalinka team recognized the exciting experimental opportunities in the natural (i.e., non-laboratory) setting that this fire-walking ceremony presented. After earning the trust of the village leadership, the local townspeople, and the fire walkers themselves, they were permitted to introduce into the ceremony a number of controls and measurements that were unobtrusive and unproblematic from the standpoints of all involved. These included video recording equipment for the purpose of memory research (Xygalatas et al. 2013a and b), but, most importantly, for my purposes here, twelve fire walkers as well as twenty-six spectators volunteered to wear heart-rate monitors. All of the volunteers wore the monitors during the roughly thirty minutes it took for the twenty-eight fire walkers to traverse the coals as well as during a thirty-minute interval one to three hours before the event (in order to obtain baseline heart-rate measures on all of the participants). Nine of the twenty-six spectators who wore the heart-rate monitors were either relatives or friends of one or more of the fire walkers, while the other seventeen, recruited at random, were unknown to the fire walkers.

Studying the correspondences between the fire walkers' and spectators' heart rates permitted the researchers to tease apart the synchronization of arousal in ritual from the synchronization of bodily movements. Considerable experimental evidence (e.g., Cohen et al. 2010) indicates that synchronized bodily movements create striking and similar responses among group members. Prolonged synchronous movements serve to align group members' cognitive states and levels of emotional arousal, which, in turn, are presumed to result in an elevated sense of group solidarity.

The empathetic projection hypothesis, at least in part, undergirds the second half of that story. This hypothesis holds that "it is the imagined responses of participants to focal events of the ritual that align their relevant cognitive states, without any strict need for orchestrated motor coordination" (Xygalatas et al. 2011: 735). The suggestion is that synchronization of bodily movement is not necessary for such outcomes but is only a particularly popular means for orchestrating arousal in common among group members. However it is achieved, it is the common *arousal* that is the underlying mechanism for the sort of empathetic responses that build prosocial feelings among members of a group.

The study offers more fine-grained scrutiny of the dynamics underlying that process. The Xygalatas-Konvalinka team suspected that, for people *affiliated* with one or more of the fire walkers, simply observing a fire walker might suffice to produce similar arousal. The findings they obtained from their study furnished stunning corroboration for that speculation. Crucially, the data they obtained from the heart-rate monitors "revealed striking *qualitative* similarities during the ritual between the heart rates of fire-walkers and heart rates of relatives and friends, with no apparent similarity to nonrelated spectators" (Konvalinka et al. 2011: 8515, emphasis added).

These findings are significant on at least two important fronts pertaining to the character of many people's religious experiences. First, they corroborated the empathetic projection hypothesis. Only the fire walkers walked across the bed of hot coals, yet the heart rates of the spectators who identified themselves as either a relative or friend of a fire walker did not differ significantly from the heart rates of the fire walkers on all four of the heart-rate dynamics that the researchers measured. These affiliated spectators who participated in the study had no physical contact with the fire walkers during their walks, but their heart rates tracked those of the fire walkers not only during their walks but also throughout the entire ceremony and, it turns out, even to some extent during the baseline epoch as well (Konvalinka et al. 2011: 8516–17).

Second, the effect is, at least in part, a function of *social* relationships. It does *not* turn exclusively on the brain's mirroring capacities, which have attracted so much attention over the past two decades (Rizzolatti et al. 2011). Even though all of the spectators witnessed the fire walking, it was only the heart rates of the spectators who were associates of a fire walker that exhibited those qualitative similarities to the fire walkers' heart rates. The nonaffiliated spectators who participated in the study were no less capable of mirroring the fire walkers' levels of arousal, as measured by their heart rates, but, in fact, they did not. Their heart rates on all four of the heart-rate dynamics that the researchers measured differed significantly from those of the fire walkers.

The Xygalatas-Konvalinka team's study of the fire walking at San Pedro Manrique provides a glimpse of the kinds of experimental controls, quantitative measures, descriptive precision, and *rich insights* that the theories and methods of CSR can supply for the study of at least some varieties of religious experience. On the one hand, their work contributes fundamentally to substantiating what are, in effect, theoretical proposals about religious experience from religious studies and the social sciences. On the other hand, this work and Xygalatas and his colleagues' subsequent work in

Mauritius (e.g., Xygalatas et al. 2013a nd b) also models for experimentalists in the social, psychological, and brain sciences how to carry on fruitful experimentation utilizing physiological measures in the field.

Afterword

Nothing seems more fitting than ending with this brief survey of some (but, by no means, all) of the most exciting experimental studies that now animate the field and lead it into new areas of endeavor. In the light of such work, neither CSR's successes nor its promise—as an ever-growing body of methods, theories, and findings for illuminating what we regard as religious phenomena and for enriching their study—are any longer a matter of debate. *Sometimes*, and the case of the cognitive science of religion is one of those times, it is a good thing that science does not get waylaid by religious studies scholars' philosophical worries.[12]

3

Twenty-Five Years of CSR: A Personal Retrospective

Harvey Whitehouse

In the nineteenth and early twentieth centuries many great scholars speculated about the causes of religion: Sir Edward Tylor argued that religion originated in an attempt to make sense of puzzling observations and experiences; Sigmund Freud argued that religion resulted from repressed feelings of guilt; Emile Durkheim argued that religion was a symbolic expression of the social order; Karl Marx argued that religion emerged as a tool of class oppression. All these undoubtedly brilliant theorists, however, suffered from a common problem: all were seeking a *magic bullet explanation*, a single cause that could explain religion in its entirety.

When we think of religion as a single monolithic entity, a natural kind requiring a general definition and explanation, we are making a fundamental error. It is no more a natural kind than the big dipper as seen in the night sky from Earth is a natural kind. It is just an illusion resulting from taking a particular earthbound perspective. What we commonly think of as "religion" isn't really a single coherent phenomenon. It is actually a loose bundle of things, such as supernatural agent concepts, afterlife beliefs, creation myths, signs and portents, altered states of consciousness, rituals, and so on. Although people readily group these things together, they are founded on very different psychological systems and are sensitive to very different triggers. We need to break up religion into its various building blocks. Each of these fractionated elements should be the focus of discrete research programs.

Perhaps the single most important insight of the cognitive science of religion (hereafter CSR) is that religion is shaped and constrained by implicit, panhuman intuitions about the way the world works and these intuitions derive from many distinct psychological systems rather than just one. Fractionating religion is not an obvious way to proceed, however, because particular religions present the fractionated elements as inextricably bound together. That is, the cultural traditions that we call religions (not only the world religions but the many thousands of smaller cults and regional movements that are studied by religion scholars and anthropologists) tend to present notions of deities, ancestors, creation myths, sacred laws, and many other aspects of religious belief and practice as bound together into systems of interlocking elements. Nevertheless, these connections owe their existence to historically contingent cultural beliefs and practices, rather than to anything written deep into human nature. CSR has provided compelling evidence that notions of supernatural agency are indeed undergirded by a variety of panhuman cognitive systems (e.g., Guthrie 1993; Boyer 2001; Barrett 2004a; Bering 2006; McKay and Whitehouse 2015), but these systems are quite distinct from the equally universal cognitive processes giving rise to creationist beliefs (e.g., Kelemen 1999b; Evans 2001) even though beliefs in supernatural agents and creationism are commonly assumed to be related. Particular religious traditions may postulate moralizing gods that are *also* creator beings, but the link between mind-body dualism and teleological reasoning about the natural world is culturally learned rather than intuitive and innate. Of course, the same point could be made conversely, that atheists not only repudiate god concepts but also tend to be extinctivists and skeptical of magic even though the denial of deities does not necessarily imply that there is no afterlife or that spells could not possibly work.

This chapter seeks to demonstrate the merits of this general approach to religion by charting in semi-autobiographical fashion my recollections of CSR's early days and by summarizing our collaborative efforts over the past twenty-five years to understand the causes and consequences of just one fractionated element of religion: *ritual*.

A little more than twenty-five years ago

In the 1980s I was a fresh-faced student of anthropology at the London School of Economics and Cambridge. Along with Oxford, these institutions were generally credited with having created the discipline of social anthropology. Within this Golden Triangle, there was a strong tradition of cultural relativism and cognitive constructivism, handed down from venerated intellectual ancestors: Rivers, Malinowski, Radcliffe-Brown, and Evans-Pritchard, among others. Steeped in this relativistic tradition, many of my undergraduate teachers believed that cultural systems shaped our psychological development to such a degree that there was nothing at the core of human nature that wasn't culturally constituted—and if there was, it wasn't interesting. Later I would discover a very different view of human nature grounded in the experimental methods of the cognitive and evolutionary sciences. But the perspectives of the harder sciences were not usually taught to social anthropologists in those days, and my private skepticism regarding the dogma of relativism led me to seek some kind of proof of the psychic unity of humankind. I wanted to go somewhere as culturally remote and historically detached as possible from the society that had nurtured and raised me. I wanted to discover the limits of human nature by encountering them somehow "in the wild."

From my undergraduate readings in anthropology, I learned that the peoples of Papua New Guinea (PNG) had never known the blessings (or curses) of metallurgy, science, literacy, or state formation and, for thousands of years, had remained beyond the reach of more complex societies evolving in other parts of the world. This extraordinary isolation, until the arrival of Western colonists mostly in the twentieth century, made the case of PNG profoundly alluring. Many of the accounts of anthropologists in the region quoted informants who recalled a time when all artifacts were made from materials gathered from the natural environment and fashioned using stone age technologies. Scattered settlements in the forest were often at war, capturing brides from their neighbors, killing and allegedly even eating people from enemy groups. At the time, it would have been (and probably still is) considered ethnocentric, or at least poor taste, to single out these rather striking facts and far more fashionable to fabricate more abstract (if spurious) idiosyncrasies of Melanesian cultures, such as their allegedly "fractal" conceptions of personhood (Wagner 1991). But what really interested me about PNG were the potential consequences of having been virtually untouched by the many waves of cultural evolution that had swept across the rest of the world for millennia.

For my doctoral research at Cambridge, I decided to travel to the island of New Britain in search of a remote cultural group known to the outside world mainly for three things. First, they were called the "Baining," which in the language of their coastal-dwelling neighbors meant something like "forest dweller" (with pejorative connotations of savagery). Second, many of the Baining rejected the teachings of Christian missionaries and instead followed a little known cargo cult. And third, one of Britain's finest social anthropologists, Gregory Bateson, had tried to study the Baining but had given up, claiming that their culture was impenetrable. These patchy and unreliable snippets of information lured me to the Baining and I ended up living with them for two years, studying their language and culture, and becoming adopted into the community.

I had been taught that a good way to establish rapport while struggling to learn an unwritten language was to acquire the skills valued locally, which in this case included gardening, foraging, fishing, and hunting. As my abilities to communicate improved, however, it became clear that what people really wanted to talk to me about were their religious beliefs and practices (Whitehouse 2005b). I soon learned that much of daily life was devoted to enticing the ancestors back from the dead, ushering in a sort of Paradise on earth. The Baining village in which I ended up living turned out to be just one of scores of villages scattered throughout the forest, together forming a movement known as the *Kivung* (which means a "meeting"). Kivung followers performed a vast array of rituals, mainly focused on three types of temples where food offerings to the ancestors were laid out and lengthy public speeches were made. One of the striking features of these practices was that they conformed quite rigidly to an established canon that was standardized across the movement, encompassing thousands of followers. This uniformity of belief had managed to cross several language barriers, binding together peoples with otherwise quite different histories and cultural systems. And yet, as Kivung members they all espoused the same elaborate ideology and performed the same repertoire of daily rituals. Given that most followers of the Kivung could not read or write, the maintenance of an elaborated orthodoxy seemed to depend on the frequent repetition of doctrines and narratives in public settings.

Although Kivung members had a strong sense of common identity, especially in opposition to their coastal neighbors and the various Christian missions, an even stronger sense of group cohesion was evident in the individual villages comprising the movement. In the village of Dadul where my research was based, I witnessed the formation of a breakaway sect that sought to be the first among all the Kivung villages to be reunited with its ancestors and showered with wealth. The people of Dadul, consequently, established new rituals to welcome back the dead, involving massive feasts and the destruction of property. Men discarded their clothes in favor of loincloths and women removed their sarongs, baring their breasts in defiance of missionary teachings. A mass marriage was conducted and other grand ceremonies performed inciting strong emotions. Eventually the community built a large roundhouse capable of housing the entire village and they joined together in all-night vigils waiting for the returning ancestors. Some of the feelings aroused in participants were powerfully dysphoric. For example, a decision was taken to remain within the communal temple from dusk to dawn and not to venture outside to answer the call of nature. Due to overconsumption at feasts during the hours of daylight, some people inevitably needed to defecate in the temple and this also triggered vomiting, adding to the hellish atmosphere of the nocturnal vigils.

When the ancestors failed to show up and the community's resources had been depleted to a critical level, the specter of starvation loomed and the splinter group was disbanded. People then returned to daily life in the more routinized Kivung movement. A striking consequence of this splintering process was not disillusionment or despair, as one might have imagined but, on the contrary, renewed conviction and social solidarity—especially within the local community that had undergone these ordeals together. Perhaps even more surprisingly, I learned that such splintering events were not unique to the village I was living in—in fact, most Kivung villages underwent experiences like this every five to ten years and the overall effect seemed to be to amplify commitment to the group rather than to undermine it.

These experiences in PNG suggested to me that rituals tend to fall into two basic clusters. There are those (like the ones I witnessed in the local splinter group) that evince strong emotions. Other examples from this region included rites of initiation—often involving terrifying ordeals. I called these imagistic practices because they make a strong impression on people and leave a lasting image in their minds. Many imagistic rituals are rare or once-in-a-lifetime events. These may be distinguished from a second cluster of rituals that are decidedly more sedate or mild, such as those observed in church on Sundays or the mosque on Fridays. These much more frequently performed rituals are usually accompanied by highly elaborated religious teachings. I called these doctrinal practices.

Doctrinal and imagistic practices constitute divergent modes of religiosity. Imagistic practices are very effective at binding local networks of people into tightly knit, emotionally bonded communities. They seem to create a sense of family connection based, not on bonds of kinship and descent, but on the sharing of life-changing ritual experiences. Doctrinal practices, by contrast, tend to spread rapidly and become standardized across much larger populations. The frequent repetition of doctrinal rituals—from daily prayers to weekly Holy Days through to all the events that fill up religious calendars—serve to cement the social identity of potentially enormous social groups, like a tribe or a nation, but not with the same intensity as imagistic practices.

Following my two years of living with the Baining, I returned to Cambridge and buried myself in the library. Reading the cargo cult literature with fresh eyes, I began to realize that the two modes of religiosity I had observed in the Kivung were apparent in other cults in the region, and perhaps more widely still. After writing up my field notes and submitting my thesis, I began to explore the applicability of this distinction between doctrinal and imagistic practices to other religious groups in Melanesia—and that is how I became involved in the establishment of CSR.

The early nineties

Twenty-five years ago I took up my first academic post as a research fellow at Trinity Hall, University of Cambridge. By that time my skepticism toward most forms of relativism and all forms of postmodernism had become difficult to hide. I had started reading books about cognitive science and evolutionary theory, and my growing interest particularly in experimental psychology was gently encouraged, at times gleefully, by several mentors, foremost among them, Ernest Gellner and Alfred Gell. But it was actually one of my peers, rather than any of the distinguished elders living in Cambridge at the time, who most inspired me to develop a cognitive approach to the study of culture: Pascal Boyer. At the time, he was, like me, a research fellow, but he came from a different and (to me) more exotic intellectual tradition under the tutelage of Dan Sperber in France. Boyer encouraged me to make frequent visits to Paris to visit his friends, especially Carlo Severi and Michael Houseman. And it was also through Boyer that I came to meet E. Thomas Lawson and Robert N. McCauley—as well as their students, including Justin Barrett and Brian Malley. I was amazed and delighted to find that we not only shared many frustrations about the way religion was typically being studied at the time but also had strikingly convergent ideas about how to improve the situation. As the frequency of our group meetings intensified, we began to describe ourselves smilingly as the architects of a "cognitivist conspiracy" bent on reforming the study of religion and setting it on a truly scientific footing. But more seriously, and even more remarkably, I discovered that Lawson and McCauley had been developing a theory of ritual that bore uncanny resemblances to my recently published ideas about doctrinal and imagistic modes (see McCauley 2001; McCauley and Lawson 2002; Whitehouse [1992] 2014, 1995, 2000). This coalescence of interests, among others, helped foment a wider interest in CSR.

Although the story from this point on could be told in many ways, involving many people and projects, it is hardly surprising that what I remember best about the burgeoning CSR field are the activities focused on the study of ritual, since that was my primary area of interest. We all knew that in order to make serious progress testing our theories of ritual would require the involvement of scholars of religion, classicists, historians, and archaeologists. I recall some gloomy conversations with my fellow conspirators about the improbability of sparking significant interest among experts in those fields. But then McCauley introduced me to Luther H. Martin.

Martin firmly believed that the study of religion was ripe for change and that it would be eminently possible to find scholars interested in CSR—especially its refreshingly precise conceptual frameworks, hypotheses, and methods of testing them. And he turned out to be right. Martin helped and inspired several of us to organize conferences to which we successfully invited many distinguished historians of religion studying

Mithraism, Gnosticism, and medieval and reformed varieties of Christianity—but also people who studied prehistoric rituals stretching back into our hunter-gatherer past (Whitehouse and Martin 2004a,b, 2005; Whitehouse and Laidlaw, 2004; Whitehouse and McCauley 2005a,b; Pachis and Martin 2009). Through these discussions many of us became convinced that there was something universal and ancient about the patterns of ritual behavior we were interested in.

Ritual and causal opacity

The first thing to note about rituals is that they are everywhere. All human societies have them and as far as we can tell they always have. Like the broader category of "religion," ritual can be fractionated into a number of different aspects including things like synchronous movement, symbolic qualities, concerns about contamination, and associated concerns about threshold and entrance, exactness and symmetry, separating and boundary marking, and so on (Boyer and Liénard 2006). Some rituals are such ingrained habits that we scarcely notice them. People often associate rituals with religion, but rituals are a pervasive feature of all aspects of our lives, not just contexts of worship. Even atheists perform a vast array of rituals everyday—by following conventions for greeting each other, handling food, deciding what to wear and how to style their hair, and so on.

One of the most interesting features of ritual from an anthropological perspective is its "causal opacity" (Whitehouse 2011a, b). That is, rituals don't have a rational causal structure—it's simply a matter of following a convention, observing the proper or the "done" thing. When we think of ordinary instrumental behavior, we assume it has a knowable causal structure. For example, when we observe the behavior of an angler, we naturally infer that the rod, reel, line, hook, and bait are all arranged so as to assist in the extraction of fish from the water. This even goes for elements we don't fully understand—like the cogs and other mechanisms in the reel, which we assume are there somehow to facilitate the process of casting out or reeling in. By contrast, when we observe a bottle of champagne being swung on a rope so that it shatters against the hull of a ship, we don't interpret this action in purely instrumental terms. It would be absurd to imagine that the bottle is going to help push the ship down the slide into the water. If it assists in the process, it's assumed that it won't be in terms of ordinary physical causation. For many, sacrificing a bottle of bubbly is just a traditional entailment of launching a ship. If it's also thought to do something useful (e.g., bring good luck in some unknown way), such outcomes are never adequately explainable in physical-causal terms.

The reason for homing in on causal opacity is that it picks out a very special feature of our evolved psychology. Humans are the only primates that engage in extensive imitation of causally opaque behavior. Chimps couldn't care less about table manners; they don't clasp their hands in prayer or do the goosestep. It's true that chimps and other primates sometimes learn things from each other but they only pick up skills that are technically useful—skills that you can see actually work. Since cultural rituals don't have any causal structure, most sensible animals wouldn't bother copying them.

But what those other animals don't realize is that rituals contain a powerful secret: by slavishly conforming to arbitrary social conventions, human groups are binding themselves together into cooperative units—and thanks to our rituals we are able to cooperate in ways that none of the other higher mammals have managed to accomplish.

One of the many little clues suggesting that ritualistic behavior is written into our evolved biological makeup is the fact that it emerges very early in human development. Even infants seem to be fascinated by causally opaque behavior and will try to copy it (Gergely et al. 1995). Indeed, the willingness to copy arbitrary conventions is essential to acquiring language—you have to accept that certain funny utterances refer to stable features of the world around us, not because there's a causal relationship between the sound and the thing it refers to but simply because that's the accepted convention.

Our tendency to copy causally opaque behavior is sometimes called "overimitation." Psychologists have known for some time that if you show children an unnecessarily complicated way of retrieving an object from a box, they will copy not only the causally necessary behavior but also imitate the pointless elements. In fact, even if you tell them to avoid reproducing any "silly" actions that don't really help with getting the object out of the box, they still copy them (Lyons et al. 2007). Apparently they can't help themselves. For some time psychologists have been thinking that overimitation evolved to help children acquire complex technical skills before they could actually understand how they work. But around 2010 I teamed up with developmental psychologist Cristine Legare at the University of Texas to explore a very different interpretation. Perhaps the function of overimitation, we reasoned, is to transmit arbitrary group conventions—rituals in other words. And perhaps what motivates this behavior is the desire to belong, rather than to learn anything technically useful.

To test this idea, we designed a study in which young children (4–6 year olds) were shown a novel action sequence using objects they had never encountered before (Legare et al. 2015). The children were split into two groups, receiving slightly different treatments. In one condition the modeled action sequence ended with all the objects back where they started. We called this the "ritual" condition because it made no sense to interpret the actions as having a causal structure leading to an outcome. In the other condition, all the objects were handled in exactly the same way as in the first treatment but with one crucial difference—the last object ended up not where it began, but in a box. We called this the "instrumental" condition because even though it was undoubtedly an odd way of putting an object into a box, there was at least some kind of causal structure buried in the action sequence leading to an end goal. Children in each of these two groups were then given the objects to handle themselves. Even though no instruction was given to copy what they had seen, all children imitated the modeled behavior to some extent. Crucially, however, children copied more faithfully and were less prone to inventing novel behaviors of their own, in the ritual condition.

Humans seem to interpret behavior in two very different ways—either as instrumental (aimed at manipulating the causal structure of the world to achieve an end goal) or as ritualistic (aimed at sharing a set of conventions distinctive to their group). One of my PhD students, Rachel Watson-Jones, helped us replicate the findings of our original experiments while also extending the paradigm. What we found is that if you prime the children with an ostracism threat, then levels of imitation associated with the ritual condition go even higher (Watson-Jones et al. 2014; Watson-Jones et al. 2016). The ritual stance seems to be all about trying to belong, to affiliate with an in-group. When you are insecure about your status as a group member, you become even more anxious to conform to group conventions—basically, you overimitate even more. In some of our recent studies, we have used verbal framings to activate the ritual stance—we told children that the model "always does it that way," suggesting that this is a conventional preference rather than an instrumentally motivated action. We have found that even this mild encouragement to adopt the ritual stance gets children copying opaque behavior more faithfully (Herrmann et al. 2013).

Ritual as social glue

Collective rituals play a crucial role in binding groups together. Based on my fieldwork in PNG back in the 1980s and subsequent investigations of more than 100 detailed case studies, it appeared that rituals tended to fall into two basic clusters or "modes of religiosity," previously described (Whitehouse 1995, 2000, 2004). Imagistic and doctrinal rituals entail quite different types and intensities of group cohesion. Imagistic rituals are very effective at binding small groups of people into tightly knit, emotionally bonded groups. It's almost as if they create new family units, connected not by shared genes but by their shared experiences in sacred rituals. Doctrinal rituals work differently. They are generally standardized over much larger groups of people than imagistic rituals because they are often linked to rigid belief systems that can be exported intact to new people. The frequent repetition of doctrinal rituals cements the social identity of potentially enormous social groups, such as nations or world religions.

Imagistic rituals build groups by creating a sense of family connection among participants. They do this by making us feel like we share something basic and essential about our innermost personal identities. Every one of us has a personal self—a set of traits that make us who were are. A lot of these key features come from our past experiences, events that have shaped our lives—our personal autobiography. The most self-shaping experiences can be quite negative ones—ordeals that we've overcome, making us stronger or wiser. Bad experiences tend to be remembered better than good ones, prompting more intense reflection afterward (Whitehouse, Richert, and Stewart 2005; Russell et al. 2016). For nearly two decades we had linked these kinds of experiences to the imagistic mode and the remarkably intense social cohesion it produces. But we lacked a precise conceptualization of this type of cohesion or a way of measuring it. That's when I discovered the work of William B. Swann and his colleagues on "identity fusion" (Swann et al. 2012).

Identity fusion may be concisely described as a visceral sense of oneness with the group (Swann et al. 2009) whereby one's social identity becomes an essential aspect of one's personal self. Strongly fused persons report intense family-like connections to other group members, high levels of personal agency, and feelings of invulnerability in their group. They also exhibit strong prosocial tendencies, for example, a willingness to make personal sacrifices to aid in-group victims of attacks (Buhrmester et al. 2015). Perhaps the exceptionally intense form of social cohesion we had been observing in imagistic traditions was captured by the identity fusion construct. Perhaps when self-shaping experiences are felt to be shared with other people—when we feel like they've been through what we've been through—the boundary between the core personal self and the social self becomes more porous, leading to fusion.

Psychologists have shown that wherever you go in the world people are highly fused with their families, even if with no other group (Swann et al. 2014). It makes some evolutionary sense that sharing tough experiences should serve as a way of fusing kin groups—after all, in ancestral conditions, the people with whom you shared life's struggles would typically have been your kin. But what may be happening, at least with some rituals, is that they hijack this fusion mechanism (Whitehouse and Lanman 2014). Painful or frightening initiation rituals, for example, serve as life-changing experiences that we never forget—and because they are also causally opaque, we reflect deeply on their meaning and significance. Initiations shape our autobiographical selves but they also make us feel that we share these experiences with others who have gone through the same rituals. This bonding mechanism has been used for thousands of years in small-scale societies, especially ones that needed to bind together young men so that they'd stand by each other on the battlefield or when engaging in other high-risk pursuits like hunting large and dangerous animals (Whitehouse and Hodder 2010).

By contrast, doctrinal rituals are all about creating social identities that are separate from our personal identities. Imagine that the most important rituals for your group are conducted on a daily or weekly basis—like calls to prayer or Sunday services. What this means is that your knowledge about the group's beliefs and practices is stored in your semantic memory—it's part of your general knowledge of the world. You couldn't remember every single call to prayer or Sunday service as a distinct experience; instead you have a set of prototypes in your head telling you how those things should be done. And those prototypes are essentially depersonalizing—they specify who does what in terms of roles and functions rather than actual people. (The priest does this and then the congregant does that—but not Fred does this and Wilma does that.) And so we enter the world of large-group thinking and identification with groups (Whitehouse and Lanman 2014).

Quantifying ritual

Although much had been learned about modes of religiosity by poring over case study material, an obvious problem with such an approach was that of selection bias. To avoid the charge of cherry-picking case studies that fitted the theory, we needed to come up with a more objective way of testing our hypotheses. About ten years ago one of my postdoctoral researchers, Quentin Atkinson, spearheaded the construction of a database of 645 rituals taken from seventy-four cultures around the world (Atkinson and Whitehouse 2010). For each of the rituals in our database, we coded for approximately 100 variables—allowing us to test the principal predictions of the modes theory statistically, and without any danger of cherry-picking convenient examples. This not only confirmed some of our core predictions; we also discovered something else that turned out to be important *archaeologically*—namely that as rituals become more frequent and less dysphoric, agricultural intensity increases. This pointed to the possibility that the transition from foraging to farming in the early Neolithic may have been linked to the rise of the doctrinal mode.

In an effort to find out whether a basic shift from imagistic to doctrinal dynamics did indeed feature in the rise of agriculture and the evolution of social complexity, Atkinson and I began working with archaeologists Ian Hodder and Amy Bogaard to construct an archaeological database covering a large chunk of Western Asia from the end of the Paleolithic to the beginning of the Bronze Age. What seems to be happening with the invention of farming is a gradual transition from imagistic to doctrinal patterns of ritual and group formation (Whitehouse and Hodder 2010; Whitehouse et al. 2013). But we could never have known this without a huge amount of labor to reorganize the archaeological evidence in a way that could be analyzed statistically. The archaeological record is great for looking at patterns over very deep time, but the available evidence from prehistory is incredibly patchy. So we've also been building databases on the *recorded* past.

Soon after we began building our archaeological datasets, I met Peter Turchin at a cultural evolution conference. Not only did he share my excitement about the idea of building longitudinal databases to study the origins of social complexity but also he had the statistical and modeling expertise I lacked. Together we applied for funding to hire historian Pieter Francois to help us create SESHAT: Global History Database (Turchin et al. 2012, 2015; Currie et al. 2015). Over the past five years SESHAT has been growing at an astonishing rate and earlier this year passed our first major milestone of 100,000 data points. SESHAT is now enabling us to quantify various aspects of ritual behavior and relate these to the evolution of social complexity. Our database will eventually be global in reach and go back in time as far as possible for each region coded. The basic idea is to assemble what we know about human history in the same way that GenBank has enabled the biosciences to organize and store our knowledge about gene sequences. So in the end, we'll have a vast storehouse of information about the evolution of social complexity that can be searched using statistical tools.

In an effort to understand how people become fused with the group, we have also been running experiments in our Oxford lab, building on early studies in which we attempted to manipulate fear in artificial rituals (Whitehouse, Richert, and Stewart. 2005). Obviously, ethics boards tightly circumscribe what we can do to induce pain and fear in our human subject pools. So now we're also increasingly taking our lab measures out into the field (Whitehouse 2012b). For example, we are going to places where people naturally go through extremely dysphoric experiences together—in regions as far apart as Japan (where some rural communities subject their young people to ritual ordeals by ice and fire) and Mauritius (where thousands of Hindus, as in other regions of the world, pierce their skin with hooks, often attached to chains used to drag heavy objects behind them, and parade for hours in the searing heat). More recently still we have been studying the agonies of defeat on the football pitch. We have found that the worst performing football teams in the UK Premier League have a more tightly bonded support base than the more successful teams and that this bonding seems to be an outcome of shared dysphoria in highly ritualized settings, involving a host of causally opaque identity markers and conventional behaviors. In all these studies we measure not only fusion with the group but also people's willingness to fight and die for group causes and to harm members of opposing groups in various ways. So at the same time as we're learning about the psychology of group bonding, we're also gaining new insights into the drivers of intergroup conflict.

From fusion to fanaticism

When people are highly fused, they will seemingly stop at nothing to protect their fellow group members from external threats. For some years, Bill Swann and I have merged our research teams (comprising lead researchers Michael Buhrmester, Oliver Curry, Jonathan Jong, Jonathan Lanman, Ryan McKay, Brian McQuinn, Miriam Matthews, and Valerie van Mulukom) to investigate the imagistic pathway to fusion and its consequences for extreme progroup action (Whitehouse and Lanman 2014). Our core hypothesis is that painful or frightening ordeals are especially memorable and personally transformative, all the more so if such experiences are causally opaque (Whitehouse [1996] 2014). When people reflect deeply on shared experiences of this kind, they shape not only the personal self but also the group, rendering the boundary between them more porous and producing identity fusion (Jong et al. 2015). Thus when the group is threatened, it feels like a personal attack. If so, maybe when jihadi fighters and other religious extremists blow themselves up in suicide attacks, it is not so much an act of unprovoked aggression against innocent civilians but an act of defense against an out-group threat.

To investigate these psychological processes in more depth, one of my doctoral students, Brian McQuinn, set off for Libya in the middle of the 2011 revolution to study the role of imagistic practices and other forms of shared suffering in the formation and cohesion of revolutionary brigades. Many thousands of volunteers were gunned down by Gaddafi's forces and yet the survivors battled on. Once established in the field, McQuinn managed to spirit me into Libya and together we surveyed 179 surviving members of four battalions (Whitehouse et al. 2014). Our goal was to measure fusion with various groups: family, members of one's battalion, all fighters in the revolution, and those who supported the revolution but didn't participate in it. Roughly half the sample comprised frontline fighters and the other half were providers of logistical support within the battalion (e.g., they drove or repaired ambulances). Our results showed that the overwhelming majority of revolutionaries were fused with their families, with their battalions, and with the members of other battalions. Only a very small proportion (less than 1%) of all revolutionaries was fused with supporters of the revolution who didn't take up arms. In other words, simply being on the same side ideologically (sharing the same beliefs and goals) didn't predict fusion. What really seemed to matter was having gone through the intense fear and pain of warfare by virtue of being in a revolutionary battalion.

Our study went further, however. We asked all participants in the survey to say which group they would choose as their primary fusion target if they could only choose one of them. In other words, we used a forced choice question—they had to say which group they were the most fused with. And here we found a striking difference between our two samples. Nearly half of frontline fighters chose their battalions over their families as the primary fusion target. By contrast, only 28 percent of those who provided logistical support chose battalion over family. One possible interpretation of this finding is that frontline fighters were more fused with each other because they had undergone more intense, self-shaping experiences together.

The insurgents we spoke to in Libya were willing to lay down their lives for each other without a moment's hesitation. It seems that when people fuse with each other and with a cause, they become emboldened by a sense of invincibility. If that process involves religious beliefs, then it may seem to the casual observer as if extreme *doctrinal* commitment is motivating extreme behavior—when actually the "deeper" underlying mechanism is one of fusion via *imagistic* pathways.

Summing up

Over the past twenty-five years, CSR has sought to disambiguate various patterns of thinking and behavior loosely associated with the label "religion" in order to establish more suitable objects for scientific study. CSR has gone on to establish a set of theoretical foundations and methodological procedures for explaining the fractionated elements of religion in ways that have led to cumulative advancement rather than merely a succession of fashionable interpretive frameworks. In the process, CSR has established new and potentially durable bridges between the study of religion and other fields of scientific research spanning the cognitive, behavioral, and biological sciences.

CSR represents a radical departure from the idea that religion is something unique and monolithic—a "natural kind" worthy of study in its own right. Scientists interested in explaining religion nowadays mostly focus on bite-size features of religion rather than religion as a whole. In this chapter, we have considered ritual as an example of this approach. We have observed some strikingly recurrent patterns of ritual behavior across all areas of human life and throughout the ages: from ancient cults to medieval churches, from revolutionary brigades to modern armies, and from martial arts groups to football fans.

Ritual is popularly misconstrued as an exotic, even quirky topic—a facet of human nature that, along with beliefs in supernatural agents and magical spells, is little more than a curious fossil of prescientific culture. Nothing could be further from the truth. Humans are as ritualistic today as they have ever been. Even the most secular political systems ever devised, for instance those under the sway of historical materialism and its vision of a communist utopia, were as devoted to ritual as any in human history. In part, this is because rituals play a vital role in bonding groups and pitting them against each other. Understanding how these processes operate isn't simply a scholastic exercise. It is vital to fostering more cohesive and successful societies—and also, perhaps even more urgently, managing the destructive and seemingly intractable conflicts that drive us into wars, riots, and revolutions.

Acknowledgments

This is a revised and extended version of a paper first presented at the Engelsberg Seminar on Religion, hosted by the Ax:son Johnson Foundation (Whitehouse

[2015], Explaining Religion and Ritual. In K. Almqvist and A. Linklater (Eds.) *Religion: Perspectives from the Engelsberg Seminar 2014* [261–70]. Stockholm: Axel and Margaret Ax:son Johnson Foundation). In reflecting back on my work over the past quarter century, I would like to acknowledge my intellectual debt to the many doctoral students and postdocs I have worked with during that period. Although there are too many to list, I would like to mention in particular those supported by our EXREL and ESRC projects: Quentin Atkinson, Nicolas Baumard, Michael Buhrmester, Emma Cohen, Oliver Curry, Pieter Francois, Michael Gantley, Patricia Herrmann, Gordon Ingram, Ben Johannes, Jonathan Jong, Christopher Kavanagh, Florian Keissling, Justin Lane, Jonathan Lanman, Ryan McKay, Brian McQuinn, Miriam Matthews, Camilla Mazzucato, Daniel Mullins, Valerie van Mulukom, Adrian Murzac, Martha Newson, Veronika Rybanska, Yvan Russell, Tara Tasuji, Rachel Watson-Jones, Claire White, Dimitris Xygalatas. Much of the work summarized in this chapter was supported by a Large Grant from the UK's Economic and Social Research Council (REF RES-060-25-0085) entitled "Ritual, Community, and Conflict" and a STREP grant from the European Commission's Sixth Framework Programme (project no.043225) entitled "Explaining Religion."

4

The Beautiful Butterfly: On the History of and Prospects for the Cognitive Science of Religion

Uffe Schjødt and Armin W. Geertz

Like the beautiful butterfly struggling out of its cocoon, the cognitive science of religion (CSR) is coming of age. After twenty-five years of struggle and excitement, CSR is on the verge of becoming what it set out to be. It grew from a protest movement in the academic study of religion, to groundbreaking theories and hypotheses, and on to become an intellectually stimulating multidisciplinary and multidimensional endeavor. We may not agree on everything but that is part of what makes CSR a science. Whether designing experiments in the lab or field, or mining ancient texts, or doing deep history, the success of the CSR revolution depends on a vibrant environment of open debate with little regard to senior/junior hierarchies or cherished hypotheses. Science is all about following wherever the evidence leads us. It is about testing our theories and developing hypotheses that can be checked, challenged, replicated, and refined by others. And in the end, slowly but surely, we will establish some actual facts about elusive issues in religious thought and behavior.

Despite all the excitement and commitment and sense of identity, we are still a small, esoteric group of people in a world dominated by big brothers and sisters. So we need to stick together and help each other learn and thrive.

Where did we come from?

CSR was part of a protest movement stemming from the first cognitive revolution during the 1940s and 1950s. Everyone knows the story of the Artificial Intelligence Revolution and Noam Chomsky's groundbreaking criticism of black box behaviorism in American psychology (Gardner 1985: 11). More positively, it was argued that there are in fact cognitive processes that function like a computer based on input/computation/output behavior. Key theoretical insights came from a wide variety of sciences: mathematics and computation (the Turing machine), neuronal network theory, cybernetic synthesis, information theory, and neuropsychological studies of cognitive incapacities. The new cognitive science was multidisciplinary from the beginning: artificial intelligence, education science, linguistics, neuroscience, philosophy, psychology, and anthropology—in other words, the very disciplines incorporated in the logo of the *Cognitive Science* journal founded in 1977 by the Cognitive Science Society.

Although we have witnessed a convoluted development of cognitive approaches in anthropology, linguistics, and psychology ever since the 1950s, with seminal insights from anthropologists such as Stewart Guthrie (1980) and Dan Sperber (1974), CSR came into force during the 1990s signaled by scholars of religion E. Thomas Lawson and Robert N. McCauley's groundbreaking *Rethinking Religion* (1990) and by anthropologist Pascal Boyer's *Tradition as Truth and Communication* (1990). What exactly where they rethinking? What discipline or disciplines, in other words, were they trying to redesign? Obviously, they were challenging concurrent tendencies in religious studies, theology, the history of religions, and the anthropology of religion. But, as we will see later in this chapter, one of the most important challenges facing CSR today—besides exploring and explaining religious cognition and behavior—is to encourage other scholars of religion not only to use the results of CSR research but also to join in on the debate, funnel valuable, sophisticated knowledge into CSR and contribute to theoretical developments. It is, as it were, a give-and-take deal.

What were CSR scholars protesting? They were protesting the blank slate assumptions of anthropologists and the mind-blindedness of religious studies. Furthermore, CSR scholars were providing an alternative to the postmodern, antiscience approaches in cultural studies that also were impacting the study of religion and theology. As Lawson and McCauley noted:

> In the study of religious phenomena (and in the history of religions particularly) research has leaned heavily in the direction of interpretive pursuits at the expense of explanatory theory. Our project aims to compensate for this imbalance. It is orthogonal to much previous research, since in this area scholars have done so little detailed theorizing about religious systems. Indeed, as we have indicated above, many have denied its very possibility! Obscurantism in spades. (1990: 2)

The most important achievement of CSR is that it has shown us how to focus analyses on causal mechanisms rather than remain enmeshed in surface features, an approach clearly inspired by Chomsky. Lawson and McCauley were interested in identifying the cognitive constraints of symbolic-cultural systems, that is, "studying the (usually) unconscious representations of cultural and social forms (and their underlying principles) which participants share" (1990: 3). A well-known claim by Pascal Boyer is that there can be "no theory of what happens in cultural interaction without some strong hypotheses about what is happening in the actors' minds" (1990: viii). Another important achievement of CSR is that it has once again raised classical questions, only now, with different methods.

What is cognition?

This approach to human cognition, however, came at a price. Inspired by the computer analogy, cognition was reduced to mental representations based more or less on logico-mathematical rules in the brain. Pascal Boyer, for instance, argued that mental representations are organized in "if ... then" rules that compete with other rules in mental computations (1994b: 78). Although the computational model has led to breakthroughs, none of the various approaches mesh well with our current knowledge of the brain (Thagard 1996: 107ff; Edelman 1992: 234). Uffe Schjødt has argued that the computational approach exposes scholars to the Binding Problem and the Homunculus Problem (2007; Barsalou 1999). Armin W. Geertz has argued that cognition is much more than mental representations because human cognition is enmeshed in somatic, emotional, social, and cultural networks. Thus, cognition is the sum total of bottom-up and top-down processes (Geertz 2010a; Donald 2001; Frith 2007; Bar 2009; Jensen 2016). Applying stringent scientific paradigms to this more complex model, however, raises challenges in both the laboratory and the field.

What have we achieved so far?

There are basically six "foundational hypotheses" in CSR. Whether they constitute "explanations" is another matter. They have, however, inspired several generations of scholars in a variety of sciences and disciplines. Although the list is somewhat arbitrary, most CSR scholars would accept the following list as representative of the early CSR revolution:

- Epidemiology of representations (Dan Sperber)
- Animism and anthropomorphism (Stewart Guthrie)
- Hyperactive Agency Detection Device—HADD (Justin Barrett)
- Ritual representations (E. Thomas Lawson and Robert N. McCauley)
- Counterintuitive ideas (Pascal Boyer)
- Modes of religiosity (Harvey Whitehouse)

Anthropologist Dan Sperber argued during the late 1960s and early 1970s that individual brains are filled with ideas that determine behavior. These ideas are transmitted from brain to brain and can end up invading a whole population. Sperber claims that "culture is made up, first and foremost, of such contagious ideas. It is made up also of all the productions (writings, artworks, tools, etc.) the presence of which in the shared environment of a human group permits the propagation of ideas" (1996: 1). In order to explain culture, Sperber argues, we must explain "how and why some ideas happen to be contagious. This calls for the development of a true epidemiology of representations" (ibid.). Another anthropologist, Stewart Guthrie, also worked on cognitive theory during the late 1970s. He is best known for his anthropomorphism hypothesis which claims that human beings universally attribute anthropomorphic features to ambiguous phenomena in the world. Religious ideas arose out of this tendency (1980: 185). In fact, Guthrie defines religion as "systematic anthropomorphism: the attribution of human characteristics to nonhuman things or events" (1993: 3). Anthropomorphizing, he argues, is a good bet in terms of survival in a dangerous and often ambiguous world. It is a spontaneous reaction and, thus, ultimately selected for in the evolution of our species. Similarly, we attribute animation to inanimate things and events (5). This ability and bias is a cognitive predisposition that guides our perception in daily life (91). Psychologist Justin Barrett picked up on Guthrie's insights and developed them further. In a series of experiments conducted together with Frank C. Keil, Barrett demonstrates how university students theologically incorrectly anthropomorphize God in their comprehension of narratives (Barrett and Keil 1996). In 2000, Barrett posited what he called a "hyperactive agent-detection device" (HADD)— reformulated more broadly in 2004 as "hypersensitive agency detection device" (2000; 2004a). In drawing on Guthrie's insights concerning the human bias toward an over-attribution of intentional action as the cause of ambiguous events, Barrett argues that HADD is an evolved cognitive mechanism:

> The implication for religion is that the HADD might lead people to posit agents, perhaps of a counterintuitive sort, that are then well-transmitted because of their easy fit within intuitive conceptual systems. Similarly, counterintuitive-agent concepts would be more likely to receive attention and be transmitted than non-agent concepts, because agent concepts are more likely to resonate with agents posited by the HADD. . . . In this way, selective pressure of the HADD might contribute to the prevalence of religious-agent concepts over other counterintuitive concepts. (2000: 31–2)

In his book, *Why Would Anyone Believe in God?* (2004a), Barrett developed his theory on the origins of god concepts to include the evidence on Theory of Mind (ToM). Thus their origins are to be found in the mental tools that evolved in our species: HADD, minimal counterintuitiveness, and ToM.

One of the more complex CSR hypotheses was presented by scholars of religion E. Thomas Lawson and Robert N. McCauley in their groundbreaking book *Rethinking Religion* (1990). In drawing on Noam Chomsky, Lawson and McCauley analyzed religious rituals as formal action representation systems in believers' mental representations about their rituals, their so-called ritual competence (6). This analysis serves as a means of identifying the universal principles of religious ritual. In a second book, in which they deal critically with Harvey Whitehouse's Modes of Religiosity Hypothesis, Lawson and McCauley add the further elements of ritual pageantry to ritual frequency (McCauley and Lawson 2002). Justin Barrett has nicely summarized their hypothesis as a series of sub-hypotheses, namely primacy of agency, right intentions, repeatability, potential reversibility, sensory pageantry, and system centrality (2004a). In a recent article, Barrett has attempted to develop a "cognitive typology of religious action" that touches base with Lawson and McCauley's typology (Barrett 2007b).

In his book, *The Naturalness of Religious Ideas* (1994a), Pascal Boyer argues that in order to explain religious ideas, we must examine the ideas that actually occur in the minds of people, that they store in their memories and then pass on to other people (51). More importantly, he was concerned with the "cognitive constraints" that evolutionary selection imposes on the types of representations that survive and are passed on trans-generationally (11). He argues that many ideas and actions are based on "intuitive unnaturalness" and the use of categories "with seemingly inscrutable characterizations and inferential gaps in their implications" (83). Boyer argues that religious representations are constrained by schematic assumptions from other domains of cognition. He posits four such domains (also termed "repertoires"): the ontological, the causal, the episodic, and the social categories repertoires (42–3). He then posits the ways in which these domains are violated and/or transferred from one to the other and, more importantly, are stimulated by default in religious concepts. The concepts that survive selection through generations are minimally counterintuitive; in other words, they violate or transfer between domains in a minimal way, called "tweaking," which are easier to remember because they draw on default assumptions without being overly bizarre.

In drawing on the past 100 years of anthropological and sociological literature, anthropologist Harvey Whitehouse has attempted to explain the well-known existence of two major types or "modes," as he calls them, of religiosity. They consist of the *doctrinal mode*, which is repetitive and low-arousal, on the one hand, and the *imagistic mode*, which is infrequent and high-arousal, on the other (Whitehouse 1995; 2000; 2002). Whitehouse argues that these types of ritual appeal to two different types of memory, namely the semantic and the episodic. The former is routinized and generalized and the latter is specifically related to personal experiences that lead to spontaneous exegetical reflection. The former leads to more orthodox forms of organization and the latter to more local, nonorthodox forms of organization (2002: 303ff, 309ff). Whitehouse's modes theory has been particularly inspiring to historians who also note these divergent modes in their historical sources (Whitehouse and Martin 2004a; Pachis and Wiebe 2010; Martin and Sørensen 2011).

Then what happened?

These and other CSR hypotheses have proved to be fruitful and the CSR community has grown into a vibrant, however esoteric, body. CSR scholars have been welcomed by psychologists, cognitivists, and neuroscientists, but the reception in their home disciplines has remained low-key.

Both Justin Barrett (2004b) and Ilkka Pyysiäinen (2013) have argued that many of the previously mentioned and subsidiary hypotheses have been supported by experimental evidence, but a persuasive consensus is far from present. CSR suffers from the same problems as the disciplines it draws from and there is a serious need to further debates on the theories, concepts, hypotheses, experimental paradigms, and methods employed as well as on what exactly cognition is. Besides taking into consideration current methodological debates in behavioral psychology on priming (Bargh 2006; Doyen et al. 2012), self-reports (Baumeister et al. 2007), operationism (Green 1992), methodological biases (Chambers et al. 2014), the need for controlled behavioral experiments and non-Christian samples (Musick et al. 2000), the need for global surveys that are not simply restricted to educated students in India and the United States (Henrich et al. 2010), and the crucial need for replication and so on, can experiments based on the assumption of cognition as simple mental representations actually tell us much?

We also have disagreements on what religion is. Most CSR scholars refuse on methodological grounds to attempt defining what religion is. The result is that although religion is "fractionated" to more fundamental mechanisms that allow experimentation, the fractionated elements are nevertheless part and parcel of unspoken, default assumptions about what religion and religious behavior is (Geertz 2016).

Meeting psychology and doing experiments (good and bad things about it)

Many of us have overheard the gossip at study of religion conferences hosted by the American Academy of Religion (AAR), the European Association for the Study of Religion (EASR), the Society for the Scientific Study of Religion (SSSR), and the International Association for the History of Religions (IAHR) that CSR scholars are the new bullies in town. Meeting psychology and embracing its methods is not for everyone. Still, we cannot help feeling that much of the hostility is somehow our own fault, or at least, due to common misunderstandings that we need to explain better and a few unfortunate tendencies that we should fight to get rid of.

Experimentation is a common target. To experimentally demonstrate the causality between central theoretical variables may sound noble, but attempts are often perceived as superficial and arrogant. For some reason, scholars of religion, who have spent entire careers teasing apart historical and ethnographic contexts to understand the complex nature of religious beliefs, get provoked when everybody's attention is drawn to one shot observations in a minimalist and highly artificial laboratory setting.

Very illustratively, the journal of *Religion, Brain & Behavior,* which is spearheading the publication of experimental religion research, has now surpassed long-celebrated journals of religion in terms of impact based on citations, including the *Journal of the American Academy of Religion, Religion*, and *Method and Theory in the Study of Religion* (according to Google Metrics for journals as of September 6, 2016).

The sheer impact of experimental research may be cause for jealousy, but our impression is that the indignation runs deeper, and that it is, at least in part, anchored in stereotypes and prejudice. Consider the following statements. Experimental psychologists with very limited specialized knowledge about religion use the impact of experimentation to buy themselves the right to speculate and make unjustified claims about complex theoretical issues that have been debated in our field for centuries. The evidence itself is based on experiments in which theoretically uninterested participants (university students) are exposed to very general religious stimuli (words like "God" or "prophet") in completely artificial contexts. The results are overemphasized in presumptuous titles, inferential promiscuity, and overly bold interpretations. Add to this stereotype that most high-impact experimental findings have proven difficult to replicate (Harrison et al. 2013; Calin-Jagerman and Caldwell 2014; Gomes and McCullough 2015). No surprise, then, that experimentation is a target of annoyance and ridicule.

Such stereotypical prejudice, however, is erroneous and grossly oversimplified. Experimentation is not a single well-defined approach. Experimental logic is by definition applied to experimental research, but experimentalists do so in widely different ways. Certainly, many researchers run highly abstract experimental paradigms on university students, but others apply the experimental logic in more authentic research environments examining actual religious practices in particular groups. The diverse experimental landscape can be divided into at least three different approaches.

Experimental psychology studies: Standard paradigms

Most studies use abstract religious stimuli in adapted versions of standard experimental tasks and paradigms (e.g., Norenzayan and Shariff 2008; van Elk 2013, 2015; Inzlicht and Tullett 2011). This line of research is focused on experimental control that is a cardinal virtue in experimental psychology. The goal is usually to isolate and test the true effect of one particular variable by controlling for other variables including social psychological confounds such as demand characteristics (i.e., cues in experiments that help participants work out what is expected of them) and social desirability (i.e., tendency of participants to respond in socially desirable ways). For this reason, researchers often use manipulations and measures outside of participants' conscious control, for example, implicit priming, Implicit Association Test (IAT), or the Stroop task (demonstration of interference in the reaction time of a task). While this may sound odd from the perspective of study of religion scholars, who usually find subjectively experienced phenomena to be more interesting, social psychologists tend to ask other questions: How does religion affect human behavior and attitudes? Does religion improve performance? Does it make humans more social? The scope is instrumental and applied value and not just for the sake of understanding. Observed effects therefore have to be generalized to larger populations and not just to a small group of believers. This fits well with the scope of CSR scholars, who are interested in cultural evolution, because general population effects of weak and generic religious stimuli can reveal theoretically important insights on universal human cognitive biases that have contributed to the evolution of particular types of religious ideas (Barrett 2000), behaviors (Liénard and Boyer 2006), and societies (Norenzayan 2013).

The highly abstract and artificial nature combined with its minimal and often subconscious manipulations, however, makes this approach less attractive for scholars interested in specific religious phenomena. For instance, fascinating titles such as *God is Watching You* (Johnson 2016) may sound highly relevant for researchers interested in religious experience, but reading this technical report leaves some readers disappointed and frustrated. The study is not about the feeling of God's presence, the religious stimuli are not recognized as religious by the participants in the study, and the experimental task does not resemble real-life behavior. Abstract social psychological studies contribute well-controlled findings, but what is gained in terms of experimental control is often lost in terms of authenticity and ecological validity. This does not mean that the experimental approach is a dead end. It just means that one should look to other kinds of experimentation.

Religion-by-proxy studies

A second line of experimentation attempts to study religious practices and experiences indirectly, for example, by using hypnosis and illusion paradigms. The idea here is to study religion by proxy or analogy, examining psychological phenomena that resemble theoretically important religious phenomena such as trance, possession, felt-presence, suspension of disbelief, mind reading, anthropomorphism, and out-of-body experiences (e.g., Deeley et al. 2013, 2014; Blanke et al. 2014; Ali et al. 2014). Research questions include: What cognitive processes support and elicit trance states? What is the perceptual basis of out-of-body experiences? Under which conditions do humans experience apparitions?

Religion-by-proxy research is not limited to the study of extraordinary experiences. It has also proved valuable in the study of rituals. Kristoffer L. Nielbo and Jesper Sørensen (2011; Nielbo et al. 2013), for example, have analyzed the core behavioral constituents of ritualized behavior, for example, elements of goal demotion, non-functionality, rigidity, and so on, to experimentally study how these basic building blocks affect perception in observers. Their observations point to specific pathways by which ritualized behavior acts on practitioners' perceptions, experiences, and interpretations.

Religion-by-proxy research enables researchers to experimentally investigate phenomena associated with religion in well-controlled paradigms and, notably, without having to deal with the serious issues of demand characteristics and social desirability that haunt studies in which religious participants practice their religion while they are being observed (see the following "Authentic religion studies"). Yet it fails to study the real deal. Whether participants deem practices and experiences as relevant to their religious worldview or not is likely to affect how they are processed cognitively. The complex patterns of religious beliefs, attitudes, values, and expectations must somehow influence the way practices and experiences are processed by believers.

Authentic religion studies

A third approach attempts to examine actual religious practices and experiences directly in the laboratory (e.g., Schjødt et al. 2008, 2009, 2011; Andersen et al. 2014) or in the field (Konvalinka et al. 2011; Xygalatas, Mitkidis, et al. 2013; Xygalatas, Schjødt, et al. 2013). Laboratory studies allow for comparatively more control and more sophisticated measures, for example, eye tracking and neuroimaging, whereas field studies sacrifice control and measurement possibilities in favor of a higher degree of ecological validity.

Conducting controlled studies of practices and experiences deemed religious by participants is difficult and requires extensive work far beyond the experiment itself. Recruiting from particular religious groups necessitates active field work and rapport with participants. Creating a context in which participants can have authentic experiences involves interviews and collaboration with informants. Figuring out ways to measure theoretically interesting aspects of behavior, physiology, and cognition in concrete religious practices may require advanced equipment and the concomitant skills to use it.

These challenges call for interdisciplinary research teams. Primary investigators typically come from traditional comparative religion research (history, anthropology, and psychology of religion) with an academic interest in a particular group of people, beliefs, or practices. Collaborators include experts on specific psychological measures, technicians to handle specific equipment, and sometimes statisticians to analyze complex data sets.

The fact that this approach is the only one that examines actual religious practices and experiences may make it look superior to religion scholars, but the approach is seriously hampered by issues of demand characteristics, social desirability, and lack of experimental control. Contrary to the more abstract approach, what is gained here in authenticity is often lost in experimental control. Yet, it enables researchers to apply experimental logic, control, and measurements on culture-specific aspects of religion. Perhaps it is best understood as a new tool in the study of religion that supplements traditional research on distinct groups of believers, ideas, and practices.

A guide for CSR scholars and adversaries

These three kinds of research show that experimentation in the study of religion is not a single well-defined approach. Each of the three approaches suffers from varying degrees of ecological validity, demand characteristics, social desirability, and experimental control. Specific points of criticism rarely generalize to all three approaches. If critics dislike the abstract designs because they seek authenticity and cultural specificity, they have a good case against standard experimental psychology but a weaker case against authentic religion studies. On the other hand, if one's interest is in the evolutionary history of human societies, subtle cognitive biases revealed in highly controlled experimental psychology may offer important explanatory models. If critics are skeptical about the effects of social desirability in authentic religion studies or simply have a general interest in the psychological nature of religious experience, studies that examine religious phenomena by proxy may be more attractive. Yet, if the aim is a detailed understanding of the practice of a particular group or practice, authentic religion studies may be the best way forward.

The approaches complement each other. One solves the methodological problems of the other. As is probably the case in most scientific methods, no single approach is superior. The ability to distinguish between them enables CSR scholars, as well as their adversaries, to identify the kind of studies that match their interests, but also to deliver qualified criticisms of specific types of experimental research, rather than lumping experimentation together in a general but misunderstood criticism.

What's next? Multidisciplinary integration

For scholars who do accept the potential value of experimentation, other questions arise. One of the biggest challenges of the coming decade will be to successfully integrate CSR research within the broader history of religions community. Pioneering attempts include the work by Edward Slingerland on early Chinese conceptions of mind and body (2013) and recent ambitious projects that test cognitive theories on large-scale historical databases (e.g., Turchin et al. 2012; Watts et al. 2015, 2016). But how can historiographers use insights from experimentation to understand particular historical data? Are they incompatible disciplines? We think not. Take, for instance, the case of Christian healing testimonies.

Christian healing testimonies is a traditional genre in religious literature, which is most often held by historiographers to be highly biased healing accounts published for the purpose of proselytizing. Anthropologists have added to this assumption the social functionality of testimonies, for example, for identity making, bridge burning, and conversion (McGuire 1977). Interestingly, the concrete healing experiences reported in testimonies are often ignored.

Cognitive research on expectation modulation has led to experimentation on charismatic intercessory prayer that represents an important context for reported healings in testimonies. In one study by Schjødt and colleagues (2011), the prayer experience and neural response to charismatic healers were found to be strongly modulated by participants' beliefs and expectations (ibid.), a finding that corresponds with experimental insights on other kinds of complementary and alternative medicine, suggesting that intercessory healing in testimonies may in fact point to very concrete healing experiences, but only in so far as the diseases reported to be cured are susceptible to expectation modulation, for example, diseases with highly subjective symptoms but little or no physiological correlates (Kaptchuk 2002).

Motivated by this insight, Paldam and Schjødt (2016) examined a large collection of healing testimonies to see if they could find evidence of expectation effects in participants' healing accounts. Well aware of the proselytizing biases and other biases particular to the genre, reading through almost a thousand cases of healing, their prediction was that most accounts would report experienced health effects on the kind of diseases that are most susceptible to expectation modulation. Support was indeed found for this hypothesis. Biblical miracles of the more dramatic kind were almost (but not completely) nonexistent, while back pain was by far the most frequent ailment reported to be cured in the testimonies. Thus, even in highly biased proselytizing materials such as healing testimonies, testifiers tend to report rather mundane effects of intercessory prayer, supporting the idea that testifiers in fact report concrete healing experiences rather than echoing the much more powerful miracles reported in the Bible.

Together, this small cluster of studies illustrates how experimental research can inform historiographical analysis by advancing new hypotheses. In this case, by focusing attention on testifiers' concrete healing experiences rather than simply discarding them as mere genre biases. The example is tailored to make a case for collaboration, but integration of historiography and experimentation, however, is not always feasible (see the methodological discussion in Houston and Frith 2000). Direct experimentation on a particular cultural-historical phenomenon often requires that it has a contemporary counterpart. Christian healing testimonies and prayer meetings have been around for centuries and still exist to be examined, but many ancient practices no longer exist. Viable experimental approaches are then limited to religion-by-proxy paradigms in which contemporary practices are used as proxies.

Conclusion

It is wonderfully safe to mature in a cocoon without too much influence from the outside. But CSR has grown to a point now where it has to interact creatively with its surroundings. The study of religion community can be a harsh and unforgiving habitat that does not always recognize obvious advances over previous ones. A mature field, now competing with other disciplines, CSR needs to persuade the broader community of its applicability.

Triumphalism is tempting for a butterfly this beautiful, but arrogance and inferential promiscuity is a real threat to its existence. We should praise other scholars of religion for being immune to overinflated promises of quantitative data. CSR must avoid getting dragged down with other psychological sciences by the current replication crisis (Bulbulia et al. 2016).

Multidisciplinary integration is necessary in order to make CSR truly relevant for scholars in the humanities. In the case of experimentation, perhaps we need to change the way experimentation is conceptualized. The ability to generalize findings to larger populations is not necessarily the main goal for scholars in the humanities, who often attempt to gain a richer understanding of a very distinct group. Experimental research in the humanities does not have to comply to the same ideals as social psychologists because our questions are often different. Rather, we may adapt the experimental approach to our needs. Authentic religion studies, as presented here, may represent such an adaptation. They only become possible when anthropologists combine their ability to describe and recruit distinct groups of people with the skills of experimental psychologists to examine the cognitive processes involved in concrete practices and experiences. By accepting the uniqueness of a group rather than treating it as a sample of larger populations, perhaps, theoretically interesting groups of participants should be understood as populations in their own right. Humanistic experimentalists, then, may conduct a kind of census experimentation that has less to do with generalization and more to do with understanding the participants.

This is obviously part of a much broader discussion, but it clearly illustrates how CSR is bound to transform itself as it integrates with the study of religion. Fly, fly, beautiful butterfly!

Part Two

State of the Art

5

Religion Explained? Some Variants of Cognitive Theory

Stewart Elliott Guthrie

Introduction

The cognitive science of religion (CSR) is widely considered the growth field in religious studies. Still relatively new at some three and a half decades, it has an international association and journal, two more journals largely devoted to it, centers or programs at a number of major universities on several continents, and a freshet of papers and books.

Several laudable features characterize the field. Scholars in CSR, more than those elsewhere in religious studies, treat religion naturalistically. They typically assume, for example, that it is in some way a product of natural selection. Most also agree that generalizations about religion must recognize cross-cultural variation. Further, they often use experimental and statistically controlled methods, which have helped prove useful even for this humanistic and relatively elusive topic. These scholars, moreover, appear confident about prospects for their field (Whitehouse 2008; Rydving 2008; Pyysiäinen 2013).

Despite its rapid growth and other marks of success, however, the claim that CSR is a science invites scrutiny. Such scrutiny suggests that it is (in the sense of Kuhn 1970) pre-paradigmatic in that its theories and definitions are diverse and contradictory (Guthrie 2007, 2016a,b). Some critics accordingly describe it as a "heterogeneous group of loosely interrelated … perspectives" (Rydving 2008: 73) that offer a "raft" (Dawes and Maclaurin 2013: 21) of theories. Although its scholars largely assume, for example, that explanations of religion should be compatible with evolutionary theory, they do not agree on *what* evolves or why (Bulbulia et al. 2008).

Most important, CSR, like religious studies more broadly, has no general agreement about either a definition or a theory of its distinguishing term and central explanandum, religion. Neither does it have a consensus that this term corresponds to any objective phenomenon (Schilbrack 2010), much less a culturally universal one. Without some such agreements in something like a standard model, CSR as a science must be called promising but nascent and inchoate.

A fundamental question for CSR, as yet not clearly answered, is whether the concept "religion" is cross-culturally coherent. Many scholars of religion (Smith 1982, 1993; Asad 1993; McCutcheon 1997; Fitzgerald 1997; Bloch 2008; Martin 2008) think it is not. If not, then should we even bother with religion, or instead choose to study some recurrent human phenomenon that *is* coherent? Bloch (2008) advocates the latter, recommending abandoning religion as a topic in favor of what he considers an *actual* human universal, the "transcendental social," that is, "essentialized roles and groups" (ibid.: 2056). Other scholars (Saler 1993; Nongbri 2013) agree that the concept of religion is culture-bound but think it can be rehabilitated by defining it in terms of family resemblance or current usage.

Facing the ongoing definitional problem, many CSR scholars (e.g., Mithen 1999; Barrett 2000; Boyer 2001; Bulbulia 2004; Whitehouse 2008; Pyysiäinen 2013; Norenzayan 2013), who often seem to assume that religion *is* a cultural universal, define it (usually implicitly) by invoking other Western terms, especially "supernatural." But "supernatural" itself is culture-bound (Durkheim [1915] 1965; Hallowell 1960; Lienhardt 1961; Horton 1967, 1993; Luckmann 1976; Saler 1977, n.d.; Klass 1995). It also is vague, often meaning little more than "false." Thus, using it to identify religion merely bases one culture-bound and dubious term on another.

Four kinds of theoretical approaches—the relationship of which is indeterminate and uneasy—in CSR stand out. Two of these may be called "strongly" cognitive, in holding that the central task for CSR is to explain the existence of religion by invoking universal features of human cognition. I shall return to these two.

A third approach is cognitive only in what seems a weak sense. Its proponents seek to discover not so much what features of cognition produce religious thought and action as, given the existence of such thought and action, what benefits to society direct and sustain them. Because of its emphasis on such benefits, this third approach could as well be called functionalist—or, as its proponents usually prefer, adaptationist—as cognitive. (It has also been called the social- or structural-functionalist, symbolist, or social-glue approach.) Its adherents (e.g., Bulbulia 2013a; Norenzayan et al. 2008) often accept a spandrel account of religion's genesis, but their main concern is neither that genesis nor its cognitive basis.

Instead, social functionalists, or adaptationists, maintain that religion is a means to create social solidarity. They have long precedent in this. Recognition that religion can unify is old, appearing in China as early as the turn of the first millennium BCE, when the new Chou Dynasty announced to the people of the conquered Shang that its conquest showed it had acquired the Mandate of Heaven. In the West, Polybius wrote in the first century BCE that religion is the chief foundation of social order, a claim repeated to the present, recently, for example, by sociologists and some anthropologists.

The claim has some truth but also problems. One problem is that communities of religion may raise conflicts of loyalty with other forms of community such as those of kinship, residence, politics, or culture. In these cases, religion is not unifying but divisive. Another problem is that ethics and morality are not always connected with religion, as the thesis requires. Gods in unstratified societies, for instance, often lack morality and do not expect it in human relationships.

More fundamental is another problem, common to all functionalism: showing how a society identifies, creates, and sustains institutions that are useful, as adaptationists think religion is. Most adaptationists imagine institutions already in existence and assume that whatever persists does so because it is useful, an assumption that may stem from an implicit further assumption that evolution is a designer. The notion of design, explicit in Darwin's writing, was criticized as anthropomorphic even in his time. Darwin responded that it was not meant literally, but was mere shorthand.

Although some anthropologists and others continue to explain religious belief by what it does for society or culture, they face two further problems: that many other beliefs perform the same functions and that functionalism "tends to leave out religion in modern industrial societies, where [secularization has] eroded all the culturally unifying functions" (Douglas and Perry 1985: 416–17). Penner (1989), specifically addressing religious studies, calls functionalism "bankrupt." Functionalists and adaptationists might do well, then, to heed Francis Crick's (1990: 138) recommendation that biologists "constantly keep in mind that what they see was not designed, but rather evolved."

Other theorists (e.g., Whitehouse 2008; Jong 2013) portray religion as a cultural universal, but one that can be explained only by fractionating it and accounting separately for each fraction. This view conflicts with the standard forms of the two strongly cognitive approaches alluded to previously, in that the latter tend toward globalism regarding religion, and their proponents aspire to produce general theories. Whitehouse (2008: 27), for example, writes that "[a]ll human populations share a common set of religious concepts and behavioral patterns" but denies that any single theory is applicable to these concepts and behaviors.

The set of features Whitehouse proposes as fractions of religion is representative. It includes afterlife, beings with special powers, signs and portents, creationism, spirit possession, rituals, ritual exegesis, the sacred, deference, moral obligations, punishment and reward, revelation (2008: 32). As a group, most of these do seem consistent with "religion" in current Western usage.

Yet most non-Western languages lack this term, and most cultures lack the concept (Nongbri 2013). Westerners and non-Westerners thus often misunderstand each other. Pre-Western Japan, for instance, lacked both word and concept for religion (Josephson 2012) and responded to Western demands for laissez-faire toward "religion" by translating it as *shūkyō*, meaning "sectarian teachings." The Japanese government deemed this phrase to include Buddhism but not Shinto, a distinction that still is popular in Japan but is at variance with Western understandings. Thus, confidence in the cross-cultural applicability both of the concept of religion and of its nominal constituent features seems unwarranted.

Another problem with trait lists like the preceding is that many of the features named—ritual, deference, punishment, signs, morality, even afterlife beliefs (Bering 2002) —occur equally in secular life. Thus if cognitive scientists are to address religion, they still need to say what they mean by it. Yet Whitehouse (2008: 19) writes that "[p]roblems of defining 'religion' can be set aside" and that whether certain traits "happen to be classified as 'religious' … is of little importance" for explaining them. But if classification as "religious" is of little importance, what is our justification (as Rydving

[2008] asks) for using this Western concept cross-culturally? And if the same or similar features *do* recur cross-culturally in sets that somehow cohere, why is this?

Here I address primarily the two approaches I have called strongly cognitive, with attention to what seems their incompatibility. I view their present partial alliance (e.g., in Barrett 2000; Boyer 2001; Atran 2002) as paradoxical and affirm one over the other.

Current theory

The two strongly cognitive theories of religion now prevalent in CSR, and often combined although seemingly contrary, may be called "intuitivism" (in the sense of Sperber 1996: 89, in which intuition comprises "spontaneous and unconscious perceptual and inferential processes") and "counterintuitivism," usually called "minimal counterintuitivism" or MCI. Intuitivism, whose pedigree dates at least to de Spinoza ([1670]) 1951) and probably to Bacon ([1620] 1960) or even Xenophanes (fifth century BCE), holds that religion consists of conceptions of, and actions directed toward, other-than-human persons. (The latter commonly are glossed, imprecisely, in CSR as "intentional agents.") It holds further that these conceptions and actions stem from intuitive predilections and are conceived and performed largely unconsciously and automatically.

Intuitivist theory is concerned largely with how religious belief and behavior is generated. It emphasizes the uncertainty of perception, our need to discern any persons present, and the chronic difficulty of distinguishing persons from nonpersons. In the face of this difficulty, it holds, we involuntarily err on the side of caution. Because the presence of persons usually is the most important possibility and because persons often operate from behind the scenes, we have evolved a high sensitivity to such presence and a low threshold for finding it. Thus we imagine persons of endless sorts, human and other-than-human, and find them everywhere.

The second strongly cognitive theory, "minimal counterintuitivism" or MCI theory, currently is regarded by many writers in CSR (e.g., Murray 2009: 460; Pyysiäinen 2013) as the standard model in the field. It is much younger than the first theory, dating back not some four centuries but only about two decades. The theory derives from "cultural epidemiology" (Sperber 1985b, 1996), which proposes that culture reproduces and spreads much as does contagious disease. As do germs, cultural concepts are said to reproduce randomly. Unlike germs, however, their spread and retention depend largely on their memorability. Religious ideas are especially memorable because they combine a universally counterintuitive element (such as invisible agents) that makes them salient, with a universally intuitive element (such as agency) that makes them understandable.

A leading advocate of MCI theory, Pascal Boyer, adds that counterintuitive is a "technical" term (2001: 65). It does not mean "strange, inexplicable, funny, exceptional or extraordinary." Instead, it means "including information contradicting … ontological categories" (2001:65) or "templates." Ontological categories (which, according to Boyer [2001: 78], comprise animal, person, tool, natural object, and plant), or templates, are intuitive and universal. They entail features that, according to the theory, also are universal (and thus at least maturationally normal) human intuitions.

For example, the templates—animal, plant, and person—entail being alive, needing nourishment, growing, becoming old, and dying. Any concept of an animal, plant, or person that does not display these features, such as a plant that does not die, is said to be counterintuitive. Moreover, the same features (being alive, etc.) intuitively belong *only* to these three templates, so that a concept that belongs to some other template, such as object, but that includes living, eating, growing, or dying, also is counterintuitive. Concepts are easiest to remember if they violate only one or two requirements of their template; hence the name "minimal counterintuitivity" or MCI.

Despite relatively wide acceptance in CSR during its two decades, MCI theory has drawn criticism (Carey 1995: 275–82; Beit-Hallahmi 2002; Guthrie 2002: 41ff; 2007: 53–5; 2016a, b; Näreaho 2008; Smith 2009; Rydving 2010; Russell and Gobet 2013; Emmons 2016; Purzycki and Willard 2016,). Critics note problems at varied levels. These include the foundational questions of what "counterintuitive" means and whether the ontologies and entailments claimed by the theory are in fact intuitive and universal. Purzycki and Willard (2016), for example, point out that some writers in MCI theory find it counterintuitive to postulate persons who are omniscient, or a cow that swims, while others do not. In another contradiction, preschool children think people are immortal (Slaughter et al. 1999), yet MCI theory holds that immortality in living things is counterintuitive. Similarly, Carey (1995: 278) doubts that knowledge that animals necessarily grow is intuitive, noting that young children do not grasp it, even concerning humans.

Carey also notes a related issue, "'the problem of perception.' The issue is: On what basis does the input module categorize entities as animals? [Scott Atran's] folk-biology module will be useless unless the cognitive system can identify the animals in the world" (1995: 278). Carey points out that no formal features are reliable indicators of animacy and that infants accordingly try to establish social relationships even with mobiles, because these respond contingently to their kicks. She concludes (279) that rather than formal features, what is salient for perceiving agency is intentionality: "A plausible solution to the perception problem is that animals are identified by the module that picks out beings with intentional states, i.e. beings capable of attention to ... goal-directed activity." Such a module, she adds, would belong not to folk biology but to folk psychology.

If distinguishing animals from the inanimate environment depends on detecting intentionality, then distinguishing *kinds* of intentionality would be necessary to the theory. The theory does not address this necessity, yet the distinction is subtle. It apparently is learned and not universal. For example, numbers of developmental psychologists (Carey 1985, 1995; Carey and Spelke 1996; Inagaki and Hatano 1987; Hatano and Inagaki 2002) write that preschool children do not have this distinction but rather see animals as persons. Adults also, among foragers and horticulturists (Hallowell 1960; Harvey 2014)—and probably all adults, unconsciously—typically see them as persons. Yet MCI theory maintains that the distinction between persons and nonhuman animals is intuitive and universal. If it is not, then animals that talk, for example, are merely odd, not counterintuitive.

Another important MCI claim is that attributing mind to lifeless objects such as artifacts is counterintuitive. However, researchers across a number of disciplines again contradict this claim. Anthropologists (Gell 1998), cognitive neuroscientists (Farah and Heberlein 2007; Heberlein 2008; Schilbach et al. 2008; Cullen et al. 2014), linguists (Cherry 1992), psychologists (Gigenrenzer 1997: 275; Wegner 2005: 22; Lillard and Skibbe 2005: 282) and philosophers (de Spinoza [1670] 1951; Hume [1757] 1957; Nietzsche 1966; Dennett 1987) have written that people ascribe mind to objects involuntarily, promiscuously, and automatically.

Yet one more claim (e.g., Pyysiäinen 2013: 7) important for religious studies—and pivotal for MCI—is that agents without bodies, which include most gods, ghosts, and demons, are counterintuitive. This too is contradicted by an array of scholars: anthropologists (Lohmann 2003; Cohen et al. 2011), linguists (Lakoff and Johnson 1999), philosophers (de Spinoza [1670] 1951; Hume [1757] 1957; Nietzsche 1966; Leder 1990; Johnson 2006), and psychologists (Bering 2002; Bloom 2004; Wellman and Johnson 2008; Roazzi et al. 2013) have written that mind-body dualism, in which mind has both priority and separability, is ubiquitous and intuitive.

If so, then the independence of mind from body displayed by gods is more parsimoniously explained as the same independence of mind that we attribute to ourselves and other humans. In light of inconsistencies such as those previously mentioned, Russell and Gobet (2013: 741) suggest that most MCI writers mistake their own personal intuitions about ontology for universal ones; and Beit-Hallahmi (2002: 309) writes, in an otherwise favorable review, simply that "Boyer describes religious ideas as 'counter-intuitive,' but their universality shows that these concepts are actually natural and intuitive."

MCI theory not only disagrees in these ways with much other scholarship but also says nothing of how or why ideas about other-than-human persons such as gods and ghosts arise (Guthrie 2016b). Rather, in MCI these ideas seem to arise spontaneously and unpredictably, and survive (or not) depending only on their memorability. Ethnographers, however, describe other-than-human persons as differing from humans in only a few ways, such as having no body (or having one that is nonhuman) or being invisible, immortal, or more or (as are many goblins, imps, and demons) less knowledgeable or powerful. Their beliefs and desires otherwise resemble those of humans. This regularity cannot well be explained by random generation. Yet (as Anttonen [2002: 27] puts it) scholars of religion have no "more serious task [than] identifying the mechanisms that generate" religion, to which such persons are central.

Perhaps for this reason, such advocates of MCI theory as Barrett (2000), Boyer (2001), and Atran (2002) combine MCI with a theory that *is* generative, namely that of the "hyperactive agent detection device," or HADD (Barrett 2000). This phrase, which Barrett (e.g., 2000) says is based on Guthrie (1980, 1993; but see Boyer 2001, who attributes the core argument to Barrett), represents a compressed version of the theory I propose. HADD holds that in the face of phenomena of uncertain origin, the "device" in question assumes that these are caused by one or more unseen agents, on the principle, better safe than sorry. It does so "reflexively" (Barrett 2007a: 6)--that is, automatically. HADD and MCI thus both attempt to account for the same thing—belief in the unseen agents of religion—but in opposing ways: the former appeals to intuition, but the latter to counterintuition.

HADD itself, though widely accepted in CSR, seems unnecessarily modest in its claims. Most important, it posits belief only in a conceptually minimal "agency" rather than in the rich and complex *persons* of religious belief and behavior. People do not pray to mere agents, but more specifically to persons. Furthermore, developmental psychology and linguistics (Meltzoff and Moore 1977; DeLancey 1990; Cherry 1992; Carey and Spelke 1996; Inagaki and Hatano 1987; Hatano and Inagaki 2002; Meltzoff 2007) support Hume's ([1757] 1957) and Nietzsche's (1966) assertions that the prototype of "agent" is the human person, specifically the personal mind. If so, then we can profitably substitute the inferentially rich concept "person" for the minimalist concept "intentional agent."

A last apparent deficit of MCI theory is that, although it belongs to a field that acknowledges the importance of Darwinian evolution for understanding human cognition, it proposes no selective advantage for counterintuitive beliefs. (One might grant that, if true, such beliefs would be informative and thus advantageous; but by definition these beliefs are "counter-ontological" [Boyer 2001: 65] and thus presumably false and disadvantageous.) Moreover, the memorability of unusual information already is accounted for by information theory, so the counterintuitivity hypothesis seems redundant.

A cognitive theory revisited

I turn finally to the second strongly cognitive theory mentioned previously (Guthrie 1980, 1993, 2002, 2007, 2016b; Koch 2009; Saler 2009; McKay 2015), which builds on four centuries of scholarship identifying religion as anthropomorphism. More originally, this theory offers an evolutionary explanation of the latter. My explanation assumes that religion is a culture-bound concept and that "religious" and "secular" thought and action are, as human phenomena, indistinguishable. In contrast, anthropomorphism—the attribution of human (and primarily mental) characteristics to nonhuman things and events—is a cultural universal. More, it is a *human* universal, the kind that Bloch (2008) recommended studying instead of religion and that Durkheim ([1915] 1965) said would be needed to explain religion as an illusion.

The argument has grown stronger over time, but its basic claim remains the same: that natural selection has favored our interpreting a mysterious world as personal, because doing so risks little when wrong and gains much when right. What has been selected in us (and in other animals; Darwin 1871; Ristau 1998: 139; Foster and Kokko 2009) is a cognitive bias that follows the logic of Pascal's Wager: when in doubt about what confronts us, bet first on what matters most. For most higher primates, certainly including humans, that usually is the presence, traces, or features of a conspecific. Thus we see, for example, faces, signals, and purposes everywhere.

This argument is not the whole story of anthropomorphism. The human imaginative focus on self and other persons has deep roots in both nature and nurture. We form our notion of agency as infants, through intimate communication with caretakers, and we conceptualize ourselves and them simultaneously (Meltzoff 2007). But how can our resulting bias toward personal models be adaptive in a largely impersonal world?

Much of my argument has precedent. Philosophers of religion from at least de Spinoza ([1670] 1951) to Nietzsche (1966) theorized religion as anthropomorphism, as have anthropologists from Tylor (1871) to Horton (1967, 1993) and Lévi-Strauss (1966), and psychologists from Freud ([1927] 1975) to Spiro (1966) and Kelemen (2004). Most called anthropomorphism intuitive and largely unconscious. They gave, however, different explanations for it, none of which seems complete. Moreover, most scholars in CSR have scanted it, and Boyer (1996b) and Mithen (1998) were brief and omitted virtually all scholarship on the topic.

Because religion as a category is culture-bound, it is better considered a concept than a phenomenon. It *can* usefully be defined by family resemblance (Saler 1993, 2008, 2009), with the latter's "more-or-less" category inclusion and open-ended set of features, none of which is held necessary or sufficient for membership. I adopt family resemblance's more-or-less approach and its open-ended set. The features in the set do not seem equal, however. One of them—relations with "other-than-human persons" (Hallowell 1960)—seems more important than the others, for two reasons. First, relations with *any* persons produce further features associated with religion, including ritual, moral norms, reward and punishment, doctrine (or at least exegesis), signs, and symbols. Thus, stipulating the single feature, person, entails some six or seven others. The second reason persons are important features is usage: the term "religion," both in CSR and among Western laypersons, now commonly requires "gods or other superhuman beings" (Nongbri 2013: 157), who of course *are* persons.

Conceptions of religion in CSR, too, usually invoke agents that are "supernatural" (as noted, a term generally left undefined), but otherwise are diverse. In my view, religion is distinguished from non-religion *only* by its broader and more systematized postulation of other-than-human persons. (Buddhism, be it noted, is *not* a problem for this view, any more than are the existence of demythologized Christianity and Judaism. Each of the three "religions" is a concept that encompasses myriad traditions, some more "religious" and others less so.)

As noted, my argument meets Durkheim's stipulation that any explanation of religion by an illusion "must identify an illusion of equal generality," as well as Bloch's (2008) recommendation that we study, not something culture-bound, but something universal. The illusion identified here—that the world is personal—is even more general, as it shapes our understandings of limitless things and events. Daily life is teleological, for example, through and through. As Nietzsche wrote (1968: entry 550), "I notice something and seek ... an intention in it, and above all someone who has intentions, a subject, a doer: every event a deed ... this is our oldest habit. Do animals also possess it? As living beings, must they not also rely on interpretations based on themselves?" But why is this so?

In current CSR, this question is passed over lightly. Some writers on religion, such as Freud ([1927] 1975), say anthropomorphism attracts us by its comfort and that we see "what we want to see." This may have *some* small truth, but much anthropomorphism is far from comforting. In a different view, others (e.g., Boyer 1996b) have called it a "projection." But projection is (as even Freud admitted; a muddled concept, a "metaphor in search of a theory" (Harvey 1996). It begs the questions both of motive—why *would* we project?—and of optics: what and where is the surface on which we project?

(For example, since "projection" here is meant psychologically, is that surface somewhere in our minds or brains?)

The only papers by CSR writers (Boyer 1996; Mithen 1998) other than mine, to my knowledge, that address the question why we anthropomorphize say that we do so because it is counterintuitive. But evidence to the contrary is multidisciplinary and massive (Guthrie 1993, 2002, 2007, 2016b; Norenzayan et al. 2008). Instead, almost all scholars of anthropomorphism, beginning with Bacon ([1620] 1960), find it spontaneous and intuitive. Hume wrote that even philosophers cannot avoid it, and Nietzsche wrote that it is fundamental in, and inalienable from, human thought.

The question why we do it was posed and partly answered by Hume ([1757] 1957). We anthropomorphize, he wrote, because we find the world complex, frightening, and chronically uncertain, so we rely on the models we know most intimately, ourselves. Despite his answer, Hume was puzzled by teleology, our sense that things and events have purposes, which he said feels like a direct apprehension or sensation. He asked the source of this feeling and, in a letter of March 10, 1751, to Gilbert Elliot, Hume wrote: "A Theory to solve this would be very acceptable" (Greig 1932: 153–7).My theory offers Hume a theory by building on him with Darwin as well as Pascal in that our evolved cognitive biases reflect our interests as organisms. Thus our default models of the world stem not merely from what is familiar, as Hume wrote, but also from what is *relevant*. My account amounts to a brief evolutionary epistemology. It foregrounds our uncertainty, as did Hume, our biological interests, and an evolved strategy in five propositions. First, to perceive is to interpret (Wittgenstein [1953] 2001; Gombrich 1969, 1973; Arnheim 1969, 1974; Greco 1998; Gregory 1997; *pace* Gibson 1986). What we interpret is sensations, such as colors, sounds, smells, and touch, each of which may have any of a number of causes.

Second, we interpret sensations by means of templates or models, choosing one of which constitutes the process of perception. Wittgenstein ([1953] 2001, pt. II, xi) wrote, for example, that people do not "just see" but "see as," and Gombrich (1969) wrote that perception is a "bet." Is the tickle on our neck a loose thread or a spider?

Third, we make our perceptual bets based on our interests, crucial to which are persons. Finding persons has priority, and we find them, signs and signals from them, and traces and likenesses of them on every hand (Hume [1757] 1957; Nietzsche 1966; Wegner 2005; Heberlein 2008; Schilbach et al. 2008). The search for such features is old, as a small stone known as the Makapansgat Pebble suggests. This pebble was carried into a cave in South Africa about three million years ago, evidently by an Australopithecine and evidently because it had features resembling those of a hominin. Our search still continues, revealing faces in flowers, rocks, and artifacts, and personal agency, purpose, and meaning everywhere.

Fourth, perceptual and cognitive ambiguity is primordial and enduring (Gombrich 1969, 1973; Arnheim 1969, 1974; Gregory 1997; Methys et al. 2014). This condition is general, but especially acute in, for example, observing animals in the wild, where cryptic patterns and coloration are pervasive. These and allied means of deception doubtless have been major factors in heightening human sensitivity to animacy. This sensitivity is not only human, moreover, but also as Darwin ([1871] 2004: 118) observed, widespread among animals. Indeed, Darwin wrote that other animals similarly over-ascribe agency and that the higher mammals may "imagine that natural objects and agencies are animated by spiritual or living essences," as illustrated by his dog's alarm at a distant parasol moving in the wind. The threat displays with which wild chimpanzees respond to storms, waterfalls, and wildfires (Guthrie 1980, 2002) also seem to illustrate Darwin's thesis.

Fifth and last, our sense of agency is stimulated by an indefinite range of features. These include anatomy, such as eyes and faces, but more importantly behavior, especially when it is goal directed or contingent. Our sensitivity to goal direction is high and our threshold for judging its presence is low (perhaps because animals come in many and cryptic forms and thus are hard to see, while their goals are relatively few and their movements relatively visible.) Thus we find goal direction, and with it agency, even in geometric figures moving on a plane surface (Heider and Simmel 1944), in a group of dots moving together (Scholl and Tremoulet 2000), and in weather (Lillard and Skibbe 2005: 282). Infants find it even in a plain, rectangular box if it avoids obstacles in reaching an objective (Csibra 2008).

The centrality of goal direction to agency detection, then, helps answer Hume's question, why teleology has such force (a force described also by Bacon 1620; de Spinoza [1670] 1951; Kelemen 1999a,b, 2004; Kelemen et al. 2013). It therefore occurs not only in humans including infants but also in other primates (Jolly 1996), and likely in other animals as well (Darwin 1871; Guthrie 1980, 2002; Foster and Kokko 2009), for all of whom the detection of agency is vital.

Summary and conclusion

CSR is pre-paradigmatic, because its theories and definitions of religion are diverse and contradictory. Of its two most strongly cognitive theories, one stresses religion's peculiarities and finds it counterintuitive, while the other stresses its continuities and finds it deeply intuitive. Unlike religion, anthropomorphism is a human universal and has parallels in other animals. Broadening our focus from religion to anthropomorphism clarifies what is central to religion, which may be described as beliefs and practices concerning other-than-human persons. These beliefs and practices stem from an enduring human need: to discover and deal with whatever persons may be present. In betting that they *are* present, we follow a strategy that predates our species.

6

Cognitive Attractors in the Evolution and Diffusion of Religious Representations

Pascal Boyer and Nicolas Baumard

The publication of *Rethinking Religion* (Lawson and McCauley 1990) ushered in the cognitive program in the study of religion, which like other scientific research programs includes a set of agreed assumptions about the problems to solve, the kind of evidence that would be relevant, and the methods to be used (Lakatos 1978). Specifically, the many scholars who joined the program assumed that religious representations and behaviors are natural outcomes of brain function, that experimental evidence is required to argue for models of their emergence in human minds, and that the two central questions in the field were, why religious representations appear in human minds at all, and why they include so many recurrent features the world over. Here we argue that the success of the program so far should lead to refining these questions and considering how an evolutionary cognitive perspective can make sense of historical variation in religious systems.

The standard cognitive model: Religious representations

What could be called, perhaps with undue grandiosity, the "standard model" of religious representations is a set of empirically grounded models of the kinds of representations usually called "religious" (Barrett 2000; Boyer 1994a; Pyysiäinen 2001). In cognitive terms, there is nothing special about concepts of, for example, spirits, ancestors, gods, and so on, which are part of the repertoire of fantastic imagination, combining salient, explicit information that violates intuitive expectations with tacit assumptions delivered by evolved intuitive systems. The concept of spirit or god is a prime example, combining as it does counterintuitive physical properties and intuitively expected mental functioning. So religious concepts are just those fantastic concepts that people happen to take more seriously than everyday fantasies or fiction and can therefore play an important role in social interaction. This deflationary account of religious cognition, studied as just another variety of human imagination, of course, runs against the grain of traditional religious scholarship. It was present from the beginnings of the research program, as *Rethinking Religion* for instance treated representations of rituals as a product of standard action psychology (Lawson and McCauley 1990).

For both religious concepts and rituals, a cognitive approach implies that people are unaware of the computations that give rise to specific intuitions or beliefs. In the same way as we do not have access to the way we parse sentences, we have no conscious access to what makes particular religious concepts or beliefs salient and compelling. This is true for most aspects of religion and is made most salient in the phenomenon of "theological correctness," when people's explicitly stated beliefs (e.g., that a god's cognitive powers are infinite) are associated with opposite intuitive assumptions (e.g., that a god attends to events serially) (Barrett and Keil 1996; Slone 2004). In the same way, underlying cognitive processes explain why concepts of religious *agents*, expected to have humanlike cognitive function, are vastly more widespread than any other type of religious concept (Boyer 1996b; Guthrie 1993). The point is worth emphasizing, as even in recent scholarship, that many have found it difficult to resist the temptation to take explicit religious beliefs at face value, for instance, in describing the "prosocial" aspects of religious prescriptions.

No such thing as Religion

One domain where the nascent cognitive program should have been more attentive to scholarship from history and anthropology is the delineation of the domain. For most anthropologists in particular, it is clear that the term "religion" is of little use, as it does not really denote a domain of social and cultural phenomena with common and unique features (Bloch 2008; Saler 1993). Indeed, the term is seriously misleading, as it tends to obscure profound differences in social interactions related to religious concepts. Most producers and consumers of the cognitive science of religious representations are familiar with a situation in which people interested in a specific variety of superhuman agency can have access to an explicit doctrine, maintained by formally trained specialists and clearly construed as different from other possible doctrines. They are also used to the idea of gods with a large or even cosmic jurisdiction, and to notions of salvation. By contrast, for most of human prehistory and history, and to this day in many communities the world over, religious concepts were and are activated in the context of cults directed at local superhuman agents, whose powers are relevant to pragmatic concerns, like prosperity, illness, and misfortune. Relations with these agents are managed by informal specialists like healers, mediums, and shamans. There is no fixed and common doctrine, and no formal training for the specialists.

Applying the term "religion" to these different kinds of situations has caused grave confusion in the study of religion, and especially in the search for the evolutionary processes that led to human interest in superhuman agency. In particular, the notion that people have beliefs, that these beliefs are public, that they express some commitment, and that religious activities are costly is derived from recent organized religions, but do not really apply to small-scale informal cults (Boyer and Baumard 2016a). As most of the features we spontaneously associate with "religion" are not actually present in human cultures, except for a few recent but familiar exceptions, it is not clear that the term "religion" is of much use to cognitive or social scientists. It is not just that small-scale cults and organized religious institutions are two forms of "religion"—they are simply not forms of the same thing at all. Indeed, if we develop good explanatory models for the success of religious organizations embedded in state polities, there is no reason to assume that any of that will be relevant to the dynamics of superhuman agent cults in small-scale groups typical of our prehistory. Projecting features of familiar "religion" onto ancestral situations leads to serious distortions in evolutionary models (Boyer and Baumard 2016b).

One particular area of misunderstanding is the estimation of the success of religious doctrines. To the outsider, it may seem that some doctrines have been extremely successful, as Islam or Buddhism, for instance, are the official religious affiliations of billions of individuals. But this is mistaking political hegemony for cognitive diffusion. As historians and anthropologists have observed for decades, most members of most religious affiliations hold beliefs that are only remotely related to the doctrinal messages of the locally dominant religious organizations. It is not just that people supplement the official beliefs with intuitive content in the phenomenon of theological correctness described previously. What also happens is that people simply add many nonofficially sanctioned beliefs and practices to the official system. In all places with religious organizations, we find that people resort to other, nonofficial specialists like healers and mediums, practice divination and sorcery, and hold beliefs about superhuman agents not recognized by institutional religious groups. The fact that people hold on to "magic" and "fakirism" is a permanent complaint of official religious specialists, the world over. They generally try to use their political clout to suppress such practices and outlaw this informal competition, but informal religious activities invariably crop up again. Religions never really win.

The next frontier: Attractors and diffusion

So far, the cognitive study of religious representations has mostly addressed aspects of religious representations that could be found in roughly similar forms the world over. That is the case, for instance, of religious notions of superhuman agency (Barrett 2000; Boyer 1994a; Pyysiäinen 2001), or the organization of ritualized behaviors (Liénard and Boyer 2006; McCauley and Lawson 2002). As a consequence, this cognitive perspective may seem less relevant to understanding differences between religious systems. There are of course notable exceptions to this, in particular Whitehouse's model of two distinct "modes of religiosity," based on different kinds of interaction that can stabilize a tradition. Specifically, Whitehouse argued that traditions can be stabilized, either through exceptionally salient, but rare, events with highly ambiguous content or by frequent transmission of highly repetitive and conceptually clear material (Whitehouse 2000, 2004). Although some aspects of this distinction may need conceptual clarification (Boyer 2005), it is an example of what we consider the next frontier for the research program, namely an understanding of how changing ecological, economic, and political environments can affect the diffusion of religious representations.

To make this more specific, we must clarify what we mean by diffusion. Cultural transmission is sustained by communication—which of course is not a simple process of downloading mental representations from one brain to another. Rather, human communication works in an ostensive-inferential manner, providing cues about one's intention to communicate, which modify a listener's representations in a relatively predictable manner (Scott-Phillips 2014). That is, listeners are not decoding a message; they are producing inferences about a speaker's representations, such that the utterance produced would be optimally relevant, given mutually accessible information (Sperber and Wilson 1995). In this active process, utterances and gestures are only the starting point—the initial cues that trigger an inferential interpretation.

This has important consequences for cultural transmission, suggesting in particular that information transmitted through myriad cycles of communication should end up in ever-increasing differentiation. That is indeed the case, but the representations we call cultural are an exception—in fact, we call them cultural precisely because they are such an exception, because they seem to appear in roughly similar forms in many different minds. These recurrent forms, or cultural attractors, are more likely to appear than other representations—even when they are not clearly communicated (Claidière et al. 2014; Sperber and Hirschfeld 2004). For instance, people are likely to form an intuitively compelling representation of superhuman agents with standard mental properties, even if they receive no information about that aspect of the gods and spirits (Boyer 1994a), indeed, even if the information they received contradicts that intuitive description (Barrett and Keil 1996).

Cognitive attractors also account for otherwise mysterious, historically specific aspects of cultural diffusion. For instance, Olivier Morin's epidemiological studies of portraiture show that, inside specific artistic traditions, once the subjects of portraits started to look straight at the viewer, this particular style overrode previous norms, because of the perceptual salience of direct gaze (Morin 2013). In a similar way, norms of etiquette are driven by attractors toward the avoidance of disgust-inducing behaviors, which explains their nonrandom cultural change (Nichols 2002). More generally, the combination of cognitive attractors, with patterns of diffusion (e.g., how many chains of transmission are formed, how many individuals they include, how frequent transmission is, etc.), can explain how some clusters of representations form traditions (Morin 2016). This in our view is the next frontier for the study of religious representations.

An example: Life-history and moralizing doctrines

As an illustration of this cognitive approach to diffusion, consider the development of movements based on (1) a moralized distortion of previous religious doctrines, or the invention of new moralizing doctrines, and (2) a preoccupation with the self and its destiny, notably in the form of techniques of the self (asceticism, moderation, meditation, etc.) and salvation of the soul. These are of course the two main features of so-called Axial Age doctrines (Eisenstadt 1986; Jaspers 1953). These doctrines emerged in very similar forms in three distinct regions, the Eastern Mediterranean, the Ganga valley and Northern China. We have argued that the phenomenon makes sense in the light of an exceptional increase of prosperity in these regions at the time (Baumard et al. 2015). As the elite in these prosperous societies reached unprecedented levels of certainty in the satisfaction of pressing needs, they adopted ideologies of moderation and self-cultivation, as well as depictions of the universe as intrinsically moral (Baumard and Boyer 2013).

The connection between relative prosperity and new religious doctrines can be explained in terms of life history theory, a branch of behavioral ecology that studies how individuals, including humans, adjust their behavioral strategies to the environment through phenotypic plasticity (Ellis et al. 2009). In a harsh environment, organisms tend to adopt a "fast" behavioral strategy, pursuing smaller but more certain and immediate benefits (displaying a higher rate of time discounting, greater disinhibiting, a greater level of risk-taking). In more certain environments by contrast, organisms tend to develop a "slow" behavioral strategy, aiming for larger but less certain benefits (lower time discounting, greater inhibition, lower level of risk-taking). Early environments influence the calibration of each individual's life strategy, with consequences, for instance, on age of menarche, early reproduction, risky or safe behaviors, levels of violence, levels of trust and trustworthiness, and many other social outcomes (Ellis 2013; Nettle 2010; Nettle et al. 2007).

Life history theory could shed some light on the variability and the evolution of religious doctrines. In contemporary societies, religion is indeed often associated with a particular set of moral values such as extended cooperation, restricted sociosexuality, and delayed gratification. By contrast, hunter-gatherer societies, agropastoralist tribes, and archaic chiefdoms are characterized by religious doctrine in which religious behavior is mostly about bartering commodities with supernatural powers in order to get resources or protection from enemies (Baumard and Boyer 2013).

In our view, differences in life history may explain fine-grained aspects of the diffusion of specific moral and religious doctrines (Baumard and Chevallier 2015). Specifically, Axial Age movements and their successors in so-called world religions promote doctrines centered on three main themes. The first one is a conception of extended cooperation, a high level of prosociality. This appears for instance in the strikingly similar formulations of the Golden Rule in otherwise very different traditions, such as reformed Hinduism, second-temple Judaism, and Taoism. This was a cultural innovation, against the background of religious doctrines that emphasized parochialism and revenge (Baumard and Boyer 2013). A second theme is restricted sexuality, which, for instance, was a salient difference between early Christians and their contemporaries (Brown 1988). This included an increased emphasis on sexual moderation and on conjugal love in places as different as Rome and classical China (Goldin 2002; van Gulik 2003; Veyne 1992). A third theme was the more general notion of moderation and patience, that is, an emphasis on delayed gratification, against the temptation of the short-term profit that is typical, and in fact required, in environments of great uncertainty.

The elites of the Axial Age seem to have experienced a shift from a fast to a relatively slow behavioral strategy. As a result, a variety of doctrines that had for centuries seemed quite compelling lost some of their appeal. Instead, elite individuals either created or adopted doctrines, the explicit content of which was closer to their intuitions. For instance, typically slow life strategies allow people to engage in generous behaviors, which in the long run accrue higher benefits from cooperation than tit-for-tat exchange (Raihani and Bshary 2015). For slow-strategy individuals who intuitively grasped this advantage, versions of moral doctrines that emphasized generosity seemed to fit their moral intuitions, and therefore seemed compelling.

Different attractors and strategies

Our interpretation of the Axial Age innovations is that changing ecological and economic conditions had modified the local attractors for cultural transmission. In late antiquity, part of the Eurasian population started to live in an environment that was safe enough, predictable enough, and affluent enough to trigger investment-oriented strategies. As a result, the Eurasian elites abandoned benefit-oriented religions in which gods are thought to provide resources, protection, or social success to adopt investment-oriented strategy in which the gods favor extended cooperation, restricted sociosexuality, and delayed gratification and condemn benefit-oriented strategy founded on greed, violence, and unconstrained sexuality.

This model might explain why moralizing religion appeared relatively late in human history, and only in the most affluent societies. Conversely, it may explain the gradual decline of sacrifices and rituals of the resource-oriented archaic kind of religion (Oster 2004; Thomas 1997), and their periodic return; Oster, for instance, argues that the witchcraft craze of Renaissance Europe can be explained by the deteriorating economic conditions of the medieval Ice Age (see also Quinlan and Quinlan 2007). More generally, this case illustrates two important facts about the diffusion of religious representations.

First, changing conditions affect the range of beliefs or practices that are intuitively compelling to individuals. Although the point may seem banal, it is usually obscured in the study of religion by a widespread belief that evolutionary and cognitive perspectives can only explain the universal aspects of religious phenomena. If that was the case, the cognitive study of religious representations could explain why witchcraft or moralizing religions are catchy, but not why the former appears to be more present in small-scale societies and the latter in large-scale societies. This interpretation of the scope of evolutionary cognitive models is widespread, with exceptions of course (see, e.g., Whitehouse and Martin 2004a). By contrast, we argue that each ecology triggers specific behavioral strategies, with their underlying cognitive processes.

A second important point is that statements of belief are of course strategic. This becomes more salient in large mass societies, where different strategies and sets of values may be available and where the benefits from adopting them depend in part on their frequency in the population. Obviously, generous cooperators do not thrive in a society of vengeful scoundrels. That is why individuals who adopt slow strategies often engage in moralizing behaviors with a view to influence their social environment.

At the same time as they adopted moralizing doctrines and an ethic of generosity, classical elites tried, often through political domination, to impose these values on a populace that may have been less intuitively seduced by such themes and more attached to the traditional kind of religiosity in which gods and spirits provide a form of protection racket. In a pattern that is familiar to historians, the history of such movements as Buddhism and Christianity is rife with conflicts between the established doctrine on the one hand and periodic popular returns to tit-for-tat religiosity.

These points are also relevant to modern religious dynamics. In the economic conditions of modern Western societies, choices of reproductive strategies—between an investment in the long-term returns of conjugality as opposed to more hedonistic and "faster" strategies of immediate gratification—seem to influence a whole set of apparently unrelated values and attitudes, religiosity among them (Weeden and Kurzban 2013).

After twenty-five years of research, mostly centered on universal or very general processes, the cognitive research program may stand to benefit from a slight change of focus. Changing conditions provide quasi-experimental variations that allow us to test more refined hypotheses about the way an evolved cognitive architecture can elaborate and transmit religious representations.

7

The Long Way from Cognitive Science to History: To Shorten the Distance and Fill in the Blanks*

Panayotis Pachis and Olympia Panagiotidou

Introduction

One of the greatest developments in the (neuro)cognitive sciences in the last twenty-five years is the realization of the intrinsic connection between human cognition and culture. This conception gave rise to a new field of research: that of social cognition (e.g., Han and Northoff 2008; Chiao 2009; Roepstorff et al. 2010), which studies the ways in which both higher- and lower-level cognitive processes are formed and are being affected by the external social and cultural world.

Religious practices and beliefs comprise one of the first domains in which the interaction between human cognition and culture is being examined. Well-articulated theories of religious behaviors and transmission of religious ideas throw light onto the cognitive underpinnings as well as cultural and social processes that mediate religious practices and precepts. Theoretical hypotheses and laboratory findings from the cognitive sciences are being applied to modern religious groups. Simultaneously, the first attempts to employ modern cognitive theories to the study of past religious practices have sketched out the opportunities arising from a coordination between the cognitive sciences and historical studies. Despite the criticism, the involvement of historians in cognitive projects seems to be imperative for both deepening historical knowledge and advancing the diachronic understanding of the human mind and its development.

History, cognition, and culture

Although the discipline of history is mainly interested in the human past, historical research tends to focus on the results of human actions rather than on the human

* We would like to thank Professor Luther H. Martin (Emeritus, University of Vermont) and Professor Roger Beck (Emeritus, University of Toronto) for their valuable contributions regarding suggestions, and criticisms.

agents themselves. This is quite reasonable if we take into account the nature of the available historical evidence. Written testimony and material remains constitute the primary sources for historical study. These sources are the products of their composers' and constructors' thoughts and actions and are directly amenable to inquiry.

In the study of literary and historiographical works produced in past eras, historians first intend to reconstruct the sequences of past events and to give an account of them. Archaeological findings enable them to delineate natural and architectural topographies. Material remains provide evidence for manufacturing techniques, cultural practices, social institutions, and ordinary activities that comprise the wider cultural, social, and economic contexts of different historical periods. However, articulating a narrative of the past and describing the available evidence are not enough for historical study. Historians inevitably seek insights into ordinary people's lives, attempt to trace changes and transitions in the course of history, and then proceed to suggest psychological interpretations of past people's behaviors, thoughts, and actions.

The History of Mentalities (e.g., Huizinga [1919] 1924; Lefebvre 1932; Granet 1934; see Hutton 1981; Burke 1986), arising within the Annales School at the beginning of the twentieth century, drew attention to past peoples' systems of beliefs, perceptions, collective attitudes, psychological expressions, and modes of thinking. This approach pointed out that in order to study different cultural and social groups throughout history, historians should transcend their own way of thinking and understand those people's beliefs, categories, symbols, and worldviews in terms of the latters' mindsets (Burke 1986: 442). Despite the weaknesses of this approach because of its focus on collective mentalities as homogenous systems of ideas and beliefs, and of its evolutionary conception of cultural change (see Burke 1986: 443–5), the Mentalities School nevertheless inaugurated a study of culture and human thought in a long-term scale.

From then onward, some scholars have been interested in the ways human thought is patterned and organized in "schemata" and "mental or perceptual categories" (e.g. Gombrich 1960; Kuhn 1970; Foucault 1966; Cohn 1975) that structure and interpret both individual and collective experiences in the world (Burke 1986: 446). In parallel, the increased interest in the human mind (e.g., Simmel [1905] 1977: vii, 39, 43; Starr 1965: 27) introduced terms and theories from psychology into historical studies. The progress in psychological research and psychotherapy gave rise to two related fields: those of psychohistory (e.g. deMause 1975, 2002) and psychobiography (see Schultz 2005). These two fields of historical research have been criticized by conventional historians. However, even classicists and historians of religions who do not explicitly adhere to psychological interpretations of history use psychological terms and suggestions when they articulate their historiographical views (e.g., Dodds 1951: esp. 106, 114, 116, 120, 123, 134, 151–2; Burkert 1985: esp. 3–4, 58–9, 142, 190, 199, 220; 1996; Versnel 1981; 2011: e.g., 148, 170, 217, 259; Bremmer 1994; Parker 1998).

Nowadays various psychological notions—suggested by psychologists and used by historians—are studied by cognitive scientists as the result of specific mental and emotional processes. The rapid developments in cognitive psychology, neuroscience, and biology along with advances in neuroimaging techniques during the last decades provide valuable insights into human cognition. A wide range of laboratory experiments run by cognitive researchers investigate and reveal the foundations and underlying mechanisms of multiple cognitive processes—like consciousness, memory, perception, reasoning, and mental imagery. In addition to research into cognitive universals, many cognitive scientists have recently become interested in the ways human cognition interacts with and is influenced by the external material and social worlds (e.g., Han and Northoff 2008; Chiao 2009; Roepstorff et al. 2010). Humans are conceived as complex "embrained and embodied" entities, "deeply dependent on culture (encultured)," and their cognition is "extended and distributed beyond the borders of individual brains" (Geertz 2010a: 304). Such a conception is connected to the realization that experimental findings should be tested in the realm of people's everyday lives. To this end, cognitive anthropologists are conducting a growing number of studies in the field in order to investigate the cognitive processes within specific social and cultural contexts.

However, since human culture not only exists in the present but also extends to the whole history of the human species, the study of modern people's cognitive operations and capacities provides only partial insights into human cognition. At this point, the contribution of historical studies seems to be of crucial importance both for advancing our modern knowledge of the human mind and for deepening our historical understanding of human agents.

The first endeavors to apply cognitive theories to past cultural practices were undertaken by prominent historians of religions. Luther H. Martin was the first researcher specialized in the study of Greco-Roman cults to describe the potentials provided by cognitive approaches for historical research. In many of his articles, he suggested cognitive approaches to his specific historiographical example, the Roman cult of Mithras (e.g., 2004a, 2015), and articulated convincing arguments about the value of cognitive theories for the study of past religions (e.g., 2005a, 2007, 2014c). Roger Beck produced the first comprehensive monograph (2006) in which he applied modern cognitive theories to the ways whereby the Mithras cult could have been apprehended by initiates in the mysteries. From then onward, discussions began regarding the collaboration between historians and cognitive scientists, which led to the emergence of a new subfield of historical research, that of "cognitive historiography."

However, the application of cognitive theories to historical material is still met with skepticism by the majority of classical historians. Yet their theoretical considerations and objections to cognitive approaches are undermined by the intrinsic mental and affective aspects of human history. Despite their theoretical and rational skepticism, historians inevitably use terms and notions closely connected with mental processes, cognitive abilities, and the emotional reactions of past people in order to study history. And their historical study deeply depends on their own cognitive equipment, since the notion of history has inherent cognitive roots.

History as a cognitive enterprise

We are interested in the past because we are humans. Autobiographical memory enables humans to have self-consciousness—to be aware that they are unique individuals, distinct from others and their surroundings, "entities" that change over time while retaining the sense that they remain the same persons throughout their lifespans (see Damasio 2000: 176). Thereby, humans are able and want to narrate stories about themselves, the world, and the events they experience. They compose these stories by recalling and (re)structuring their memories, recognizing causes and results, tracing relationships, filling in the blanks, and striving to give coherence and meaning both to their lives and to perceivable reality. Beyond self-awareness, humans intuitively categorize themselves as belonging to the same species and are able to recognize conspecifics among other beings and entities (Boyer 2001). In the long term, they look for human origins, and they are interested in their predecessors' stories (Claus and Marriott 2012: xiv). What personal stories are for individual humans, history is for the human species. In this light, historical thinking is a cognitive task.

In principle, the cognitive sciences legitimate the study of the past and the need for history. At the individual level, historical thinking is natural (Martin 2007: 38), and narrative constitutes a valuable cognitive tool for organizing and making sense of human life and the world (see Geertz 2010b: 29–30). When history develops into an academic enterprise, however, it becomes a complex and highly demanding endeavor. Historians are bestowed with the task of narrating the history of the human species and the world in which humans live. However, this very task is predetermined, restricted, and structured by the processes and capacities of human cognition. The cognitive sciences offer historians the possibility of becoming aware of these restrictions and the predispositions of their own mind and thereby free their work from their own preoccupations (cf. Whitehouse 2005a: 308–11; Martin 2007: 39–40).

The human past comprises innumerable events, infinite ideas and thoughts, countless actions, practices and beliefs, social interactions, and cultural expressions unfolding every second. Individuals are not able to remember and recount every single moment of their lives, and, similarly, historians are unable to record and narrate every single instant of the human past. Much as persons do when they narrate their own life stories, historians are called on to understand the gist of past events and to organize their knowledge according to logical patterns and sequences. While individuals use memos and memorabilia (e.g., photos, diaries, notes, etc.) in order to remember their past, historians use historical sources generated by other people. Personal memos operate as anchors (see Geertz 2010b: 30) that trigger an individual's own memory and recall their own representations. Historical sources are the first piece of information on which historians can rely in order to reproduce mental and public representations produced and shared by others at some time in the past. Studying the sources, historians gain some access to the minds who created the data on which they draw, and they then generate their own mental representations and simulations of past events, which they organize following the logico-structural principles of narrative (cf. on the narrative structure of historical thinking, see Martin 2007: 39–40; Whitehouse 2005a: 312–15). Thus, history does not consist in recording a farrago of unrelated events and practices randomly developing in space and time. Rather, historical narratives fit the structures of human cognition that tends to recognize causal relationships and logical correlations in both natural and social events and to look for consistency and coherence in scattered knowledge and information. Given that historical sources provide fragmented evidence and not a complete account of the past, the cognitive capacities of historians enable them to envision a wider picture of a particular historical era and to deduce more extensive inferences and conjectures. In this light, historians are historiographers who write a history of the past based on the available evidence that is, in turn, organized according to cognitive principles (Martin 2005a: 12–13).

In studying the sources, historians may choose to focus on different aspects of the surviving evidence and to suggest various interpretations of this evidence. Their focus, choices, and decisions are largely determined by their own interests as well as the wider needs, demands, perspectives, and preferences of their era (cf. Blake 1955: 61). In all cases, however, their work is governed by common principles of human cognition. Classifications of different types of evidence (e.g., primary and secondary sources, written testimony, material remains, etc.) reflect the universal tendency of the human mind to categorize various aspects of the phenomenal world. Conceptions of history and separations of different historical periods depend on metaphorical perceptions of time. Notions like era, culture, and society constitute conventional terms used by historians to organize their material. While these arguments may sound self-evident, such principles—generally understood as common sense—are widely used by historians in order to fill in the blanks of their historical accounts. However, the cognitive sciences have put forth well-articulated theories, testable and supported by scientific methods, in order to investigate what is widely perceived as common sense but that is deeply rooted in the way humans think, conceive, and reason about themselves and the world. In addition, historians take for granted and use a wide range of cognitive tools (e.g., metaphors, categories, concepts, narratives) in order to structure their historical accounts. Cognitive studies investigate the construction of these tools in terms of the common features of human cognition. Their findings throw light onto the processes and mechanisms that shape and modulate human thinking and reasoning, providing historians the means to examine and to understand more deeply the various forms and contents shaped by common cognitive tools in different historical contexts (cf. Martin 2005: 12).

Further, the common attempt of historians to embed their evidence in its specific historical and cultural contexts is insufficient for understanding this evidence. As Lawson and McCauley (1993) point out, these contexts themselves demand explanation, since they do not constitute independent entities or hyper-institutions that externally determine people's lives (Martin 2007: 42–3). Instead, they are the output of continual interactions between human agents and their external natural, social, and cultural worlds—interactions that constantly affect both cognition and the world (cf. Geertz 2010a; Malafouris 2010; Roepstorff et al. 2010).

The main features of human cognition that were developed over human evolutionary history were prompted by the need to adapt to a constantly changing environment. However, evolution and cognitive development are not only triggered by changes in the natural world. As psychologist Margaret Wilson (2010) suggests, various cultural contexts provide infinite stimuli that not only generate different "contents of cognition" but also further modulate and modify cognition. Simultaneously, alterations in the ways people think and act are imprinted on the external natural, cultural, and symbolic milieu that, in turn, affects people's thoughts and perceptions. In this framework, cognitive theories provide crucial insights into those universal cognitive and affective proclivities that enable humans to construct specific images, assumptions, and mental categories of perceivable reality as well as into those processes that form different contents of these perceptions in different historical contexts. In this way, the employment of cognitive theories and models for interpreting historical sources may throw light on the processes of historical transitions and cultural diversities with direct reference to the human agents who mediated and generated changes and alterations in the external natural, social, cultural, and symbolic world (see Martin 2014c: 353; McCorkle 2014: 143).

In parallel, historical contexts provide a new field of research for the cognitive sciences in which the laboratory and experimental findings can be tested in the "real life" of past eras (Eidinow and Martin 2014: 8; Martin 2014: 351–2). Although experimental methods cannot be applied to people who lived in the past, historians can investigate whether the common cognitive abilities of the human species underlay and generate the historical material they have at their disposal. In this way, they can assess—confirm, refine, or even reject—the theoretical assumptions about cognitive evolution and the universalities of the human mind and brain claimed by the cognitivists. Historical studies can further point out which cognitive processes may be sensitive to cultural influences affecting the ways people perceive and conceptualize the world as well as what kinds of external stimuli can "re-engineer" the human mind (Wilson 2010: 180–1).

Given that the notion of history and of historical awareness is an output of human cognition as well as that human cognition has an intrinsic historicity, the (con)junction of cognitive sciences and historical studies seems—explicitly or implicitly—inevitable.

Conclusion

The cognitive sciences and historical studies may differ in terms of subjects of research, available sources, and methodological tools. What they share, however, is that they are both conducted by researchers who are situated in a specific cultural and social world that affects their thoughts and choices (Heintz 2010: 13). In addition, they share a common interest in understanding human beings.

Based on these premises, the lines that separate these two fields of research become less rigid. Taking into account contemporary developments in scientific methodologies, collaboration between scholars from different disciplines has proved to be extremely valuable for making progresses in their research. Researchers with different expertise join their methods and knowledge in order to approach shared goals of inquiry and to reach more pluralistic insights and conclusions. In this wider milieu, historians appear to be among the last to abstain from this kind of collaboration (see Xygalatas 2014b). This hesitation may derive from the unique features of historical data that determine the limits and potentials of historical research. However, cognitive approaches to historical material do not claim to replace traditional historical approaches and classical studies. They are mainly addressed to those historians who share with cognitive scientists a common interest in the human mind throughout history (see ibid.: 195). In this framework, cognitive findings and models could embellish historians' theoretical considerations and assumptions by contributing a scientific grounding for historical theorizing. At the same time, historians could provide cognitive scientists with further empirical data for supporting, ameliorating, or modifying their theoretical hypotheses or even to determine further questions for research.

Collaborations between historians and cognitive scientists do not entail the demolition of either's methods, goals, or much less, existence. Yet willful neglect of knowledge about human mental processes can lead to obsolete approaches and interpretations of historical agents' mental representations, psychological reactions, intentions, choices, and hidden agendas. Simultaneously, cognitivists' claims of universal proclivities and capacities of the human mind will remain theoretical assumptions, if there is no historical investigation to validate these universalities from the "real life" of "real people." It remains for scholars from both fields to realize the mutual benefits of their collaboration in order to achieve their common goal of advancing knowledge.

8

The Indispensability of Cognitive Science for a Genuine History of Religion

Anders Klostergaard Petersen

The title of this essay may appear contradictory, verging on a Tertullianic oxymoron: What has cognitive science to do with history of religion? Do they not exclude each other by being located at different analytical levels and in conspicuously different academic cultures?

A majority of colleagues, I surmise—and that pertains to both sides of the spectrum—holds that cognitive science and the study of religion embody the divide of what C. P. Snow famously dubbed the two cultures (1959). Only by acknowledging the difference pertaining to the gap between natural science and the humanities can representatives of the two disciplines fare well. Personally, I think differently. I espouse the view that the emergence of cognitive science in the study of religion twenty-five years ago not only brought about a new intellectual impetus for thinking in novel ways but also, even more importantly, led to a partial departure from the time-honored, but with respect to the quest for truth, no less detrimental humanities vs. natural sciences dichotomy. This is potentially the most advantageous gain of cognitive science in the study of religion, since it is beneficial for a development that could ultimately transform the discipline into a truly scientific study of religion. By expressing myself in this manner, I foresee a number of "humanist" colleagues on their guard against such scientific usurpation of the discipline. Let me be frank. I am a proud and stubborn humanist myself who could not do without the philological competences I have spent years to attain. Yet, it is regrettable that we have come to a point in the academy at which the ways between science and humanities have almost inevitably parted. For pragmatic reasons (life is short and our competences limited), we can no longer pursue Goethe's and so many other enlightenment thinkers' ideals of encyclopedic knowledge; but retaining mortal soberness should not be conflated with justifying scholarly parochialism and academic myopia. I am not a scientist, but I as a humanist would find it difficult were I to ignore the insights provided by colleagues who can bequeath me with scientific frameworks needed in order to confront questions that I otherwise would have had to leave aside.

In this essay I shall by means of one example only set forth how I, as a historian of religion, have benefited considerably from cognitive science. I shall exemplify it by having recourse to the moot question of ritual efficacy and magic in ancient religion. I could also have concentrated on questions pertaining to continuity, development, transitions, discontinuities, and so forth in the history of culture and religion that we also have had serious problems in coming to terms with. In my view, such puzzles can only be adequately tackled when posed in the wider context of cognitive science and within a biocultural evolutionary framework known from, for instance, Merlin Donald and Robert Bellah (Donald 2001; Bellah 2005, 2011; Petersen 2013a,b, 2015, 2017a).

Due to constraints of space, however, I have decided to focus on the matter of ritual efficacy. In terms of both history of scholarship and popular understanding, it is among the most unresolved and thorny questions in the field. Yet, this is only one showcase to palpably demonstrate how historians may benefit from cognitive science. It can provide us with sound theory that may ultimately help us not only to get a grasp on elements about which we were previously forced, out of embarrassment, to beat about the bush but also to progress in our penetration of religion as intrinsic to human culture. In conclusion, I push the argument even further to the extent of endorsing the view that a genuine history of religion cannot be without the dependence on insights from cognitive science. Initially, however, I shall present some fundamental considerations in terms of philosophy of science with respect to history of religion.

History of religion as vertical anthropology

I think of the subject of history of religion as being similar to foundational epistemological and hermeneutical problems confronting anthropologists. History of religion is a form of anthropology by virtue of two facts. First, similar to the study of modern civilizations we are, as humanists, committed to preserve cultures—be they old or novel—for future generations. As elegantly and somberly formulated by Paul Valéry in his first letter of *La crise de l'esprit*, civilizations are perpetually prone to annihilation and disappearance.[1] Therefore, we are obliged to maintain them in memory by transmitting their worlds to future generations. Secondly, were we only engaged in the task of preserving worlds threatened by eradication and oblivion, we would be reduced to be caretakers only. The perpetuation of tradition is only one, albeit a crucial, part of anthropology. The second aspect is just as vital—that is, the endeavor continuously to question our world by comparing it with different worlds. The hermeneutical challenge does not only involve a faithful rendering of other worlds but also an attempt to preserve them in their alterity. By retaining the otherness of different cultures—contemporary or past—in their dissimilarity from our world, we may establish a bulwark protecting us not to fall victim to prejudicial thinking. Obviously, the alterity I speak of is of a relative nature only. It cannot be absolute. First, we would not be able to acknowledge the alterity of the other culture were it not for the fact that resemblances exist that, in the first hand, enable us to identify differences. Second, in as much as we strive to retain the otherness of the other cultures scrutinized, we necessarily make use of our own interpretative schemes. That is the inevitable problem of presentism so profoundly captivated by de Certeau (1975: 40.56–76). We have access to the otherness of the other culture only by reconstructing it in familiar terms.

Be that as it may, the argument will suffice to illustrate my understanding of history of religion as vertical anthropology. Similar to anthropologists working with the alterity of other cultures and societies different from their own, we as historians concentrate on the otherness of past cultures. We do it for two reasons. The first one concerns the need to preserve the pluriform richness of human culture by also studying cultures long extinguished. The second involves the continuous questioning of the manner in which we conceive of the world. In order not to reify our behavioral patterns and essentialize our thinking as the only natural one, it is necessary to confront other worlds, which by virtue of their difference may serve to correct misguided views.

If the history of religion is mainly concerned with the cultural and societal differences that exist between our current world and the worlds examined, this looks different when one supplies historical studies with insights from cognitive science. The latter may help us keep the first perspective at balance by providing us with an awareness that enables us to place emphasis on elements that connect people across time and space. The inclusion of the cognitive science perspective in historical studies is beneficial by not falling victim to overstressing the foreignness or radical alterity of other cultures. In the same manner, the balance—preserved by supplementing a cognitive science approach with sound historical studies—is a key for not exaggerating the similarities found to exist between human cultures and thinking. Thus, the relation between cognitive science and the historical study of religion presupposes a serene but tense relationship between two trajectories reciprocally that keep each other at fruitful balance and distance. On the basis of this perspective, I shall proceed to exemplify *in concreto* how I, as a historian of religion, have not only benefited from but also could not be without the inspiration and insights from cognitive science.

Why resuscitate terms like "magic" and "ritual efficaciousness"?

When working with the ancient world, one is continuously confronted with the question of what to do with sources that blatantly defy a world that subscribes to a scientific "universe of contact."[2] Once again, I underscore that this is no different from the epistemological problems confronting any anthropologist. Needless to say, however, when working with past worlds, we have the additional and intensified hermeneutical problem that we are playing a twofold game by acting both as an anthropologist as well as the informant. We are also the ones giving voice to the past sources and, thereby, bequeathing them with presence.

Admittedly, we face severe problems when studying ancient texts that endorse strong notions of ritual efficacy that we have related to by tradition. For example, the history of research on ancient miracle stories and magical texts is well known. Indeed, we can focus on the social function that such texts may have had just as we—in accordance with a recent and vibrant scholarly trend—can deconstruct traditional terms (see, for instance, Nongbri 2013 and Otto 2011). Sometimes such exercises may be useful, although in the long run the latter in particular proves tiresome and philosophically naïve. First, to embark on deconstructing scholarly terms is ultimately nothing but piggybacking and parasitism. Second and more seriously, the enterprise involves the subscription to two philosophically contradictory understandings. The scholar in question wants to have her or his cake and eat it too by simultaneously endorsing philosophical nominalism and realism. On the one hand, it is taken for granted that the traditional scholarly view identified with the use of particular terms such as "magic" and "religion" as fallaciously ascribing confined and definite meanings to the words used. This is the nominalist attack on the scholarly time-honored terms. On the other hand, the deconstructive enterprise is founded on the view that no singular, monolithic meaning can be assigned to the terms used, since there is no indispensable relationship between sign and referent of the sign. Hence, it attributes a credulous positivistic position to the scholars adhering to the traditional nomenclature. Thus, the enterprise philosophically undermines itself by endorsing a realist position that, by virtue of its nominalist stance, it cannot ascribe to. To put the argumentative delusion blatantly clear: We think that magic does not exist; but simultaneously we acknowledge that magic exists as a fact referred to by a particular term used by the scholars whom we need to deconstruct.Third, there is in such scholarship an astonishing inability to distinguish between second- and third-order language, since emic and etic levels of analysis are conflated with each other. Additionally, second-order categories are by necessity taken to be second-order categories without acknowledging that they may be transferred to third-order categories, which obviously makes a world of difference. Take magic, for example. It is in dire contrast to such philosophically gullible ways of conducting scholarship that I turn to a resuscitation of the term "magic" and related nomenclature as well as the problems involved. Alternatively, it is difficult to conceal the fact that whether we as historians examine the social function of these texts ("my ritual efficacy is the witchcraft and magic of the others") or simply have recourse to deconstructing previous uses of the term, we are hopelessly incapable of rendering the texts meaningful in relation to the aspect that it all comes down to, namely the question of the alleged efficacy of the powers invoked.

It was for me a groundbreaking experience to read the work of my colleague Jesper Sørensen, *A Cognitive Theory of Magic* (2007a). Sørensen courageously faces the core of the matter: what to do with magic? Rather than pursuing the easy way of discarding the term, he makes it clear that this will not help us. We may designate magic by the traditional term or we can call it "ritual efficacy," or even "x" or "y"; but that is not going to get us anywhere unless we can come up with a clear definition undergirded by an explanation of the phenomenon that so palpably defies our intuitive understanding of how the world works.[3] And, ultimately, a scientific worldview is nothing but a more complex and advanced systematization of our intuitive experiences of how the world functions (although that is already a lot): a world in which people do not walk on water or hover in the sky, use endless strings of gibberish to attain specific goals, touch inanimate objects with their fingertips whereby the objects are transformed into living entities, and so forth. That said, however, there may be different situations in which humans for various reasons are likely to subscribe to such representations.[4]

Accounting for magic from a cognitive science perspective

Sørensen essentially returns to a founding text in anthropology and the study of religion: Frazer's *Golden Bough*. But rather than doing as the majority of scholars have done, namely ridicule Frazer for having the mind-set of a parochial Victorian scholar,[5] Sørensen takes him seriously. At the very least, Frazer, in contrast to later generations, understood that in order to account for religion and to explain it, we better be able to account for ritual efficacy. To render Frazer understandable to the present academy, which has considered him the scholar par excellence to be castigated and ridiculed, but also to account for the thorny problem Frazer dauntlessly strove to tackle, Sørensen makes two moves. First, by using cognitive blending theory, he translates Frazer's ideas about contamination into the linguistic categories of metaphor and metonymy. Hence, we are in a better position to acknowledge what Frazer was speaking about when using similarity and contagion as the two key concepts for understanding magic. Whereas the one phenomenon works on the basis of likeness between two entities related by an overarching principle (metaphor), the other is founded on physical participation between the two entities involved (metonymy). In case one, we may think, for instance, of ritually simulating a bath in which we habitually cleanse ourselves of dirt. By virtue of the likeness, a similar effect may be ascribed to a metaphorical bath such as baptism in which the *initiand* is understood to have been cleansed of impurity. In the second case, we may think of inflicting harm on an enemy. If we can get access to something belonging to that person, such as a piece of cloth, we may engage in noxious magic by performing destructive actions on that piece of vestment. The linguistic figure, pars pro toto, functions, similar to indexical signs in the strict Peircian sense, in the same manner. The one partakes in the other by virtue of belongingness. Sørensen's second move is to transfer the semiotic categories into a cognitive science framework.

To illustrate the argument, take a prototypical example in magic that really takes the bull by the horns: piercing a doll vested with elements originating from one's foe and, thereby, inflicting pain, not to say death, unto one's mortal enemy. Causal expectations about the relationship between the pre-ritual situation ("I want to injure my enemy"), the ritual (the needle piercing of the doll made to look like my enemy), and the post-ritual situation accomplished by the ritual ("I succeeded in damaging my enemy") are weakened in favor of symbolically weaker basic cognitive representations built on iconic and indexical relationships. It is the contrast to ordinary expectations about causality and intentionality that permits attribution of efficacy to the ritual. The patent absence of common notions of causality (we do not go around injuring persons by inserting needles into figures in effigy) and the decoupling of intentionality (I am not directly responsible for the efficacy of the act, since some culturally postulated superhuman power has to perform the actual injury to that other person) may, in particular situations, trigger representations of the people partaking in the ritual in order to explain the ritual action in terms of other intentionalities at play. In this way, the amalgamation of iconic and indexical signs serves to undergird and stabilize the exchanges taking place at the symbolic level: "I have an antagonist whom I want to cause harm." The similarity between doll and actual person works iconically to suggest the result hoped for. So does the insertion of the needle penetrating elements de facto belonging to the adversary.

The argument rests on three insights from groundbreaking works in cognitive science. The first concerns our proclivity to attribute agency to a context in which it is not necessarily warranted, the so-called hyperactive agent-detection device, acronymically formulated as HADD, as it has been highlighted by Guthrie, Barrett, and Boyer. Humans are evolutionarily inclined to search for meaning in order to protect themselves from potential dangers or to improve their situation. When no apparent meaning can be assigned to semiotic "noise," humans are—based on their cognitive equipment—prone to search for other channels of meaning and intentionalities: for faces in the clouds, as Guthrie poignantly put it.

The second point rests on the observation that a constitutive feature of ritualization is the disconnect between the ritual participants' own intentions and the causal representations invoked by the ritual. Probably this disconnect is responsible for the ritual participants' urge to ascribe culturally postulated intentional superhuman agency to the accomplishments of the ritual. The fact that no obvious causality is at stake triggers ritual participants to search for alternative ways of instantiating causality, since we as humans intuitively take for granted that any action entails agents responsible for its performance.

The third point closely relates to the previous ones. It is the conspicuous lack of strong causal relationships between the individual sequences of the ritual and the expected or hoped for results of the overall performance that is a prerequisite for the stimulation to cognitively search for other forms of agency and intentionality. It is evident that these three points reinforce each other, since they underscore elements that reciprocally enhance representations of magical phenomena or ritual efficacy in a very straightforward manner. These points not only help us get a firm, scientific grasp on a traditionally protean concept in the study of religion, namely magic, but also enable us to take up the question of ritual efficacy in a non-enigmatic and conundrum-like manner. In light of the type of thinking briefly presented here, there is nothing strange about ritual efficacy. It is all plain and natural, and we can, therefore, explain it. We may even push the argument and studies further, when the technology allows us to do so, by actually measuring states and periods of effervescence when oxytocin and endorphins are secreted in high numbers, thereby contributing to the creation and enforcement of the self-conviction that some "real" superhuman agent is at play. At the present, it is difficult to do so with the brain scanners available, but technology is bound to enhance our chances of pursuing such studies in the future.

Personally, it was the work of Sørensen and colleagues on whom he relied that helped me in two significant ways. First, I was given the theoretical means that enabled me to work with a corpus of texts, that is, the Greek Magical Papyri, which I had previously found difficult to approach with respect to that overall and glaringly question-requiring problem: ritual efficacy. How could one possibly utter strings of gibberish and, thereby, obtain very concrete goals such as a cure from disease? In Sørensen, I came to understand why, since he theoretically paves the way for an understanding in which the palpable lack of causality between the nonsense expressed in the context of the ritual (endless strings of abracadabra) and its aim would have enabled people to ascribe efficacy to it and to relegate it to a superhuman agent. Needless to say, the same phenomenon occurs in the Gospel of Mark when Jesus suddenly turns to Aramaic as a magical language (5:41; 7:34). The intended audience begins searching for special agency and intentionality at play.

Even more so, through my increased engagement with cognitive studies over the years, I became familiar with a theoretical framework that enabled me to approach the question of ritual efficacy in general without falling prey to nonscientific mumbo jumbo about alleged effects. Such theorizing has also provided me with the means to return to Greco-Roman and Jewish miracle narratives (including early Christ-religion) that recount highly counterintuitive stories involving strong elements of a culturally postulated superhuman presence and its ability to act contrary to laws of nature. Far from explaining everything, the use of cognitive science for casting light on such narratives has helped me work with them in a considerably more informed manner. Biblical scholars usually beat about the bush with respect to such narratives in order to avoid the problems that really come to the fore both in the texts and in popular readings of them. What should we do with Jesus walking on the See of Galilee? Such stories are attention grabbing by virtue of their instantiation of highly counterintuitive elements. Yet they, thereby, trigger the listening audience to ascribe efficacy or reality to the mythical world depicted. And needless to say, this becomes even more prominent in a ritual setting (for more thinking along these lines, see my 2017[b, c, d] essays).

The indispensability of cognitive science for a history of religion worth its name

A number of other examples could have been adduced but I hope to have shown how I, as a historian, have found it immensely helpful to engage with insights from cognitive science. In general terms, the emergence of a cognitive science of religion twenty-five years ago has been extremely important by partly contributing to overcoming the far too great a gap between the humanities and the sciences. Cognitive science has been crucial in helping us bridge that ghastly gap.

As regards my own work, the emergence of cognitive science has led me to whole new ways of looking at the world, not least those ancient worlds with which I am mostly concerned. Cognitive science has enabled those of us who like to pose grandiose questions with respect to culture and religion to return to problems that have for far too long been ignored, if not even forgotten, in the humanities. Bellah's return to the old Durkheimian question of cultural evolution in the beginning of this millennium is a telling example. He already had the interest as a young scholar (cf. his 1964 article) but came to acknowledge that, were he to pursue a career in the academy, he would have to abandon this quest.[6] It all changed when Bellah began reading the works of Merlin Donald (1991) in the late nineties and realized that his interest in cultural evolutionary questions that he had missed in the early sixties was now available within a "safe" scientific framework that avoided the pitfalls of previous forms of evolutionary thinking.

Due to constraints of space, I have focused on one particular case that illustrates the help historians of religion may gain from cognitive science. I have concentrated on the traditionally knotty question of the ritual efficacy of magic that we have not been able to account for. Out of academic embarrassment, we have typically focused solely on the social dimensions pertaining to magic or have simply striven to explain away the phenomenon by engaging in terminological deconstruction. Yet, cognitive science provides us with a theoretical framework that enables us to take the bull by the horns. We may in fact explain magic in such a way that its enigmatic character disappears. It is no longer a conundrum, but constitutes an all too human cognitively explainable phenomenon. Similarly, such an approach allows us to turn to the miracle texts of antiquity in order to explain those strikingly counterintuitive aspects about which we have traditionally been beating about the bush. There is nothing strange about it. It may all be explained in a straightforward and natural manner.

Finally, I have made a plea for the indispensability of cognitive science for the study of the history of religion. The inclusion of such a scholarly trajectory helps keep the peculiarities of particular cultures, highlighted by history of religion, at bay by emphasizing human cognition in general. The relationship, however, is not one-sided. History of religion also serves or should be assigned the role of keeping cognitive science at bay, so that universal traits are not excessively exaggerated at the cost of an appreciation of the specificities of those sociocultural contexts in which they appear as distinct manifestations. In this manner, and in line with my overall argument, I conclude by subscribing to those beautifully expressed words of the encyclopedist par excellence, Goethe, in his Faust: "Wie alles sich zum Ganzen webt, Eins in dem andern wirkt und lebt!"

Part Three

CSR 2.0

9

Exiting the *Motel of the Mysteries*? How Historiographical Floccinaucinihilipilification Is Affecting CSR 2.0

Leonardo Ambasciano

> *[I]ntroducing new methods invigorates theoretical debates about basic concepts, types of data and the relation between models, data and the world. [. . .] Having embarked on an experimental study of religion, what happens to the field we originally belonged to?*
>
> Sørensen and Nielbo (2013a: 229, 226)

Introduction

In 1979, artist *extraordinaire* David Macaulay published a lavishly illustrated book with the intent to poke fun at the religious preconceptions that the archeological milieu of the nineteenth century was fond of, while targeting various (pseudo)historiographical tropes. The plot Macaulay devised was aimed at imagining present buildings, infrastructures, and paraphernalia as if they were archeological documents: 2,000 years after an unpredictable catastrophe buried late-twentieth-century North America, historians and archeologists thoroughly investigate what remains of that ancient civilization. In a bout of overzealously academic reverence, everything is primarily interpreted through religious lenses: highways become landing strips for cargo cults, a room in a squalid motel turns into a burial chamber, a bathtub is a sarcophagus, a water closet a sacred shrine, and a toilet plastic board a set of female ornamental jewelry (Macaulay 1979; Figure 9.1).

Despite the inevitable laugh (or possibly because of it), Macaulay's clever satire is a serious reminder of the deleterious impact of preconceptions on historiographical research: the more ancient the time frame under examination is, the worse the influence of wild speculations and greater the role played by absence-based inferential assumptions. A whole academic branch of history of religions (HoR henceforth) flourished, providing this kind of hazardous interpretations, and scanty remains interpreted as "motels of mysteries" proliferated along with neo-Cumontian and pro-Eliadean patterns of diffusion (cf. Penner and Yonan 1972; Wiebe 1999; Sjöblom 2005; Martin

Figure 9.1 "Harriet proudly strode around the site wearing the Sacred Collar and matching Headband. She also wore the magnificent *plasticus* ear ornaments and the exquisite silver chain and pendant" (Macaulay 1979: 36). From *Motel of the Mysteries* by David Macaulay. © 1979 by David Macaulay. Reprinted by permission of Houghton Mifflin Harcourt Publishing Company. All rights reserved.

2014b; Ambasciano 2014, 2016a). The first theoretical wave of cognitive science of religion (CSR from now on) contributed to overthrow this fideistic paradigm, already under siege by a devastating postmodernist critique. What about now?

What is CSR 2.0?

Today, some of the most vibrant and successfully funded areas of contemporary CSR are "1) the neurobiology of religion, 2) experimental science of religion and anthropology and 3) simulation approaches" (Pyysiäinen 2012: 13) or, according to another recent review, "(i) Laboratory experiments; (ii) Quasi-experimental field research; (iii) Quantitative historical research" (Bulbulia 2013b: 145), including the burgeoning field of social network theories (SNT) and computational big data approach (e.g., Collar 2013; Lane 2013; cf. Bentley and Ormerod 2012; Czachesz 2013). These vast "separate but interrelated areas" (Pyysiäinen 2012: 13) represent a major shift from the "original theoretical approaches within the human sciences," still typical of the first wave of CSR, toward a "more empirical, interdisciplinary field of study" (ibid.: 19–20; cf. Xygalatas 2014a). For the sake of brevity, these subfields are henceforth collectively labeled CSR 2.0 (sensu Xygalatas 2013).

This briefly sketched experimental turn constitutes a natural development in line with the history of other cognitive branches (e.g., artificial intelligence, (psycho)linguistics, neuroscience, etc.; see Bechtel et al. 2001). More specifically, CSR 2.0 has managed to avoid the ahistorical patterns of classic HoR, the postmodernist rejection of any kind of scientific or institutional knowledge, the poststructuralist proliferation of incomparable case studies, and the problematic gaps in our documentary record of ancient cultures and religions all at once by building on some experimental threads already traced by its CSR ancestor (e.g., Martin 2003, 2005b, and Slone 2006) and by delving deeper into present-day social and psychological experiments.

As in other biological and social sciences, the sine qua non methodological basis which underpins CSR 2.0 is the restrained setting in which the experimenters can control the signal-to-noise ratio through the manipulation of the involved variables, the reduction of potential biases via blinding procedures, the systemization of the sequence of procedures in a clear and reproducible mode, the uniform coding of the gathered data according to a shared and agreed on scientific methodology, and so forth. All variables being under control (e.g., sex, age, education, occupation, etc.), the "gold standard" of randomized comparative/controlled trials theoretically grants the "accurat[e] determin[ation] [of] causal relationships" (Xygalatas 2014b: 198). As in any scientific research, the implementation of a standardized lab protocol available to anyone in the field allows for further replication, falsification, or verification of every single step of the experimental process by the scrutinizing community of peers and fellow researchers (McCauley 2011). In order to investigate the relationship between universal cognitive proclivities and cultural particularities, crosscultural experimentation is often deployed (Henrich et al. 2010; McCauley 2011: 299n. 158). Yet, crosscultural comparison is just the first step.

The implementation of positron emission tomography (PET), magnetic resonance imagery (MRI), and functional MRI (fMRI) has revolutionized the methodological approach in cognitive science (Schjødt 2009) by shifting the focus from the theoretical "mind" to the actual "brain" and its functioning in vivo (Geertz 2009, 2010a, 2013: 45). Although naturally afflicted by external noise and by the difficult assessment of multiple variables, anthropological fieldwork carried out as "naturalistic experiment[ation]" offers a much needed correction to the artificial lab environment (Xygalatas 2014a: 344).

Finally, quantitative data gathering entails the addition of a diachronic dimension to such experimental approaches, whether by network analysis and simulation

(Xygalatas 2013), by phylogenetics and biological models (Bulbulia et al. 2013a 396), or by the "longitudinal study of religion", that is, contemporary data gathering "with a view to using these data later" in order to test the (future) historical interaction of religio-cultural covariates (Bulbulia 2013b: 150).

The diachronic extension of the sociopsychological and experimental results into the past is epistemologically supported by three of the main tenets of CSR 1.0, whether explicitly stated or implicitly assumed (Geertz 2014: 265):

1. The cognitive machinery of Homo sapiens is the result of the taxon's constrained evolution as a social primate (Guthrie 1993; Atran 2002).
2. The same cognitive mechanisms underpin both everyday and religious reasoning (Boyer 2001; Barrett 2004a).
3. Complex cultural behaviors (including religion and religions)—whether adaptive, maladaptive, or accidental by-product—are panhuman universals that emerged when the taxon developed behavioral modernity, reputedly between 100,000 and 50,000 years ago (Boyer 2001: 371; Tremlin 2006: 13–41; Martin 2014b: 349; see also Richerson and Newson 2009; Renfrew 2009; Geertz and Markússon 2010; Shea 2011; Geertz 2013).

The same basic scientific premises characterize also the adjacent fields of the evolutionary science of religion (ESR)—that is, the study of religious beliefs and behaviors as (potentially) adaptive (cf. Wilson 2002; Wilson and Green 2012; contrast with Colborne 2016)—and cultural evolution, which maps the divergence of cultural traits over time and tests competing hypotheses in historiography to detect diachronic evolutionary patterns within cultural/religious systems (e.g., Bulbulia et al. 2013b; Turchin et al. 2015; contrast with Pigliucci 2010: 45–55, and Kundt 2015).

Notwithstanding the well-known pitfalls of oversimplification, and taking into consideration the frequent theoretical and practical interbreeding between most of these experimental branches (see Purzycki et al. 2015), the following paragraphs also include a critical assessment of some of their conceptual underpinnings.[1]

It should be evident by now how rapidly sensu lato CSR 2.0 has developed since its inception, especially considering that in 2004 Justin L. Barrett, one of the pioneers of the experimental approach in CSR, lamented that the discipline still had "few experimentalists in its ranks and inadequate ethnographic work to draw upon" (Barrett 2004c: 415).

The misunderstanding of presentism

Given the aforementioned inclusion of a diachronic dimension in the ranks of CSR 2.0, it is somehow distressing to find a depreciative account of historiography in its most challenging works (an early example in Wilson 2002: 87). Basically, the idea is that "historians and scholars of religions [are] mere providers of data—essentially glorified research assistants" at the service of those scholars involved in the more prestigious scientific study of religion (Slingerland 2014b: 124; cf. Lane 2013: 167). This attitude is partly justified, given the failure of both the psychoanalytical approach of psychohistory (Weinstein 1995) and the post hoc, connecting-the-dots ideological frames of HoR (e.g., Smith [2001] 2004; McCauley 2011: 163; Martin and Wiebe 2012; Martin 2014b: 346). The problem, however, is widespread. Historiography tout court seems to be under fire.

As of late, it has become commonplace to read apodictic statements like the following: "History presents us with the puzzle, but the answers come to us through the recent work of anthropologists, behavioral economists, and social psychologists" (Norenzayan 2013: 60). Commenting on the role of narrative explanations in macrohistoriographical change, Alex Mesoudi wrote that, even though such narratives might be

> absorbing and meticulously researched [...], explanations of cultural change in history are typically limited by their informality and subjectivity. Purely verbal explanations cannot yield specific, quantitative predictions that can then be tested against historical data. And without clear criteria for testing alternative theories, historians often end up endlessly arguing back and forth with no way to determine which theory best explains a particular historical trend. (Mesoudi 2011: 123)

A recent groundbreaking and thought-provoking compendium about cultural evolution and the emergence of human ultrasociality, written by an interdisciplinary group of thirteen scholars, ends with a "wish list for future research" aimed at different disciplines with the telling exclusion of historiography, while the paragraph that intends to spur collaboration for "*Everyone!*" [*sic*] lacks a clear historiographical content (Jordan et al. 2013: 116). Apparently, CSR 2.0 indulges in throwing the historiographical baby out of the epistemological water.

The *Concise Oxford Dictionary* defines *floccinaucinihilipilification* as "the action of estimating something as worthless" (Pearsall 1999: 544). This obsolete noun might be used to describe this belittling of the historiographical dimension that, in turn, is a consequence of what historian David Hackett Fischer labeled as the "fallacy of presentism," that is, "a complex anachronism, in which the antecedent in a narrative series is falsified by being defined or interpreted in terms of the consequent" (Fischer 1971: 135). This historiographical blank slate has also led to the overzealous elimination of any poststructuralist framework to understand the social behaviors of H. sapiens (see Martin 2014b for a reappraisal), prompting the equally inappropriate outcry to defend the postmodern humanities against CSR (e.g., Day 2010). Two misconceptions seem to underpin the presentist bias:

1. Formulating hypothesis-driven research and testing predictions are the only way to do "real" science.
2. Mathematical modeling and statistical techniques are thought of as the ultimate mode to deal with the past (cf. Turchin 2008; Bulbulia et al. 2014).

Despite some previous brilliant attempts at reframing "[r]eligious studies as a life science" (Bulbulia and Slingerland 2012), I claim that both ideas are questionable and that the underlying misunderstandings have mainly arisen because of the epistemological unfamiliarity of CSR with "life sciences" as a whole and historical natural sciences in particular. In the following paragraphs, I highlight some perplexing problems that I think are detrimental to the discipline, with a caveat: as I mentioned earlier, most of the themes listed in this chapter overlap in various CSR 2.0 studies and ongoing projects, and they have been organized as follows for the mere sake of convenience.

Unbiased hypothesis-driven research?

To do science does not necessarily mean to be epistemologically objective and reliable. In the murky borderlands between natural sciences, the social sciences, and the humanities, "science" might become an empty buzzword just to gain respectability and prestige (Pigliucci 2010; Pigliucci and Boudry 2013). In this regard, perhaps, the gravest risk concerns the neurobiological study of religion, prone as it is to be exploited by "willing retheologizers" (sensu Wiebe 1989: 37) eager to find an explicitly physiological legitimization of their emically fideistic advocacy. The discipline of neurotheology (Newberg 2010: 51), thought of as the scientific way to support a "metatheology" and a "megatheology" (2005: 6494), represents the materialization of such a risk. Another related problem occurs when CSR researchers turn to pure theological and/or intelligent design-like explanations to justify the cognitive appeal of religiosity (Barrett 2009; Barrett and Trigg 2014). Both stances exit de facto scientific research and, fortunately for our brief digression, they have been already flagged, assessed, and falsified (cf. Boudry and De Smedt 2011; Martin 2014b; Geertz 2009; Wiebe 2013: 144; see Ambasciano 2016a).

More specific arguments for CSR 2.0 involve funding and confirmation biases. First of all, as it has been already noted (Wiebe 2009), the mix of funding opportunities from private and religion-friendly institutes and/or institutions and a certain idea of hypothesis-driven research are selectively promoting a pro-religious stance that affects the choice of the initial hypotheses to test and imposes a potentially discriminatory use of evidence from the documentary sources. Therefore, the first problem is the constraining pressure private funding and external referees may exert on CSR 2.0 and related disciplines, especially when they collectively act as a selective force capable of shaping and directing the research (cf. Turchin et al. 2015: 87). Consequently, the premises of the hypotheses to be tested are sometimes presented ipso facto as pro-religious instead of establishing a null hypothesis as a starting point (e.g., Bellah 2011 as baseline of Turchin et al. 2015; cf. Wiebe 1989; see also Stausberg 2014b and Ambasciano 2016b). Other recent hypothesis-driven studies in CSR 2.0 have targeted the positive resilience and beneficial support provided by religious affiliations in certain population samples against a set of variables and longitudinal data. In most cases, given the selective choice of control groups, coded characters, and variables, the outcomes usually support the initial hypothesis (e.g., Bulbulia 2013b: 151–2; cf. Jong 2013 and Galen 2015). The same occurs when comprehensive psychological meta-analyses are performed within the same religiously prosocial framework (van Elk et al. 2015). Although word limit prevents any further comment, it should be noted that the pro-religious stance in social science research has become a major issue that urgently needs to be reassessed and readdressed (Martin and Wiebe 2014; Galen 2015; see also Alexander 2013: 408–9).

The second problem concerns the potential incidence of confirmation bias. The spreading citation of Robert Wright's book *The Evolution of God* (2009) supplies an interesting case. Wright's volume has been recently cited as a sound example of the selective change through time of the Abrahamic god from a "rather limited, whimsical, tribal war god—a subordinate in the Canaanite Pantheon—to the unitary, supreme, moralizing deity of two of the world's largest communities" (Norenzayan et al. 2016: 8; passage previously included, almost verbatim, in Bulbulia et al. 2013a: 398; both versions preceded by Norenzayan 2013: 133). Although Wright's book was not intended as a research publication, the main idea of *The Evolution of God* is assumed to be a solidly scientific starting point. However, Wright's volume provides a biased account of the "evolution" of the Abrahamic god and supplies a plea for the legitimization of metatheological advocacy (Beit-Hallahmi 2016; cf. Wiebe 2005). From Wright's perspective, the ultimate roots of the cultural diachronic mechanisms of change are tautologically identified in a deus ex machina that implicitly acts as a guarantor of the moral order and as the supervisor of an orthogenetic path of directed cultural development (Wright 2009: ch. 20; see also Wright 2000; cf. Pinker 2011: 694, and Ambasciano 2016b; on the concept of Abrahamic religions, cf. Hughes 2012).

The interdisciplinary team that uses Wright's book as a starting point is especially interested in asserting the importance of moralizing/monitoring/punishing High/Big Gods for the evolution of human ultrasociality (Norenzayan et al. 2016). However, even when deprived of its prosocially biased, intelligent design-like background, Wright's thesis should not be taken sic et simpliciter as a starting point, but rather considered as an assumption to be rigorously tested through cultural phylogenetic analysis (to avoid equating correlation with causation) and carefully contextualized against the historiographical and contextual backdrop of intertwined multicausal features (e.g., a cogent falsification has been already provided by Watts, Greenhill, et al. 2015 and Watts, Sheehan, et al. 2015). This begs the question of what might count as a scientific hypothesis-driven research in CSR 2.0 and raises the issue of careless data mining in secondary historiographical sources. Other examples of a priori presentist biases are the epistemically ungrounded exploitation of contemporary hunter-gatherers as an interpretive paradigm for reimagined Paleolithic cultures (e.g., Donald 1991: 268; 2009: 101; Bellah 2011; Norenzayan 2013: 51, passim; see Ambasciano 2016a) or the oversimplification of the tangled historical roots of cultural events with the aim to favor the inborn, innate cognitive mechanisms that seemingly prompt religiosity to spring up everywhere even when banned (cf. the "Soviet experiment" in Norenzayan 2013: 172, a passage that disregards the complex history of Western policy, Christian missions, and local religious revivals during and after the era of Soviet Socialism; cf. Znamenski 2007; Ambasciano 2016b).

Moreover, the projection of present-day experimental results into the deep past of H. sapiens has sidelined the social and cumulative construction of culture (Bloch 2008; McCauley 2011; Pyysiäinen 2012; McCutcheon [2010] 2012; Geertz 2014: 267). The Early Neolithic site of Göbleki Tepe (located in the Şanlıurfa Province of modern-day

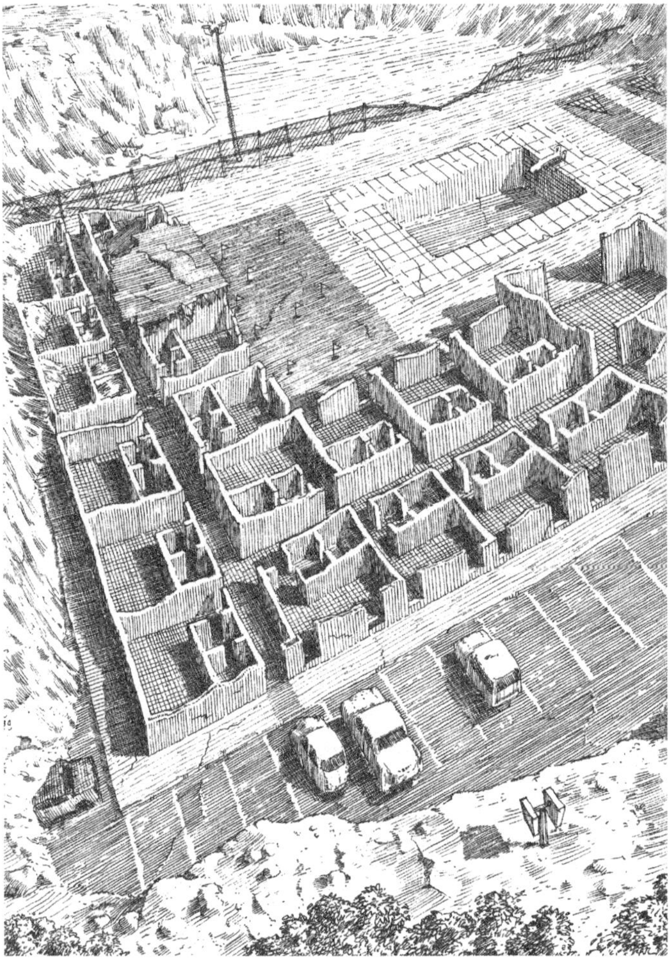

Figure 9.2 The "vast funerary complex" as recovered by future archeologists, actually a motel whose rooms are misinterpreted as being individual tombs. Seen from the same biased perspective, the structure comprises a "sanctuary" (the huge hall), a (sacrificial) kitchen, a "ceremonial pool" (more prosaically, a swimming pool), and a "specially marked funerary game areas intended to occupy the spirits of the dead during eternal life" (a miniature golf course; Macaulay 1979: 39). From *Motel of the Mysteries* by David Macaulay. © 1979 by David Macaulay. Reprinted by permission of Houghton Mifflin Harcourt Publishing Company. All rights reserved.

Southeastern Turkey, dating to ca. 11,000 years ago) provides an interesting example. This site has been cited as compelling historical evidence for the emergence of Big Gods and their temples as a social glue that preceded, and boosted, sedentism and agriculture (Norenzayan 2013: 119–20). Put bluntly, religion came first (contra the economy-came-first approach proposed in Baumard and Boyer 2013 and Baumard

Figure 9.3 Planimetry and stratigraphic units of the archeological site of Göbekli Tepe. Enclosures A–G are highlighted. Graphics: Klaus Schmidt, © Deutsche Archäologische Institut.

et al. 2015). Just as the "grandest room" of the "motel of the mysteries"—actually the entrance to the building—was mockingly identified as the "communal sanctuary" in a "vast funerary complex" (Macaulay 1979: 39; Figure 9.2), a series of outstanding enclosures in Göbleki Tepe are ipso facto interpreted as religious buildings (Figure 9.3). However, the conclusive archaeological evidence for the proposed interpretive framework is currently lacking (see Banning 2011: 630, for "architectural variability"). Even if the religious nature and use of those structures were confirmed (cf. Dietrich and Notroff 2015), the correlation would still not prove causation as to the Big Gods' hypothetical scenario of ultrasocial and political developmental complexity (Skjoldli 2015; Watts, Greenhill, et al. 2015; Ambasciano 2016b; cf. Stoczkowski [1994] 2002).

The past, mathematically speaking

"Cliodynamics," a term coined by Peter Turchin to indicate the mathematical modeling "of temporally varying processes and the search for causal mechanisms" in order to test competing hypotheses in human history (Turchin 2008: 34), has been co-opted by CSR 2.0 to tackle the problem of the origins of ultrasociality and to quantify the historical role of religion/s among other historical variables. Based on this premise, the huge, collaborative work in progress called "Seshat: Global History Databank" aims to build "a vast repository for structured data on theoretically relevant variables for any past human society, for which such data exist," from 10,000 years ago to the present (Turchin et al. 2015: 86; see the Seshat Web page dedicated to methodology available at http://seshatdatabank.info/methods/).[2]

Unfortunately, the ecological focus on the discovery of internal dynamics and patterns of dichotomous predictions, while heuristically indispensable, minimizes the incidence of the innumerable sequence of interconnected previous events that may result in complex nonrecurrent accidents. Well-known examples of such contingency are the "black swans" (Taleb 2007), a label used to define the nonpredictable nature of historical junctures as a result of stochastic interactions rationalized post hoc via hindsight bias, and the evolutionary "panda principle" (Gould 1986), according to which selective pressures favor the rearrangement and novel use of already existing physiological structures in unexpected ways. Contingency constitutes the main characteristic of all historical sciences, namely paleontology, paleoclimatology, epidemiology, evolutionary biology, historical geology, cosmology, and, of course, historiography (Gould 1989: 283). The methodological hazard is evident in the potential conflation and/or convergence of at least three different problematic threads:

1. Pro-religious hypotheses assumed as starting points (as touched on in the previous section)
2. A paucity of data affecting nearly every domain of the past: the diachronic distance from the present and the loss of fine-grained details are directly proportional
3. Our innate tendency to draw inferences, connecting the dots, and see "meaningful patterns in both meaningful and meaningless noise" (Shermer [2011] 2012: 70; cf. Martin 2014b).

A similar set of theoretical problems affects the application of SNT theories to the ancient world when adopted, but not thoroughly adapted, from the contemporary big data approach. Here, again, the gaps and the poor quantitative samples make the analysis susceptible to include a high level of non-epistemic noise, while statistical techniques suitable to control biases and normally used in big data analysis might contribute to strengthening those very biases and reinforcing, or tautologically confirming, the a priori hypothesis (cf. also Cleland and Brindell 2013: 199; pace Bulbulia et al. 2014). A complicating factor is the irrelevance of the content of the cultural message, as typically assumed by SNT analyses, that starkly contrasts with the most basic assumptions of CSR. In the end, most of the SNT analyses depend on classically descriptive, prosopographical attribute analysis (Woolf 2012).

Does this mean that trends are nonexistent or not detectable, or that the aforementioned projects are something akin to the teleological philosophies of history of old (cf. Medawar and Medawar 1983)? Not at all. Of course there are trends and patterns, ceteris paribus, and these projects, like other phylogenetic assessments of cultural clusters' change through time (Bulbulia, et al. 2013; Watts, Sheehan, et al. 2015), hold much promise for a scientifically renewed study of the *longue durée* (Braudel [1958] 1980; Gould 1986; Sulloway 1996; Armitage 2012; Smail 2008; Christian 2010; Diamond and Robinson 2010a; Martin and Sørensen 2011; Shryock and Smail 2011; Guldi and Armitage 2014). Yet, on the thin line between methodology and epistemology, an excessive trust in the strictly physics- or ecology-like structure of recurrent patterns in historiography might be misleading (Pigliucci 2010: 52; Spinney 2012). In complex cases such as historiography, with its mix of predictable patterns and unpredictable contingencies, only the continuing description and accumulation of ancient data and, most of all, the ongoing discovery of new data might settle once and for all two competing narrative hypotheses. Let me illustrate these points with the following examples.

In 1984 paleontologists David Raup and Jack Sepkosky identified a common pattern in the occurrence of mass extinctions during the Phanerozoic (i.e., the last 541 million years ca.). Using a compendium of the available paleontological record, they pinpointed a mass extinction every 26 million years starting from 250 million years ago and looked for non-terrestrial cosmological causes (Raup and Sepkosky 1984). Astronomers then proposed a whole bunch of hypotheses, and a tentative consensus emerged indicating as a possible source of the recurrent pattern the timely disturbance of a passing star (dubbed "Nemesis") through the Oort Cloud, perturbating the course of asteroids and causing some of them to periodically strike the Earth (Raup [1987] 1999). It was an elegant answer for what looked like a statistically well-established pattern, sprung from an incredibly interesting and lively research program (cf. Sepkoski 2001; Parsons 2001: 48–79). However, the alleged pattern was revealed to be a statistical artifact (Clavin 2011). Every single mass extinction event counts as a single story until proven otherwise (e.g., Alvarez 1997). In other words, mass extinctions do not form a single "species" of events, but they superficially resemble each other because of their effects.

Also, think about Lord Kelvin's (William Thomson; 1824–1907) proposal of a mathematically sound and quite recent geological origin of the Earth, already in contrast with the coeval paleontological record (i.e., evolution needs much longer time) and later disproved by the discovery of radioactivity (Rudwick 2014: 231–3).

Or, just to pick up another example, let us consider the branch-based clade *Arctometatarsalia*, nested among the non-avian theropod dinosaur lineage. *Arctometatarsalia* was originally diagnosed by a particular morphology of the foot skeleton, named "arctometatarsalian condition" (Holtz 1994; Hutchinson and Padian 1997; see Figure 9.4). Further studies and new data revealed that the arctometatarsalian condition was reached independently (at least four times) by the several lineages that were thought to share a common ancestor: *Arctometatarsalia* has turned out to be a polyphyletic assemblage (i.e., a set of distantly related organisms grouped together due to erroneous phylogenetic hypotheses) in all the recent analyses of theropod relationships. Hence, as elegant as the explanation was, the arctometatarsalian group was an artifact based on the then available paleontological record, the comparative selection of out-groups, and various miscodings (Holtz 2001). Therefore, given the highly homoplastic nature of the features supporting the clade,[3] *Arctometatarsalia* is untenable as a parsimonious evolutionary explanation—a thesis confirmed by the subsequent discovery of the ancient members of the lineages previously grouped in that clade, each lacking the arctometatarsalian condition (Figure 9.4; for homology, i.e., inheritance from a common ancestor, and analogy, that is, the results of independent evolutionary pressures on similarly constrained structures, in the historical study of cultural phenomena, religions included, see Smith 1990: 47–8).

Mathematical soundness and sheer elegance in historical sciences do not necessarily mean correctness (Gould 1989: 279); even so, errors conducted within a legitimate,

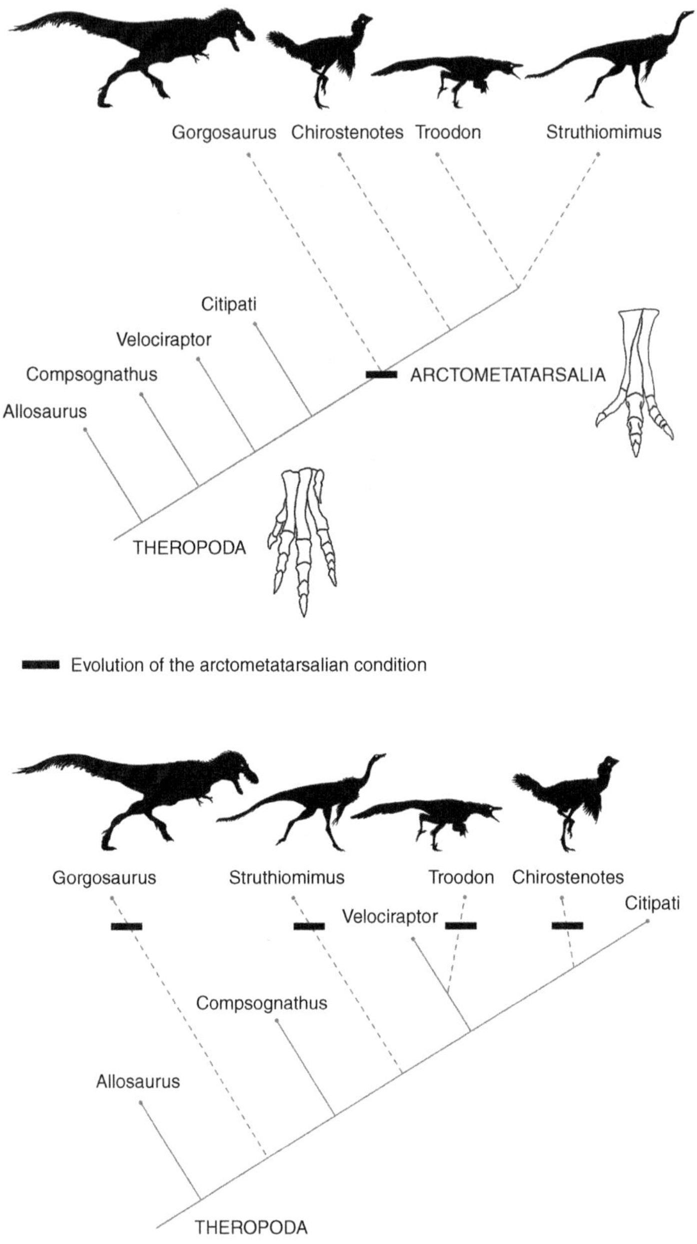

Figure 9.4 The rise and fall of the *Arctometatarsalia* clade in two cladograms. The top image illustrates the result of the original analysis by Holtz (1994) in which he proposed the monophyletic ascendance of the clade. The bottom image, instead, represents a more recent re-analysis of the original dataset by Holtz that, with the benefit of hindsight, indicates that the arctometatarsalian foot evolved independently four times (black rectangles) from the original non-arctometatarsalian condition. Reproduced by permission of Fabio Manucci. © 2016 by Fabio Manucci. Scientific supervision by Andrea Cau.

bias-controlled methodological and epistemological framework constitute the backbone of scientific progress. This is why, concerning CSR 2.0, methodology cannot do away with epistemology or descriptive analysis (Dennett 1995: 21). No matter how vast the database or how reliable the algorithms, the outcome will always be context-dependent (i.e., codings and matrices are subjected to human action and selection, while the historico-cultural dataset itself might not be trustworthy per se; cf. Lane 2013: 169; Watts, Sheehan, et al. 2015; see also Medawar 1969: 28). For instance, the assertion that the Seshat databank, originally created within a conceptual Axial Age perimeter, is in toto "theory-neutral" tends to minimize the selection of descriptive secondary sources (Turchin et al. 2015: 87). Mathematical modeling might help discern the incidence and relationships between attested historical and cultural variables, yet the paucity of data, the taphonomic bias, the incidence of contingent events, and the "hoary problem of induction," which not even Bayesian confirmation theory can truly escape, keep on affecting the resulting picture (Mesoudi 2011: 114; Cleland and Brindell 2013: 199).

If it's not dressed in a lab coat, it's not science. Or, is it?

In order to convince private institutes and/or public institutions of the direct, real-world application of such a discipline in an unfortunate politico-cultural environment that is recklessly ignoring that academic research is cumulative and its results are unpredictable a priori, and, consequently, is cutting the funds reserved for "basic scientific research that has no clear, or even hinted at, bearing on practical issues that might benefit the public" (Pigliucci 2015), CSR 2.0 seems determined to develop a neontologization of the humanities (more on this in a moment), with a pinch of a physics-like futurologist penchant for prediction (Barrett 2011: 163; Whitehouse 2013b: 281; Turchin et al. 2015: 79). Consequently, researchers in CSR 2.0 are urging their more historically oriented colleagues to join the experimental ranks (cf. Xygalatas 2014b: 198), given that the "potential represented by the sheer volume of data from ancient cultures greatly outweighs the limitations in the ways in which we can interrogate this data" (Slingerland 2014b: 123).

Are these heartfelt pleas epistemically appropriate? Yes, because given the obvious biological *trait d'union*, a middle ground can be ultimately found, and no, because three problems reinforce the floccinaucinihilipilification of historiography and limit the potential usefulness of this epistemic marriage (Lawson 2004):

1. The idea that scientific approaches must be sic et simpliciter lablike.
2. The predictive power of a scientific historiography.
3. The aforementioned neontologization.

First, the roots of the epistemic misjudgment according to which only lablike science is effective are, unfortunately, very deep (Gould 1989: 279; Mayr 2004: 32–3; Diamond and Robinson 2010b: 1–5). If historians have "often import[ed] an oversimplified caricature of 'hard' science, or simply bow[ed] to pronouncements of professions with higher status" (Gould 1989: 279), it is not uncommon for hard scientists to contemptuously regard history tout court as pseudoscientific based on their mistaken understanding of logico-mathematical evidential relations as exclusive *explananda* (Cleland and Brindell 2013: 193). However, Carol E. Cleland and Sheralee Brindell have rightly argued that the latter presupposition rests on fallacious, logically narrow assumptions that neglect the difference between two kinds of scientific inquiry—that is experimental sciences and historical field sciences. The salient feature that distinguishes the latter from the former is the over-determination of causes and the under-determination of effects. In order to overcome the experimental limitations and the sheer impossibility to manipulate the variables to unmask false positives and false negatives, historical field sciences rely on past records and on the detection of an intertwined triggering event(s) and "smoking guns" in the multicausal entanglement of historical facts (Cleland and Brindell 2013: 194, 196).

Second, for the time being there is no way to devise or implement predictive, reliable mathematical models that may allow researchers to track down the precise interaction of complex, multivariable open systems on a diachronically global scale (Cleland and Brindell 2013: 190; cf. Popper 1957; Gould 2000; Lewontin and Levins [1999] 2007; for historiographical retrodictions, see Eldredge 1999: 186n. 13). Not to mention that, even if we grant for the sake of argument that this could turn out as a workable possibility, there will always be a moment in human historiography when individual concerns and contingencies are quintessential (Pigliucci 2010: 52).

Finally, neontologizing CSR as a whole might be a grave conceptual mistake. Neontology is a branch of biological science concerned with the extant animal kingdom embedded in well-known ecological relationships, the antonym being paleontology—the discipline that studies the fossilized remains of extinct animals from the deep history of planet Earth. As far as the social sciences and the study of the contemporary social world are involved, fair enough: a "biological study of behavior" reserved to H. sapiens and akin to proper ethological analyses (Hogan and Bolhuis 2009), reinvigorated by cognitive and evolutionary sciences, is firmly within our grasp (Bulbulia and Slingerland 2012). Yet, considering all the presentist biases recalled previously, if the very same approach is thought to be feasible for past human history and historiography, here is a perfect Whiggish recipe for disaster. As a matter of fact, modern and contemporary historiography are characterized by an overabundance of private and public documents that allows us to have a much better picture of the recent past, even in its minuscule details. Ancient historians, notwithstanding Hollywood imagery and overblown assertions by some scholars, do not normally have such a possibility.

Conclusion: My map is (not quite exactly) your territory

While the unity of mind of H. sapiens as a taxon theoretically allows for psychological studies to test the heuristic value of competing retrodictions within a reliable margin of accuracy (e.g., Griffith 2014), cultural transmission did not happen in a historical void. Then again, we miss a huge amount of relevant data, and this invariably affects the very production of hypotheses to test. What to do next, then?

It is interesting to note that the belittling of historiography is strikingly similar to the harsh judgment that the triumphant genetics imposed on the Modern Evolutionary Synthesis, "denigrat[ing] paleontology as a kind of glorified stamp collecting. If you wanted to determine the mechanisms of evolution you studied genetics in the laboratory" (Switek 2010). Yet, this opinion was *en vogue* before the "paleobiological revolution" of the second half of the past century, when new theories about evolutionary trends, vigorous debates about mass extinctions, and levels of selection sprang up and reestablished the role of paleontology at the high table of academia as a historical science (cf. Raup and Jablonski 1986; Sepkoski and Ruse 2010). We still lack a similar revolution in historiography, but the resurgence of *longue durée* and related scientific approaches (i.e., cognitive and evolutionary) holds much promise for the near future (McNeill 2001). Therefore, temporarily splitting CSR 2.0 from cognitive historiography, that is, the study of "dead minds" and their ideas, has probably been the wisest decision (Martin and Eidinow 2014; cf. Martin and Whitehouse 2004 and Martin and Sørensen 2011). Yet, this is just the first step.

Religion has been explained (Preus 1987: xviii, 205–11; see Guthrie 1980; Lawson 2000; Boyer 2001; Slone 2006) much in the same way evolutionary theory explained the mind-blowing variety of organisms in space and time. Does this mean that we should give up any further descriptive study or narrative explanations? The short answer: no. Case in point: Did evolutionary theory prevent paleontology from being also a narrative, descriptive discipline? Absolutely not. Incomplete logs of facts in the whole range of the historical sciences need interpretation. Since uncertainty cannot be fully erased, interpretations require a consilience of inductions able to minimize internal errors (Shermer and Grobman [2000] 2009: 30–1). For instance, we definitely need more unbiased, qualitative, historico-cognitive, strictly scientific (re)descriptions and codings of socioreligious taxa from the past, as in basic paleontological research. Without this new and/or updated data, any quantitative modeling is miserably bound to fail, just like classic HoR failed before (Martin and Wiebe 2012; Martin and Pyysiäinen 2013; Watts, Sheehan, et al. 2015). We do not need another catalog of "motels of the mysteries," a zombie revival of Wilhelm Schmidt's *Urmonotheismus*, theologically biased Axial Age ideas, or an Eliadean (proto)shamanism in disguise (see Ambasciano 2016a).

I have argued that interested historians and researchers in CSR could greatly benefit from a deeper understanding of the history, methodology, and epistemology of other historical field sciences, especially paleontology and paleobiology, which have already devised brilliant ways to conceptually reframe these problems. Ultimately, I contend that most of the previously highlighted problems could be quite easily overcome by carefully adapting experimental sciences to the peculiar historical milieu, both theoretically (involving paleontologists, archeologists, cultural geographers, philosophers of science and, of course, historians) and practically (just an example, implementing cultural geography and spatial/innovation diffusion to improve SNT theories; e.g., Haggett 2001: 202–31, 482–503). Interdisciplinary collaboration has always been one of the most important features of the cognitive sciences (Bechtel et al. 1998). We absolutely need a CSR 2.0, but without a comprehensive understanding of historiography, the experimental revolution is condemned to remain ad libitum unfinished.

Acknowledgments

I would like to thank with gratitude the editors of the present volume, an anonymous reviewer, and, in alphabetical order, Andrea Cau, Emina Hadzifejzovic, Fabio Manucci, and Elisabeta Pop for their help during the writing process. All remaining errors are obviously my own.

10

Minimal Counterintuitiveness Revisited, Again: The Role of Emotional Valence in Memory for Conceptual Incongruity

Michaela Porubanova and John H. Shaver

Introduction

The inauguration of the systematic study of memory is generally attributed to the German scientist Herman Ebbinghaus (Bower 2000), though the possibility of a memory advantage for certain types of information has intrigued scientists since at least as early as Aristotle's treatise "On the Soul" (Danziger 2008). It was F. C. Bartlett, however, who began to investigate the effects of culture on memory, thus initiating cognitive approaches to the study of cultural phenomena (Schachter 1995). In Bartlett's groundbreaking research (1932), American students were taught a Native American folktale, "The War of Ghosts," and asked to retell the story. His study showed that the output of memory (i.e., what is recalled) can be substantially different from the input (i.e., what is taught) and that a large source of variance in output is the result of a priori cultural knowledge. In other words, Bartlett ([1932] 1995) observed that cultural experience affects how knowledge is organized and how incoming information is processed. Bartlett further suggested that preexisting knowledge is organized in schemas, or basic mental recognition devices that provide unconscious inferences about the functioning of the world. His work became the foundation for modern schema theory (D'Andrade 1995; Mandler 1984; Purzycki and Willard 2016; Rumelhart 1980; Schank and Abelson 1977).

Despite the early insights of Bartlett, for the first half of the twentieth century, cognitive psychology largely ignored the effects of culture on cognition, and anthropology largely neglected the influence of cognition on culture (D'Andrade 1995). Beginning in the 1960s, however, cognitive approaches grew in prominence across the social sciences, and roughly a quarter century ago, cognitive science of religion (CSR) emerged with claims to being a distinct approach to the study of religious phenomena. From its beginnings, CSR has placed an emphasis on the cognitive origins of religious phenomena and paid particular attention to how universal features of the mind might generate religious beliefs or behaviors (e.g., Lawson and McCauley 1990).

Early in the development of CSR, Boyer and Ramble (1991) suggested that some religious concepts—minimally counterintuitive or "MCI" concepts—are "attractive" to our cognitive architecture, and this attractiveness explains their (presumed) cross-cultural prevalence. Specifically, MCI theory posits that concepts that violate our innate, hard-wired ontological expectations (i.e., those inference sets that comprise naïve physics, naïve biology, and naïve psychology) represent attention-grabbing information and, thus, are more memorable than concepts with other structures. Such ontological expectations are "deep inferences" about the world, and violations to those expectations differ from the sorts of violations to culturally learned representations organized in schemas, that is, violations to "shallow inferences" (Purzycki and Willard, 2016). From the beginning, then, CSR scholars assumed they were assessing memory advantages for concepts that violate evolved expectations, rather than expectations of cultural schemas. This seems in part a result of greater influence of European cognitive anthropology over American cognitive anthropology in the development of mainstream CSR. Contemporary American cognitive anthropology focuses on the study of cognitive representations of cultural information as organized in generalized schemas (e.g., D'Andrade 1995), while European cognitive anthropology tends to focus on how evolutionarily specialized cognitive modules process and attract cultural information (e.g., Sperber 1996).

MCI theory has today become the staple in CSR, despite suffering from several serious deficits (Purzycki and Willard, 2016). Here we focus attention on two of these shortcomings. First, and most importantly for our concerns here, the theory relies heavily on questionable and unoperationalized assumptions. No one working under the rubric of MCI research has ever demonstrated that expectations of naïve inferential systems are being violated when people are presented with incongruous information (ibid.). Meanwhile, research has shown that violations to people's *cultural expectations* enjoy a memory advantage over what are conventionally conceived as violations to evolved ontological expectations (Porubanova et al. 2013; Porubanova et al. 2014). Thus, there may be nothing unique, or particularly memorable, about MCIs. Second, MCI research often does not examine other factors known to influence memory, specifically those that could possibly counteract/interact with a memory advantage for concepts of an incongruous nature. Here we address this latter possibility and examine the relative influence of emotion on memory for incongruous concepts.

Emotion and conceptual incongruities

CSR research (e.g., Barrett and Nyhof 2001; Boyer and Ramble 2001; Porubanova et al. 2013; Porubanova et al. 2014) as well as general cognitive psychological research (Chang and Sanfey 2009; Hirshman 1988; Schmidt 1985; Stangor and McMillan 1992) suggests that expectation violation might be one of the driving factors in the long-term retention of concepts. In other words, the attention-grabbing and surprising qualities of conceptual incongruities might underlie their potential for greater recall. Importantly, information that violates an expectation is often accompanied by an initial negative emotional response (Bettencourt et al. 1998; Hirshman 1988). Even if only slight, incongruous concepts should evoke surprise (by their very definition), and therefore, any memory advantage for incongruous concepts, if it exists, might lie in the physiological responses they evoke (Purzycki 2006) and the accompanying elaboration of the concept that stems from revision of the currently activated schema (Bartholow et al. 2001).

The prioritized processing (i.e., increased rehearsal, attention, and elaboration) of emotional information clearly has an evolutionary origin—emotions gauge incoming stimuli for potential fitness relevance and effects (e.g., Frank 1988). Automatic responses to threatening stimuli, in particular, induce rapid, adaptive behavioral responses and are already evident in newborns (Cummins and Cummins 1999). In general, we experience negative emotional states when we are exposed to threats to reproductive and somatic effort, and positive affect in response to cues that reliably enhance fitness (Hamann 2001; Lazarus 1994). Moreover, recent research suggests that there is a memory bias for fitness-relevant information (Broesch et al. 2014).

The proximate mechanisms postulated behind the memory advantage of emotional information are increased attention (Anderson 2005; Anderson and Phelps 2001), rehearsal (Hamann 2001; Talarico and Rubin 2003), novelty and excitement (Dolan and Vuilleimer 2003), and poststimulus elaboration (Christianson 1992). In general, emotional stimuli are preferentially selected by the attentional system (Anderson 2005; Öhman et al. 2001); highly emotional events are more memorable than neutral events (Anderson and Phelps 2001; Reisberg and Heuer 1992), even when they are task irrelevant (Pessoa and Underleider 2004; Schimmack 2005).

Despite the long-known influence of emotion on memory, speculation that emotion ought to influence retention of MCI concepts (Purzycki and Willard, 2016; Taves 2015) and transmission of cultural information (Nichols 2002), few CSR studies have addressed their effects (Purzycki 2010, 2011). Next, we examine the relative influence of emotion on memory for schema-congruous and schema-incongruous information.

Subjects

Seventy-three undergraduate students at the State University of New York (SUNY), Farmingdale (53 female) completed the experiment. As remuneration for their participation, all participants received course credit. Ten students did not complete all phases of data collected, thus data from sixty-three participants (48 female) remained for analysis.

Stimuli

The concept set consisted of thirty concepts created by the experimenters and colleagues. We employed two factors in concept design: violation category (conceptual incongruity [concepts with a status that is inconsistent with its conceptual prototype[1]], or congruity [concepts that are consistent with conceptual prototypes]) and emotional valence (positive, negative, or neutral). Each category combination (violation category + emotional valence) was represented by five concepts.

Each concept contained a person or an object performing an action (see Appendix). The experiment was run in SuperLab 5 and was carried out on a 17-inch color screen laptop. Participants viewed the concepts approximately 56 cm from the computer screen.

Emotional valence was determined by pretesting sixty concepts with fifty-five college undergraduates at SUNY Farmingdale, administered via SurveyMonkey in a computer lab. Students rated each concept for valence and arousal on a scale of 1–5. Based on valence ratings, we selected negative concepts (average rating: 1–1.5), neutral concepts (average rating: 2.75–3.25), and positive concepts (average rating: 3.5–5) for inclusion in the experiment. The mean valence rating of each emotional valence category was significantly different from each of the other $F\ (2, 18) = 608.46$, $p < 0.001$; negative concepts ($M = 1.22$, $SE = 0.08$) were rated as lower on emotional valence than neutral ($M = 2.93$, $SE = 0.04$): $t = -1.71$, $p < 0.001$; negative lower than positive ($M = 4.08$, $SE = 0.09$): $t = -2.86$, $p < 0.001$; and neutral lower than positive: $t = -1.15$, $p < 0.001$.

Procedure

Prior to the beginning, the participants were asked to sign an informed consent form and instructed to remember the information presented on the computer screen. The experiment involved three different experimental phases, with each phase presented twice: two study phases, two distraction phases, and two test phases. During the study phases, concepts were presented in a random order at the center of the screen (white print on a black screen) for 2,000 milliseconds with 1,000 ms interstimulus interval. A fixation cross was displayed at the center of the computer screen prior to each concept presentation to signal the beginning of a new trial. The study phases were followed by two-minute distractor tasks that asked participants to solve simple arithmetic mathematical equations. After the distractor tasks, participants were asked to write as many concepts as they could remember in three minutes. Recall of a concept was considered accurate either when the entire concept was recalled correctly or when subjects included semantically equivalent words (e.g., if the concept included "a child" performing an action but the participant indicated "a kid" performed the action). If the recalled concept was semantically similar to the presented one, a consensus of two independent raters blind to the hypothesis was necessary to deem the concepts equivalent. No feedback was provided during the free recall tasks. A week after the first experiment, participants were contacted via e-mail with a request to complete a surprise recognition survey. These surveys asked participants to indicate whether or not a concept was included in the original experiment and how sure they were about its inclusion. The recognition survey included the original thirty concepts and thirty fabricated concepts in randomized order. New items included an equal number of violations with similar emotional valence ratings as those presented in the original experiment. These surveys were conducted online through SurveyMonkey.

Results

Immediate recall

A 2×3 (violation category [conceptually incongruous, conceptually congruous] × emotional valence [negative, neutral, positive]) repeated-measures analysis of variance was computed in order to examine the effects of violation category and emotional valence on immediate recall (Mauchly's test was nonsignificant). Recall for each category was computed as a percentage of concepts in each category recalled correctly in both recall 1 and recall 2. There was a main effect for violation category: $F(1,72) = 10.16$, $p < 0.01$, $\eta^2_p = 0.13$; and also a main effect for emotional valence: $F(2, 144) = 7.82$, $p = 0.001$, $\eta^2_p = 0.10$. An interaction between violation category and emotional valence was also observed $F(2, 144) = 8.28$, $p < 0.001$, $\eta^2_p = 0.10$. Follow-up post hoc tests using Bonferonni corrections revealed that memory for incongruous concepts (M = 21.08, SE = 1.35) was lower than memory for congruous concepts (M = 25.09, SE = 1.35), $t(69) = 4.02$, $p < 0.01$. These findings are similar to studies that have shown an immediate memory advantage for congruous over incongruous concepts (Barrett and Nyholf 2001; Norenzayan, Atran, Faulkner, and Schaller 2006; Porubanova et al., 2014). Additionally, negative concepts were recalled to a greater extent (M= 26.37, SE = 1.61) than neutral ones (M = 19.66, SE = 1.31), $t = 6.72$, $p < 0.001$. However, no significant differences were found between neutral and positive or positive and negative concepts.

In terms of the interaction between emotion valence and violation category, post hoc comparisons revealed that recall of congruous versus incongruous concepts differed only in the case of positive concepts, namely recall of positive concepts was superior in case of congruous concepts (M = 28.77, SE = 1.88) as opposed to incongruous concepts (M = 17.67, SE = 1.76), $t(69) = 11.10$, $p < 0.001$. The differences in the recall of negative and neutral concepts in terms of the two different levels of violation category were not significant (Figure 10.1).

Delayed recognition

We used a 2×3 (violation category [incongruous, consistent] × emotional valence [negative, neutral, positive]) repeated-measures analysis of variance in order to examine the effects of violation category and emotional valence on delayed recognition (Mauchly's test was not significant). Memory performance for delayed recognition was measured as a percentage of concepts correctly recognized in each category (Figure 10.2). There was no effect of violation category on concept recognition; consistent concepts were recognized to the same extent (M = 70.44, SE = 2.31) as incongruous concepts (M = 71.35, SE = 2.35). However, there was a main effect of emotional valence on concept recognition: $F(2, 126) = 8.36$, $p = 0.001$, $\eta^2_p = 0.12$. The interaction between violation category and emotional valence was not significant. In terms of emotional valence, negative concepts were recognized (M = 75.78, SE = 2.44) more often than neutral concepts (M = 65.31, SE = 2.69), $t = 10.47$, $p < 0.001$. There was no difference in recognition between negative and positive concepts, but we observed a marginally greater recognition of positive (M = 71.60, SE = 2.48) over neutral concepts (M = 65.31, SE = 2.69), $t = 6.29$, $p = 0.05$.

Summary

Even though an effect of emotional valance on information processing has been widely documented, the emotional valence of concepts as a potential contributor to the

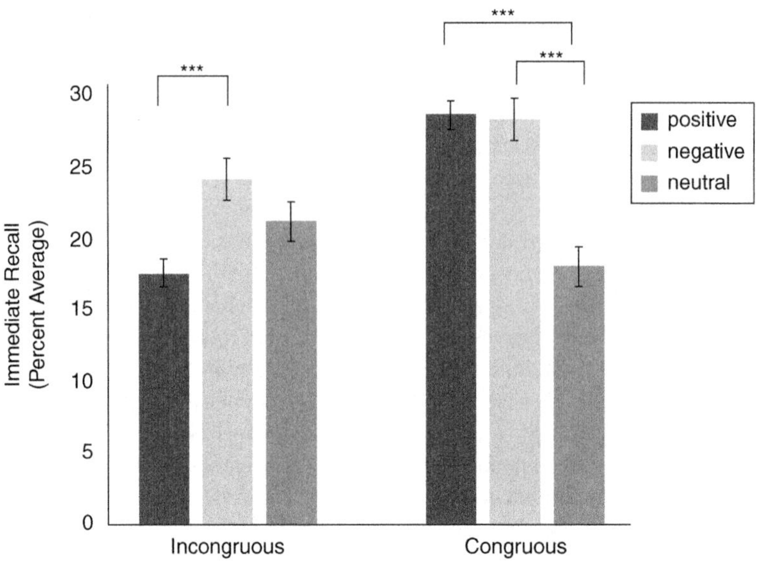

Figure 10.1 Combined recall for Incongruous and Congruous concepts that vary in emotional valence. The graph indicates percentage of items recalled correctly in each condition.

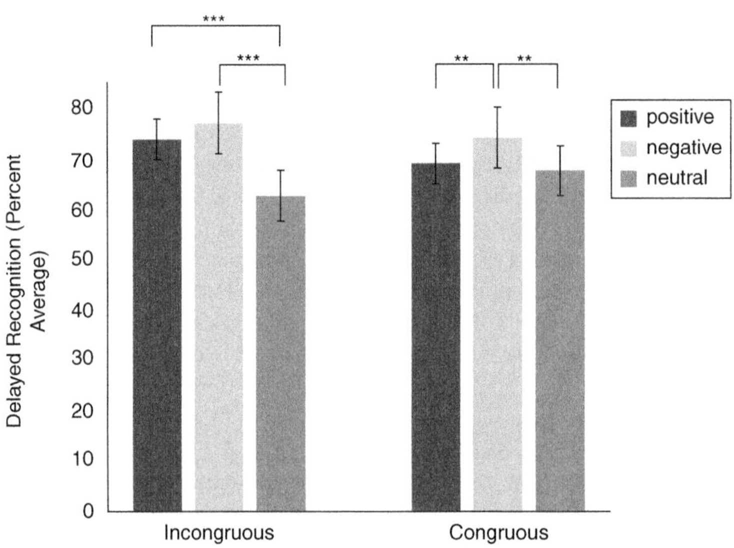

Figure 10.2 Performance in delayed recognition (average percent recognized) for Incongruous and Congruous items that vary in emotional valence. The graph indicates percentage of correctly recognized items in each condition.

memorability of conceptually incongruous information has largely been overlooked (cf. Purzycki 2010, 2011). Variance in concepts according to valence ought to influence memory for emotional conceptual information due to increased cognitive attractiveness, independent of the possibility of an alleged memory advantage of conceptual incongruities. Indeed, several studies have shown that emotionally evocative events and information (both emotional images as well as words) are remembered better than neutral stimuli (e.g., Anderson and Phelps 2001; Hamann et al. 1997; Kensinger et al. 2002; LaBar and Phelps 1998). The present work investigated the role of emotional valence in the memory for conceptual information, varying in novelty. Our results show that the emotional content of concepts is a better predictor of memory as reflected in the recall and recognition performance than incongruity itself. In terms of the emotional valence of concepts, we found a memory advantage for negative information over both positive and neutral in immediate recall. This was corroborated in the delayed recognition task; however, the advantage of positive concepts over neutral ones was only marginally significant. For both memory tasks, memory performance for negative concepts was superior to the recognition for both positive and neutral concepts. These findings are in accordance with studies that demonstrate that highly emotional (and negative in particular) events can be indelible (Brown and Kulik 1977; LeDoux 1992) and better remembered than neutral events (Reisberg and Heuer 1992).

In terms of concept violation, our findings corroborate results of other studies demonstrating that congruous concepts are recalled better or to the same extent as incongruous concepts (Barrett and Nyhof 2001; Norenzayan et al. 2006; Porubanova et al. 2014). In our case, immediate recall for congruous information was superior to incongruous information, while delayed recognition was equally good for both categories. This further challenges the view that minimal counterintuitiveness alone explains the abundance of religious ideas.

Incorporating culture into future research

CSR has adopted many of the understandings of the cognitive sciences. In its development, however, CSR seems to have neglected many of the insights of cognitive anthropology, particularly schema theory. Schema theory assumes that cultural information, including religious concepts, is cognitively arranged in schemas. The majority of CSR research, however, is based on a more modular view of the mind. For example, MCI research assumes that those religious concepts that violate innate conceptions of the world (i.e., evolved ontological assumptions) hold a memory advantage over those that do not contain any violations.

We do not know, however, whether a "flying wiener dog" violates evolved intuitions or learned associations. Research *does* show that items that violate *cultural* expectation concepts are better recalled than those assumed to violate *deep inferences.* While we are committed to evolutionary approaches to the study of mind, an application of schema theory to the study of religious concepts is both warranted and tractable. For example, when conceived as counter-schematic concepts, one can assess the degree of violation of a concept using the methods of cognitive anthropology (triad tests, pile sorts, etc.), and then examine how different degrees of violation affect recall.

Although the data we presented does not avoid the following pitfalls, we conclude with a few suggestions to allow for a greater focus on the role of culture in the transmission of religious concepts. First, research would greatly benefit from a greater ethnographic focus. Bartlett ([1932] 1995) found that the greatest fidelity in story transmission occurred when a "grandparent" generation was around to ensure that the "parent" generation was teaching its "children" a myth correctly. Such a methodology—one that approximates the way that the majority of cultural information is transmitted in real life—is a far cry from methodologies employed in MCI research. It could be that an individual's learning, from a computer or a sheet of paper, that is devoid of social and cultural context is too evolutionarily novel to adequately test human memory for cultural information. Cultural information is, by its definition, learned from group members. Recall tests should also reflect natural conditions, and perhaps future experiments should employ novel methods to reflect how information is recalled in normal life. Unexpected recall tests might be particularly useful in this regard. At the very least, a more "natural" form of experimentation is necessary.

Moreover, it is important that future research investigates more extended delays between the presentation of information and its recollection. Perhaps a more ecologically valid approach would involve incidental memory for, and spontaneous recall of, information. Frequently we hear a story that might involve bizarre or counter-schematic information, but we are rarely required to deliberately remember this information. This is remarkably different from the effortful learning involved in much MCI research. Cultural information is often encoded and consolidated incidentally without any effortful strategies for memorization. Frequently, too, information is recalled only when one is motivated to retell the story/information in a socially and culturally appropriate context.

While structural and emotional factors might be essential and important for understanding the abundance of religious concepts and folktales, culture should be also taken into account (Gervais and Henrich 2010). Context biases reduce conscious efforts to remember and transmit information (e.g., via authority and/or prestige biases) and enhance the attractiveness of cultural information to other individuals. Future research ought to examine how the source of cultural information influences the retention of concepts that vary in schematic congruity.

Additionally, the learning of religious concepts often occurs in the context of religious rituals, when people are aroused. Perhaps the emotional valence embedded in a concept is not as important as the experienced arousal state. Future research should examine the role of experienced emotion and how that affects recall of information varying in arousal, valence, and congruence.

In conclusion, we suggest further advancement by returning to the insights Bartlett employed nearly 100 years ago, and including field research, particularly those employed by quantitative ethnographers (e.g., Purzycki 2013). We also suggest that an incorporation of schema theory would allow for a greater emphasis on the role of culture in the transmission of religious concepts. Schema theory, moreover, incorporates other aspects of cognitive architecture (e.g., attention, emotion, mental representation of information, memory consolidation, visual and other imagery, and inferences) into the study of cognition.

Acknowledgments

The authors thank Jordan Kiper, Benjamin Purzycki, and Paul Swartwout for their help in the early stages of designing this study; Ashley Corso, David Brocker, and Martin Smira for their help with data collection; and Martin Lang for his comments on the final stage of the manuscript. During the preparation of this manuscript, Porubanova was partially supported by the Provost's Office of Farmingdale State College, and Shaver was supported by a Royal Society of New Zealand Marsden Fund Grant (ID: VUW 1321).

Appendix

Concept set

Violation category	Emotional valence	Concept used	Mean valence rating
CON	**Negative**	**A boy that beats smaller children**	**1.22**
CON	Negative	A child that tortures animals	1.15
CON	Negative	A man that beats his wife	1.05
CON	Negative	A man that steps on puppies	1.20
CON	Negative	A pigeon that pecks people's eyes	1.15
CON	Neutral	A bee that makes a hive	3.12
CON	Neutral	A dog that sounds like a giggling schoolgirl	3.00
CON	Neutral	A fox that dives into a hole	2.93
CON	Neutral	A goose that drinks really cheap whiskey	3.05
CON	Neutral	A rat that eats seeds	3.07
CON	Positive	A child that teaches life lessons	4.44
CON	Positive	A cow that eats grass	3.81
CON	Positive	A robin that sings songs	4.24
CON	Positive	A rose that grows quickly	4.37
CON	Positive	A skunk that sprays perfume	3.78
INC	Negative	A flying boy who throws knives at people	1.27
INC	Negative	A flying coyote that rips people's throats out	1.22
INC	Negative	A zombie that eats peoples' guts	1.43
INC	Negative	an invisible man who screams at cancer patients	1.32
INC	Negative	an invisible wolf that eats children	1.17
INC	Neutral	A girl who transforms into a statue	2.76
INC	Neutral	A hammer that listens to conversations	2.88

Violation category	Emotional valence	Concept used	Mean valence rating
INC	Neutral	A living lizard made of stone	2.93
INC	Neutral	A man who transforms into a tree	2.71
INC	Neutral	A mouse that transforms into a human	2.85
INC	Positive	A dandelion that talks to people	4.00
INC	Positive	A mouse that tells jokes about cats	3.68
INC	Positive	A thistle that gives good advice to people	4.02
INC	Positive	A tulip that listens to people	4.10
INC	Positive	A willow tree that thinks about world peace	4.35

Note: CON = congruous concepts; INC = incongruous concepts

11

Constraints on Theory-Building in the Cognitive Science of Religion with Reflections on the Influence of Physics Envy and the Principle of Sufficient Reason

Benson Saler and Charles A. Ziegler

Introduction

The words "science" and "scientific" enjoy a good deal of cachet in our society. That is certainly the case among proponents of the cognitive science of religion (CSR), who use the expression "cognitive science" in titles of professional organizations and academic programs, grant applications, and journal and conference titles. Those who do so impress us as taking pride in being associated with research that is labeled "scientific." For many of them, allegiance to science constitutes an intellectual commitment to an epistemological stance: that science is not just another way of generating and developing claims to knowledge but is a uniquely rigorous enterprise in which observations and inferences offered in support of knowledge claims are subject to replication to validate their evidential bases. This intellectual commitment includes the experimental testing of hypotheses derived from theoretical hunches.

In an essay published in 2008, the psychologist Justin Barrett maintains that "theoretical projects" in the cognitive science of religion "have outpaced empirical ones" and that it is not difficult to find "examples of the time-lag from theoretical pronouncement to empirical substantiation" (2008b: 296). Thus, for instance, while Pascal Boyer's theory respecting the transmission of minimally counterintuitive concepts was published in 1993, it was not experimentally tested for almost another ten years. For this and other reasons, Barrett laments what he perceives to be a general tendency in CSR "to attempt to solve theoretical problems through argumentation alone instead of engaging in systematic testing" (ibid.: 297). Now, regardless of what transpired in the recent past, it is clearly the case today that controlled experiments are very much in vogue in CSR—so much so, indeed, that some critics within the field (e.g., Luther Martin 2014a, to be quoted later) complain about the seeming dominance of experimental findings over data derived from historiography and ethnography.

The current popularity of experimentalism in CSR owes much, we believe, not only to its justification as an empirical arbiter but also to its linkage to the methodologies of well-established sciences. Experimental methodology in CSR is essentially the same as that adopted by practitioners in the physical sciences, numbers of whom have had considerable success in using it to provide causal explanations of observed phenomena in innumerable particular cases. In those sciences, however, the major claim to fame of experimental methodology has been the elucidation of general principles that have an enormous scope of application (they include the so-called laws of nature described in textbooks of classical physics).

Since CSR is a new field, the commensurately small body of work by those CSR proponents who have used experimental methodology leaves us with an unanswered question: To what extent can the use of this approach in CSR duplicate the kind of results attained in the physical sciences? One of the two primary goals of this essay is to answer this question. The second goal is to address some issues respecting the scientific credentials of CSR. In particular, does Ara Norenzayan's "Big-Gods theory" hint at a form of physics envy detrimental to the advancement of theory? And, if so, does the critical recognition of insufficient reason encourage us to take a different path?

We attempt to explore those questions in what follows by discussing factors such as physics envy and the principle of sufficient reason that have some bearing on possible answers.

Physics envy

Biologist Joel E. Cohen is generally credited with coining the term "physics envy" when he employed it as a metaphor in a book review published in a 1971 issue of the widely read journal *Science*. It captured the interest of many readers, which led to its use in a variety of writings during the following four decades. Its most popular use was in criticizing authors who incorporated seemingly out-of-place mathematical reasoning in their publications. In such cases, the usage-related metaphorical meaning of physics envy was that it prompted authors to incorporate superfluous mathematical reasoning in their publications solely to make the text more sciencelike. In 1971, when Cohen coined the term, its usage-related meaning differed markedly from that of the popular version that appeared later.

To understand the significance of this difference, it is pertinent to note that Cohen's version is based on penis envy, which is the English translation of *Penisneid*, the German name given by Sigmund Freud ([1908] 2014: 158) to a hypothecated stage in the psychosexual development of adolescent girls, a stage marked by their envy of boys' possession of a penis. Envy, we note, is an emotion denoting a desire for something possessed by others. This desire may be satisfied by acquisition of its desideratum, but only if such acquisition is not unachievable in principle.

By 1971, penis envy had become generally familiar to practitioners in the biological sciences who comprised the target audience for Cohen's review. So he could rely on them to realize that his use of the term "physics envy"—which, in print, visually evokes the term "penis envy"—implies a hidden comparison of these two terms that suggests that they are alike in some notable way despite being otherwise different. The word "envy" shared by both terms points to such a likeness. Thus, readers familiar with penis envy and the fact that its desideratum is unachievable in principle are led to infer that this is also true of physics envy. This description of physics envy fits one of the common definitions of "metaphor" (see Lakoff and Johnson 1980).

The usage-related metaphorical meaning of physics envy in Cohen's 1971 review is that it prompts biologists to waste time and resources by attempting to duplicate in biology the sorts of results obtained in physics and to do so by merely adopting the methodological approach used by practitioners in that discipline. In short, physics envy falsely encourages biologists to attempt to achieve results, some of which are unachievable in principle in biology.

Let us look more closely at Cohen's review and go on from there to look again at physics envy in historical perspective.

Cohen's book review (a cautionary tale)

Even a cursory reading of this review reveals that Cohen is not merely critiquing a book published in 1970 by the theoretical biologist Robert Rosen. Rather, Cohen uses his critique as the illustrative component of a cautionary tale intended to influence the development of mathematical reasoning in theoretical biology. Rosen's book describes what he claims is a new mathematical approach for discovering general principles in biology that are similar in scope to the lawlike principles of classical physics. Cohen points out that "[u]nfortunately, the grammatical form of principles that claim [such] generality does not tell how useful those principles are in particular experiences" (1971: 674). Cohen, moreover, deplores the fact that Rosen presents the theoretical aspects of his new approach in great detail without first establishing their empirical foundation. Cohen, in contrast, maintains that establishing the empirical basis of such principles is a necessary part of the first stage of the theory-building process because it enhances the likelihood that succeeding stages will remain grounded in reality.

In support of his argument, Cohen notes that successful examples of theorizing in biology, such as the extensively used models of molecular biology, were *preceded* by work that elucidated their empirical basis. Cohen claims that the failure to use this proven methodology tends to occur "[b]ecause the elaboration of language is, for some people, much easier than the labor of establishing concordance or tension between general principles and experience" (ibid.). Consequently, he avers, "there is enormous temptation to spin out language that has the sound and syntax of general principles, and to declare it scientifically satisfying without scrupulous regard to its relation to reality" (ibid.). He claims that Rosen's book is flawed in that respect and that it "is not the only recent work in theoretical biology that succumbs to this temptation" (ibid.). Furthermore, Cohen declares that such a flaw cannot be avoided by adopting the strategy of first developing a full-blown theory and then purporting to establish its empirical validation by showing that a distorted, tailored-to-fit version of that theory is consistent with some preexisting empirical data cherry-picked from the biological literature.

Cohen (1971: 675) also quotes Rosen's claim that his methodology "is a manifestation in biology of the same kind of reasoning used in physics." Cohen disagrees, and he expresses his disagreement in the form of a metaphor based on the *Cinderella* folktale. For he points out that Rosen's own account of his methodology "describes attempts to force the size-12 feet of ecological communities, cellular biochemistry, and neural networks into the glassy size-4 slippers of the formalism of statistical mechanics" (ibid.). Cohen thus equates Cinderella's stepsisters' attempts to force their oversize feet into the glass slipper with Rosen's attempt to demonstrate that elements of his overblown theory fit some biological data. As pointed out earlier, Cohen's remarks are more than a book review. In toto, we call them a cautionary tale because Cohen clearly seeks to apprise others of the danger that physics envy may lead to a dead end, and by so doing, "physics envy," he says, becomes "the curse of biology" (ibid.). But he also implies that by exposing a threat to progress in biology, the role of physics envy is not unlike that of a dead canary in a coal mine that exposes a threat to human life.

The social sciences: Some considerations

It is pertinent to begin by noting that the unprecedented success, achieved during the nineteenth century in developing the natural (i.e., physical) sciences, formed the basis for the technological advances that fueled the industrial revolution. Consequently, in the last quarter of the nineteenth century, in countries where the link between science and the economic benefits of industrialization were recognized, the prestige of science increased dramatically and physics became the exemplar of a so-called natural science ideal. A hundred years before Cohen introduced physics envy as a metaphor, many practitioners of the embryonic social science disciplines apparently experienced physics envy as an emotion. And since the social science disciplines were gradually edging into academia, their developers were motivated to facilitate the process by latching on to the prestige that accrued from being called a science. They attempted to do so by emulating the "natural science ideal." So, in the last quarter of the nineteenth century, as the philosopher Lewis Beck reminds us, researchers in the social sciences "tried to attend to observable and measurable entities and to connect these by causal or functional laws" (1949: 386).

By the turn of the century, however, it was apparent that such "laws" were not forthcoming. The "natural science ideal," advocated by many social scientists (although adamantly rejected by some), had turned out to be less clear cut and attainable than its supporters had supposed. By the middle of the twentieth century, it had become apparent (see Beck 1949) that *complexity of subject matter* is the major factor separating the natural and social sciences. In the natural sciences, contrary to what is often supposed, subject matter evinces far less complexity than subject matter in the social sciences that consists, as Beck (389) puts it, "of highly complex constellations of complex events in systems that are only poorly isolated." It is essential to note that Beck's stipulated definition of complexity is not just a diversity of parts. It also refers to the likelihood that it may not be feasible or perhaps even possible to isolate some parts from others.

Beck notes that variables in the natural sciences can be isolated from many others and understood without reference to those others, so only a small number of variables have to be known for practitioners to discover functional laws relating one to another. "Solely for this reason," Beck says, "are simple experiments possible" (1949: 387). Yet, because of the isolatable nature of variables in the subject matter of the natural sciences, such simple experiments prove capable of yielding lawlike principles of immense scope such as Newton's Law of Gravitation.

Beck goes on to point out that, in contrast, "social scientists must deal with societies of individuals of almost infinite internal complexity and variability" (ibid.). It is, moreover, impossible to deal with an infinite number of variables; so the generally unisolatable nature of the variables in the subject matter prevents their reduction to a manageable number. This, in turn, precludes experiments capable of yielding principles of great scope, as is achieved in the natural sciences. For those reasons, Beck concludes, "it is too soon to expect ... a Newton of the social sciences" (394).

Sixty-some years have elapsed since Beck affirmed that a Newton of the social sciences would not appear any time soon. But from our vantage point in the twenty-first century, our wonderment at the rapid progress in science and technology during those years prevents us from being too dogmatic in our predictions about science. Nevertheless, it seems plausible to suppose that Beck's prediction will not be negated during the further development of theory in CSR over the next decade. So in the context of that supposition and the expectation that such development will involve CSR much as it is presently constituted, we offer our answer to the question posed in the introduction to this essay regarding the adoption by CSR proponents of the methodological approach used in the physical sciences.

There is some bad news and some good news. The bad news is that the probability such adoption will enable CSR proponents to duplicate those results obtained in the physical sciences that yielded lawlike principles of great scope is vanishingly small. The reason is that cognitive sciences like CSR involve the interdisciplinary study of thought and behavior using methods and theories from a broad range of disciplines whose subject matters, as pointed out earlier, typically have very high complexity levels.

The good news is that adoption will enable proponents to duplicate those results obtained in the physical sciences that yielded great success in providing causal explanations of discrete phenomena in countless particular cases. In fact, numbers of CSR proponents have already displayed a tendency to avoid high-level generalizations characterizing science, and some (wisely in our opinion) already emphasize in their publications specific methodological procedures governing discrete studies rather than attempting to incorporate particular cases into general or "totalizing" theories.

In the sections that follow, we deal with the second goal of this essay—namely, discussion of certain issues touching on the scientific credentials of CSR.

Parsimony and sufficient reason

Scientists and mathematicians sometimes speak of "elegance" in their work. That word is hard to define, and when requested to do so, some are content to say, "We know it when we see it." Computer scientists, however, may be the most articulate (and down to earth) when they call a program elegant because it uses a small amount of code to surprising effect. In any case, elegance often reduces in part to parsimony, and the principle of parsimony is of importance in all the sciences.

The principle of parsimony, broadly put, bids us to choose the simplest explanation that fits the evidence. In doing so, we are told, one ought not introduce and multiply assumptions beyond necessity. Or, to rephrase it in the form known as Occam's Razor: all else being equal, the simplest hypothesis that explains the data should be chosen. But, of course, all this is easier to say than to do. How do we determine "necessity"? And how do we know that all else is equal? And if we know or believe that all else is not equal, what do we do about it? Such problems notwithstanding, philosophers and scientists have actually gone a long way in meeting the difficulties that these and related problems entail. In science, for instance, Occam's Razor is not an arbiter between theories nor a sovereign principle of logic nor a scientific result. Rather, it is a heuristic, a guide for the development and evaluation of theoretical models (Beck 1949). As such, it pivots on the principle of simplicity, which in this case gives expression to the notion that simple theories are preferable because they are probably easier to test than complex ones.

Now, while the development of models and theories in the sciences may sometimes produce overelaborated structures, we think that they suffer more often from inadequate data collection and/or oversimplified theories. Inadequate data are, of course, regrettable in any sort of serious research, and they can only be set right by the provision of more or better data or a replacement or rephrasing of a problem. But when data seem sufficient for the explanatory task and yet proffered interpretations and explanations fail to make convincing sense of them, we may face a different sort of setback, that of oversimplification. We may, indeed, run up against a failure to actualize the principle of sufficient reason. That principle cautions us against simplifying to the extent that our proffered explanations are inadequate for explaining the data. Ideally, sufficient reason encourages us to explain by citing sufficient causes. The history of science archives volumes of theories that proved inadequate for their explanatory ambitions and so were shelved. Should there be any doubt, moreover, of respecting the places of the principle of sufficient reason and the principle of parsimony, the history of science strongly suggests that it is more important for explanatory posits to be sufficient rather than parsimonious.

The principle of sufficient reason or, more precisely, the difficulty of satisfying it, is of great significance for theorizing in the cognitive science of religion and some of its affiliated disciplines. Indeed, a recent controversy—that of the role of so-called Big Gods in the emergence of large states or empires—is, in significant measure, a quarrel over what counts as sufficient reason.

Big Gods

In 2013, the social psychologist Ara Norenzayan published a book entitled *Big Gods: How Religion Transformed Cooperation and Conflict.* That book has attracted a great deal of attention. Thus, for instance, the journal *Religion* published a review symposium in 2014 in which twelve scholars representing different disciplines expressed their opinions. Further, the editor, Michael Stausberg, introduced the review symposium with an essay that sketches both the scholarly work of Norenzayan that preceded his book and the major criticisms that the twelve reviewers made of the book. In this essay we depend not only on the *Big Gods* book but also on the review symposium in *Religion* and Stausberg's introductory remarks.

The primary observation that Norenzayan seeks to explain is as follows: after many millennia of living in relatively small groups of hunters and gatherers, human beings in various places made transitions to larger societies, eventually to some of several thousand members and, in time, to still larger aggregates. Norenzayan poses and attempts to solve the problem: How were humans able to cooperate among themselves so that they were able to form "large, anonymous, yet cohesive" societies? (2013: 3).

Norenzayan argues that the prosocial mechanisms characteristically available to hunters and gatherers and to relatively small-scale tribalists are inadequate for developing and sustaining the "ultrasociality" required to support cities, states, and empires. Such ultrasociality, Norenzayan maintains, does, however, exist in large-scale human societies, and it is often (but not always) a religious achievement. More particularly, it often signals belief in "Big Gods"—that is, belief in "supernatural agents" that are credited with the power to intervene in human life, that are believed to punish or reward (in this life or the next) those who violate the moral code or who adhere to it, and whose interventionist inclinations are thought to produce a moralizing surveillance well in excess of what is supposedly found in small-scale hunter and gatherer societies. Belief in gods who know and care about human behavior and who act on their knowledge, Norenzayan argues, facilitates social cooperation on a relatively high level, for numbers of experiments indicate that people who believe that they are being watched are more observant of the moral code than those without that belief. The development of belief in Big Gods, Norenzayan maintains, was thus a major factor accounting for the development of large-scale societies.

Now, in presenting his main thesis (which is richer than our brief summary of it here), Norenzayan makes statements that some critics regard as subverting his argument. The most obvious are these: (1) Big Gods are a "potent cause but not the sole cause that led to the social expansion of some groups" (2013: 9), and (2) "Big God religions, without being a necessary condition, dramatically expanded the cooperative scale of societies; but other cultural institutions could have similar effects" (194n. 12). Donald Wiebe points out that Norenzayan's statements undermine the claim that the emergence of larger societies in the Holocene was a "religious achievement." That is because his admissions conduce to the conclusion that religion is neither a sufficient nor a necessary condition for the emergence and continuance of large-scale cooperative societies (2014: 679). Since we have it on Norenzayan's testimony that Big Gods are not the sole cause, what then are the *sufficient reasons* for the development of cities, states, and empires? How, indeed, does one persuasively handle the likely relevance, complexity, and subtlety of numbers of variables—ecological, economic, demographic, sociopolitical, and religious—and their multiple relationships? On our reading of his work, Norenzayan fails to satisfy us in the matter of sufficient reasons. What may at first strike some readers as being a clearly stated and straightforward set of problems turns out to be, on reflection, complex enough to cross (metaphorically) a classical physicist's eyes.

The problem of sufficient reason in this case is even more difficult to resolve when we take into consideration some additional aspects—and criticisms—of Norenzayan's argument. Michael Stausberg tactfully remarks that Norenzayan's book "failed to convince a majority of the [twelve] invited readers" who participated in the review symposium sponsored by the journal *Religion* (2014a: 604). Their negative criticisms, in our judgment, are often persuasive and sometimes harsh.

Negative criticisms include, among other things, claims that Norenzayan's thesis is not supported by properly used historical evidence. Luther Martin (2014a: 636) endorses and expands on that claim not only with respect to Norenzayan (whom he charges with exemplifying it) but also with respect to the cognitive science of religion at large. CSR, he writes, "has come to be dominated by a circle of primarily experimentally oriented researchers who talk mostly among themselves" and who, in many cases, "are unfamiliar with the 150-year literature in Western scholarship about the vagaries and ambiguities associated with the category and study of 'religion,' or about the theoretical and methodological issues of historiography with which these scholars have struggled." Additional major negative criticisms include the claim that there are theories alternative to that of Norenzayan's purporting to account for the rise of prosociality and large populations and that Norenzayan has not adequately evaluated them. Still other criticisms deal in one way or another with the thorny issue of cultural evolution and Norenzayan's possible contribution to our understanding of it. This touches on arguments and assumptions about whether or not there is such a thing as cultural evolution—and, if there is, how do we reconcile it with historiography?

The participants in the review symposium advance solutions to the various problems that they claim to encounter in their unraveling of Norenzayan's thesis. It seems that none of the symposium critics think that Norenzayan provides sufficient reasons for the rise of large social orders. But virtually all think that some of what Norenzayan says is well said and worth pursuing. We concur. Of great interest to us, moreover, is the fact that the critics sometimes expand their criticisms beyond Norenzayan to the cognitive science of religion at large. When they do so, they not only point to what they regard as weaknesses in CSR but also propose possible remedies. Among those proposals, we think, one of the best comes from the editor of the journal. "Progress," Stausberg writes, "will probably lie not in posing ever-bolder questions, but in finding research questions that are relevant within evolutionary theory but which can still speak to historical and ethnographic research" (2014a: 605).

Canaries in coal mines

Norenzayan's theory may have suffered fatal blows at the hands of its critics. But if so, and somewhat like the proverbial canary in the coal mine, an appreciation of its life and death can be useful. The well-informed and thoughtful criticisms of it show us how seductive and yet misconstrued some bold theories may be, and we are wisely warned against them if we would flourish intellectually. Yet how did someone as obviously talented and imaginative as Norenzayan come to engage in theory-building that in places is internally self-refuting and that elsewhere suggests an inadequate exploration of data and theories alternative to those championed in the "Big Gods" book? We are not in a position to answer that question with authority. But we think that it is an interesting question, and so we speculate.

We suspect that Norenzayan was motivated in part by a species of physics envy. We have in mind a generalized sort, not the early kind that extols and would mimic the mathematical precision and transparency of successful classical physics. Nor do we mean the design, execution, and analysis of experiments that test hypotheses derived from theories. Rather, we have in mind an idiom that seeks to heighten credibility by invoking terms that affirm an association with science that is not entirely warranted. To borrow words from Joel Cohen, we suspect that Norenzayan gave in to "an enormous temptation to spin out language that has the sound and syntax of general principles, and to declare it scientifically satisfying without scrupulous regard to its relation to reality" (1971: 674).

In any case, CSR is likely to benefit from ongoing critiques of physics envy, for the trappings of science are alluring even if some of the substance presented sometimes belies the trappings. And in a new, exciting, and, we think, very promising field, one can easily lose one's self in two derivative temptations: to theorize beyond present capacities to sustain such theorizing, and to identify one's work as scientific even when it may not fully qualify as such. So we will probably see a number of Big God–style theories in the years ahead. Happily, however, we are also likely to have a number of bright critics on hand, metaphoric canaries in future metaphoric coal mines, who will detect toxicity that stems from a failure to accommodate to constraints on theory-building in CSR.

12

The Effects of Relative Stable Feedback Loops: Cognitive Science and Historical Explanations

Jesper Sørensen

Should one take into account human psychology, including its brain functions, cognitive processes, and emotional response patterns, when trying to understand religious phenomena? As self-evident as the confirmatory answer to that question might seem today, for most scholars the answer was self-evidently negative in 1990, when the cognitive science of religion took off with a series of publications. "Nothing can be explained by cognitive approaches to religion that is not better explained in the sociology of religion," a professor in sociology of religion once told me, and this was not merely an expression of professional possessiveness directed against the new kid on the block. It also reflected an antipathy toward psychological explanations in general (also referred to as "psychologism" and identified with Freudian psychoanalysis, James' psychology of religion or phenomenology of religion a la Rudolph Otto) combined with a corresponding overestimation of the explanatory power of such theoretical constructs as "social structure," "culture" and "society" shared by many sociologist, anthropologists, and historians. An initial indifference was later replaced by examples of outright hostility toward cognitive approaches[1]—a reaction that was met by a, sometimes, overconfident, at other times, entirely triumphalist attitude by proponents of cognitive science of religion, who seemed to believe that *all* questions could now be addressed through cognitive studies and that a new golden dawn of scientific explanations is upon us.[2] Even if such millenarian enthusiasm perhaps forms a necessary part of all radically new research agendas—imbuing a certain fervor into the social group, instilling charismatic authority to its founding figures, ostracizing alternatives approaches, and thereby formulating strong identities available for new scholars—perhaps now the time has come for a less emotional assessment of the strengths and weaknesses of cognitive approaches to religion as well as how these can relate to more established paradigms?

In order to initiate such assessment, a first step is an estimate of the current conception of the theoretical object investigated in academic studies of religion. What do we actually study? In recent decades most, if not all, traditional categories used to investigate and explain religion have been heavily debated if not outrightly debunked, leaving the discipline in terminological mire. A cursory reading of a program in a randomly selected religious studies conference could leave one with the impression that scholars of religion are equally concerned with talking about *how* they conceptualize the world rather than actually investigating it (thereby reflecting a general tendency in postindustrial societies to talk about how we talk instead of acting on problems). The development of cognitive science of religion appeared in conjunction with this postmodernist critique of time-honored concepts in the humanities and social sciences, and there was some agreement between the two concerning the problematic status of the traditional categories of the discipline.[3] However, the answers proposed could not be further apart. Postmodernist scholars seem to have committed themselves to ever-finer genealogies of specific concepts in order to expose these as subtle instruments of power, apparently under the persuasion that this essentially Freudian unveiling would somehow miraculously solve "the problem."[4] As if, for instance, genealogically unveiling "magic" as a Western polemical concept would somehow undo the pertinent question of why people around the world pierce dolls in order to harm a non-present enemy.[5] By contrast to such guilt-ridden and ultimately parasitic attempts to cleanse the academic study of tainted concepts, cognitivists critiqued traditional concepts for failing to "carve nature by its joints" and instead proposed a reductive search for more basic explanatory concepts and principles in psychology, in brain science, and in evolutionary biology. These concepts and principles, in turn, were used to generate models and hypotheses about the origin and transmission of religious ideas and practices and, as an inevitable side effect, dissolved religious phenomena into larger classes of phenomena. "Religion" and "religious phenomena" lost their special character and were seen as just one potential (by-)product of cognitive processes and brain states ultimately explicable by evolutionary models. Together with other basic concepts, "religion" is thus conceived of as an "impure concept" that must be fractionated into more basic elements in order to arrive at adequate hypotheses and explanations.[6]

We need not further rehash this story, as it is probably known to most readers of this volume.[7] What I wish to point to is how this generally helpful development has positioned cognitive studies of religion in a specific relation to its subject matter in general and in particular to time and the potential causal role of historical configurations. Further, I wish to address the possibility of a broader conception of the role of cognition in the study of religious phenomena and, especially, how such an approach might be integrated into investigations of particular historical trajectories.

Deep history in human cognition: Macro-history

"Religion" is a shorthand, synthetic term covering a number of beliefs and practices considered to be by-products of evolved cognitive structures, some of which perhaps (opinions differ) at some point in time have been exapted to serve the adaptive function of social integration. This is a very condensed version of the strongest explanatory story framed by cognitive approaches to religion within the last twenty-five years. Inspired by the pioneering work of Dan Sperber (1975), developments in cognitive science[8] and the emergence of evolutionary psychology, scholars such as Stewart Guthrie, Pascal Boyer, and Scott Atran radically reshaped our understanding of religion by dismantling it into constitutive parts, subsequently explained by reference to activations of a number of functionally distinct cognitive modules.[9] The theory contends that these modules or, more moderately, innate learning dispositions give rise to domain-specific cognitive processing and that each of these evolved in order to address a limited repertoire of recurrent evolutionary problems.[10] Further, as the input conditions of these modules can be met by stimuli not belonging to their proper functional domain, parasitic structures emerge that effectively activate cognitive processing even if such activations in themselves have no adaptive function.

Accordingly, cultural practices and ideas, including those deemed religious, can be explained as essentially parasitic activations of functionally distinct cognitive modules. Information about unseen agents that modify or control our environment, gods or spirits that observe all we do and know what we think and feel, or rules of ritual cleansing are all examples of culturally disseminated ideas and practices, which activate systems that originally evolved to solve particular problems such as evading predators, cheater detection, and avoidance of pathogens. In itself such parasitic activation ensures the relevance of religious ideas, but their success and cultural proliferation is further explained as a function of a transmission bias that favors cultural representations that combine intuitive expectations with minimum doses of counterintuitive violations of these expectations. Thus, omniscient and omnipotent gods are represented as humans with nonhuman abilities; spirits are represented as agents with unlimited access to the same information that we normally gossip about; and ritual cleansing is often performed on things already clean. Hence, combining intuitive and counterintuitive elements in one representation ensures that they are both highly salient and thus memorable and that they activate relevant inferential processing.

In short, this epidemiological theory argues that the cultural proliferation of religious ideas and practices is the result of a repeated selection of representations combining high-cognitive relevance and attention-grabbing violations of intuitive expectations. They are simply good to think with and easy to communicate, and these two basic variables are among the major forces informing the morphology of religious phenomena at all times and at all places. Now, one of the great advantages of this explanatory approach is its scope: it explains religious phenomena by reference to universal, species-specific, cognitive constraints; it embeds these in a broader evolutionary landscape and thus helps evaluate the role of religion in "human deep history"; and it builds, so far, an incomplete inventory of basic mechanisms of religion that enables cross-cultural comparative studies. It, thereby, not only salvaged an otherwise compromised comparative phenomenology of religion (Jensen 2003); it also instigated a new empirical and often even experimental study of religious phenomena that, among other things, readdressed two questions as old as the discipline itself: what is the origin and what is the potential function of religion?

The cognitive science of everyday behavior: Micro-history

Deep history explanations subsume individual cases or data points, whether acquired through observation or experimentation, in order to test hypotheses derived from general theories. But how informative are they when the explanatory focus is not on the general mechanism but rather on the microlevel phenomenon itself? How do macro-historical settings describing the deep roots of religion in human evolution help explain concrete behavior in specific historical and social settings? In order to understand this relation between ultimate explanations derived from evolutionary theorizing and the concrete decisions and behaviors of humans observed in both "natural" and experimental settings, we need to identify what questions early cognitive approaches to religion sought to address.

With the noticeable exception of Lawson and McCauley (1990), pioneering cognitive theories of religion were all developed by anthropologists, and this fact has had an important impact on the trajectory taken by the cognitive science of religion. It introduced a certain bias toward understanding religion in terms of processes of oral transmission, memory, learning, and pragmatic relevance in small-scale societies, understood within a theoretical framework admitting little if any room for context-specific cultural history. In a certain sense, the "deep-history explanations" of religious phenomena described previously are ultimately motivated by a theoretical concern realized by the core anthropological endeavor: to understand or explain concrete human behavior observed at a particular point in time as an expression of some set of underlying regularities—in other words, the everyday pragmatics of human beings in different cultural and social settings is to be explained by reference to an underlying system or structure. Whereas most twentieth-century social and cultural anthropology found the foundations used to explain observed behavior in "society," "social structure," or "cultural systems," cognitive approaches found their explanatory bedrock in evolved cognitive systems activated or evoked by specific contextual and pragmatic cues. This is a significant step ahead as it replaces weakly defined, conceptual constructs ("society," "culture") with a more precisely specified cognitive mechanism that allows for more rigorous predictions as well as experimental testing.[11] In fact, this reductive move has changed the status of these time-honored concepts from that of *explanans* to that of *explanandum*—"culture" and "society" are what need to be explained by reference to lower universal levels, whether species-specific cognitive processes, or more general evolutionary mechanisms.

The question remains, however, how well suited are deep cognitive structures when trying to understand concrete social behavior in a full contextually rich embedding? In a sense these reductive explanations encounter the same problem as prior structural-functional explanations. Both apply stable, macro-level universal features in order to probabilistically explain fluctuating or even contingent microlevel phenomena. In the former case, behavior is explained by stable structures such as culture or social organization,[12] in the latter case by equally stable evolved cognitive mechanisms.[13] This is all well, but it comes at a price. By subsuming all individual behaviors to expressions of general principles—a move only made possible by treating data not fitting the overall statistical trend as "noise"—it effectively annihilates history, both individual and cultural, as a causal factor. It is thus at odds with historiography's attempt to explain particular events as the outcome of a prior series of (causally) related events rather than as merely singular activations explicable by reference to universal and inherently context-distant structures. And thereby it is also at odds with approaches understanding "religion" as consisting of distinct patterns of ideas and practices, that cumulatively unfold over time creating "traditions," that is, relatively stable clusters of ideas related to particular types of institutions and social structures within certain geographies. Thus we find the reinstatement of the old distinction between nomothetic and ideographic approaches, but as we shall see, the opposition might not be insurmountable.

The emergence of fairly stable feedback loops: Meso-history

Explaining microlevel phenomena as contingent expressions of stable macro-level structures is a time-honored scientific principle. The apple falling from the tree is explained by gravitation, the workings of an engine by thermodynamics, and the phenotypic expression of an organism by genetics. However, this picture is increasingly criticized in some domains for being too simplistic and failing to recognize how certain effects feedback into the system by altering the boundary conditions for future expressions. For instance, probabilistic epigenetics (both in biology and psychology) consistently points to the causal role of a broad range of environmental factors in the phenotypic expression of organisms. Phenotypical expressions themselves form part of and actively alter the environment, whereby relatively stable feedback loops are established that prevent deterministic predictions of how otherwise stable genotypes are phenotypically expressed in novel environments (Gottlieb 2007). In short, altering the environments will cause identical genotypes to have different phenotypical expressions and an organism's behavior—not to mention the potential social structures and cultural forms that the joint actions of many organisms give rise to—form part of its environment. An organism, including its genes, morphology, and behavior, is best thought of as a complex system in constant development.[14]

Human evolution is characterized by humanity's ever-increasing ability to radically modify its own evolutionary niche, both in terms of altering the physical environment by means of technology and by transferring physical behavior into symbolic interaction, as when we threaten someone rather than committing ourselves to actual violence (Odling-Smee 1995; Odling-Smee et al. 2003; Laland et al. 2010; Rappaport 1999). Both are methods of lowering an organism's energy expenditure, but we thereby, effectively, reintroduce history as a causal factor in explaining human behavior, if by history we mean the continuous and cumulative change of the physical, social, and cultural environment affected by the actions of human individuals and groups.[15] Whereas both structural-functionalist and evolutionary approaches for different reasons disregard history as a causal factor in favor of structural, synchronic explanations, the existence of such feedback loops reintroduces the *meso-level* of diachronic history back into the equation (Sørensen 2010).

Initially, at least two domains are of direct relevance when investigating such feedback loops. The first and most obvious is technology. Humans modify their environment by means of tools that both extend and augment human abilities. From the hammer boosting the power of the fist and extending its reach, over modern transportation and communication technology, to nanotechnological modifications of molecular structures, technological development continually enhances the human capacity to modify or even construct our physical and social environment and thus magnifies our ability to remove hindrances from our pursuits. However, once invented, a technology exerts a pressure on its makers by effectively demanding the acquisition of the particular skills necessary to master it. This is so in the trivial sense that all humans at any point in history are embedded in a technosphere that the individual must adapt to in order to survive.[16] For instance, we must all learn to use digital technologies and transportation systems in order to function efficiently in a modern Western world. More profoundly, even though technology is constrained by basic cognitive skills, the continuous pressure toward adapting to local technospheres makes it unlikely that these cognitive skills are stable. These skills undergo constant modifications on at least two levels: first, through the individual's lifetime, as he or she, with more or less success, obtains mastery of new technologies and thereby enhances adaptive fitness within a given system; second, a slower level of species-wide modification as the general ability to absorb new technologies in itself constitutes an adaptation that allows individuals with larger dexterity in constructing, as well as adapting to and thus fully utilizing, new technologies to increase their adaptive fitness. Individuals mastering the technologies of their time are, all things being equal, preferential mates as they can attain higher social status and ensure access to resources. A general ability to acquire technological skills should thus be favored through sexual selection.[17]

Of importance in relation to religion is the role technology plays in establishing relatively stable social institutions ensuring and controlling transmission of technological knowledge. This is the case in the formation of status groups that attain prestige due to a monopolistic control over certain technologies (e.g., scribes) and these groups' often intimate relations to religious cults, authorities, and organizations. It is also evident when it comes to the goal of religious behavior. Rituals that seek to influence aspects of the environment beyond the controls of contemporary technology seem to flourish. Thus there exists an intimate relationship between human individuals' and groups' technological control over their environment at any point in time, the goals and wishes expressed through ritual behavior, and how these are embedded in relatively stable systems of authority and in social institutions. If technological development is an instance of cumulative cultural evolution characterized by a ratchet effect (Tomasello 1999), not only does this development entail a concomitant evolution of cognitive capacities favoring the ability to acquire technologies skills and further advance technologies already present (i.e., of human creativity). It also means that other behavioral domains, such as religion, change in conjunction to technological/social coevolution. Understanding this change thus necessitates understanding religion in relation to concrete historical trajectories.

This points to the second relevant domain of inquiry, that of symbolic or conceptual structure. This domain encompasses the pertinent observation made by both anthropologists and historians: that the conceptual structures, ideas, and beliefs, as well as behaviors of people belonging to distinct societies or cultures (however conceived) differ; that it takes significant time and effort to gain access to the symbolic or conceptual structures present in any given culture or historical period; and that such structures seem to express a certain level of systematicity and stability over time. However, the problems raised by this observation are not simple. First, rather than merely stating cultural differences as basic facts subsequently used to explain differences in individuals' behavior (cultural determinism) or, more radically, arguing that these differences are in fact incommensurable (cultural relativism), a cognitive approach should seek to explain how both differences and stabilities arise, how conceptual structures relate to and potentially mold species-specific cognitive dispositions, as well as how the pragmatic behavior of concrete people both support and change otherwise relatively stable systems. Second, the degree of systematicity and coherence within any conceptual structure and its stability over time is a thing that should be empirically investigated rather than merely stated. Third, the borders of any system to neighboring systems remain contested and are likely to change over time. Cultures or social groups are not neat and easily discernable homogeneous units, and neither are conceptual systems. Rather, both are more or less loose patterns of shifting associations, dissociations, and alliances, and even though some patterns might gain prominence in certain periods, both social configurations and conceptual systems are better envisioned as consisting of a plethora of local clusters that, for shorter or longer periods and in shifting relations with other clusters, compete for predominance in their relevant domains.

In short, a cognitive historiography needs to address the ever-pertinent question of the relation between stability and innovation, as well as how stable local systems engage with other systems at the same level as well as systems at other levels of organization. Thus, an epigenetic perspective raises the degree of complexity considerably, as interactions ensue both within and between levels, and small permutations in one system potentially result in large changes in another and allow new patterns to emerge. Accordingly, should the previous speculations about how religion relates to the process of technological/social coevolution have led the reader to believe that religious phenomena are mere (epiphenomenal) reflections or products of this more basic process, he or she would have missed the point that all levels and systems potentially have causal effects and that the strength of these effects might themselves change over time as a result of their relation to other systems.[18]

Even though cultural and conceptual patterns might appear almost epiphenomenal when seen in the light of macro-historical developments, the meso-historical level remains decisive when attempting to understand or even explain concrete human behavior. The notion of cultural models is a helpful tool when modeling how such patterns achieve some degree of stability. A cultural model denotes a relation between several components, such as concepts, behaviors, and institutions, that remain stable over time, are diffused within a given cultural unit, and therefore can function as a stable referential as well as inferential device.[19] One lucid example of a pattern remaining relatively stable for more than 2,000 years is the prevalence of apocalyptic thinking in the Western world. Taking its beginning in ancient Near East, Axial Age conceptions of historical time as a narrative unfolding of a relentless battle between good and evil forces reaching a cataclysmic end in the Day of Judgment, the apocalyptic understanding of time and human destiny remains a widespread narrative structure in modern society that informs not only evangelical Christians anticipating the (ever) imminent rapture but also secularists fearing the looming end of the world whether in the form of nuclear holocaust, environmental collapse, or the rise of malignant machine intelligence (Gross and Gilles 2012). Evidently, the strength of this particular pattern has fluctuated as it competes with alternative narrative models, such as recent New Age conceptions of time as a process of eternal spiritual evolution. In order to evaluate its present prevalence in popular culture ranging from films describing life in a post-apocalyptic world, or popular fiction such as Cormac McCarthy's *The Road*, to popular science books, such as Jared Diamond's best seller, *Collapse*, we need to analyze how this cultural model relates to other models (e.g., "evolution" or capitalist notions of perpetual economic growth), how it is now entrenched in conceptual systems and metaphors (e.g., "we have reached the end of the road"), *and* how it potentially gains cognitive relevance by latching on to both evolved systems such as predator avoidance (e.g., evoked by the figure of "the Beast" or "Terminator") and pragmatic relevance by entering into argumentative structures in the everyday human interaction (e.g., in politics).

However, cultural models need not take the form of such long-lasting grand narratives. Studies of conceptual metaphors have pointed to the pervasive role of metaphorical structures in human cognition (Lakoff and Johnson 1980, 1999). We consistently see one thing in terms of another, thereby utilizing inferential potential in one domain to understand properties of another. Some metaphors might be more or less universal (e.g., "seeing is believing") grounded in basic biological structures or evolved cognitive dispositions; others are evidently confined to particular social, cultural, and historical settings making them hard to comprehend for outsiders. For instance, "time is money" can be understood as a basic metaphor that remains a key conceptual building block of modern capitalist economy. It enables the transaction of labor for money and thereby the construction of a free labor market supplanting different versions of serf and slave economies that builds on a conception of ownership of an individual or some of that individual's produce. The metaphor is not simple, and it requires the emergence of underlying concepts and technologies, such as the understanding of time as a consecutive series of identical, quantitative, and measurable units, as well as the ability to represent value through symbolic, material tokens guaranteed by reputable, social institutions. However, once in place it pervades our conceptual understanding of both time and labor (e.g., we can now "*invest* our time" and most modern employees receive an "hourly wage")—in short, it forms our expectations to specific domains, and more generally, such conceptual metaphors allow inferential structures to pass between domains and inform our expectations to the world.[20]

The formation of stable expectations is of crucial importance in this discussion. Significant aspects of the human cognitive machinery can be understood as aimed at probabilistically assessing possible outcomes in order to regulate future behavior. Searching for external regularities is thus an integral part of this process, as regularity reduces the cognitive workload allowing internal predictive models to guide behavior.[21] This is obviously the case when we are engaging our natural surroundings, but is equally the case in social interactions. Regularities in behavior facilitate cooperation, and collective behavioral idiosyncrasies can be understood as prediction-enhancing systems that become strengthened through usage, eventually resulting in stable feedback loops (e.g., a "handshake" as a greeting). A most conspicuous effect of the feedback between behavioral regularities, prediction, and usage is the emergence of language and other forms of symbolic communication. Symbolic communication is, among other things, founded on the ability to spot regularities in material tokens, and their repeated usage enables referential structures to solidify and connect with social institutions as well as social status systems. We thus not only expect words to have a stable referential core but also further expect that symbolic tokens (whether expressed in waves of sound or patterns of ink) elicit somewhat similar inferential processes in the receiver, and we judge their epistemic status based on their relation to persons, institutions, and groups. Further, as it was the case in relation to technology, the ability to process symbolic information can be understood as an adaptation. An individual is enmeshed in a semiosphere (Lotman 2005) consisting of a plethora of conceptual models, ideas, and symbolic expressions that it must master in order to navigate its social environment. Stable symbolizers have better access to resources and thus have an advantage in sexual selection. The concept of a semiosphere is not meant as an argument of semiotic solipsism according to which "meaning" mysteriously emerges as an effect of formal relations between signs. In line with the general epigenetic model defended in this chapter, it rather points to the emergence of regularities and local systemic properties as a necessary condition for language and other symbolic structures to fulfill their role as cognitive augmentations. It is through reaching some level of both referential and inferential stability that language and other means of symbolic communication can serve their cognitive functions: to align the representational space of individuals and thus enable the expansion of empathy, facilitate cooperation, and reign in maladaptive effects of competition between individual organisms.

Where do we go from here?

Elsewhere I have described the effect of the stable feedback systems underlying symbolic communication as an *immunology* of representations (Sørensen 2004, 2010). Stable conceptual clusters, which incorporate symbolic, behavioral, technological, and social institutional variables, can be seen as the cultural equivalents of white blood cells that prevent the easy access of some novel concepts and ideas into already established conceptual and cultural systems while allowing others to pass. In order to properly understand human cognition and behavior in concrete situations, we must add such a temporally protracted, cultural level of constraints to the pragmatic and cognitive selection pressures argued in a strictly epidemiological model of representations (Sperber 1996; Boyer 2001). A meso-historical perspective seeks to model the network of conceptual clusters forming a gradually changing cultural landscape on which deep-historical dispositions and well as pragmatic constraints unfold.

However, does this call for cognitive science of religion to embrace cultural historiography entail a similar embrace of a strictly interpretative method depending on the pattern-recognizing abilities of the individual analyst? This need not be the case. The digitization of significant textual corpora as well as the more easy access to various sorts of big data allows for much more rigorous quantitative testing of claims made about stability and prevalence of certain conceptual clusters as well as claims about the relation of such clusters to both behavior and social institutions.[22] Integrating these methods, broadly conceived under the heading of Digital Humanities, with a cognitive science of religion that has hereto been mostly concerned with universal cognitive constraints, would make possible a more refined study of religion or religious phenomena that acknowledges the intricate interpenetration of biological, psychological, cultural, and social variables.

Part Four

Looking Forward

13

The Road Not Taken: Possible Paths for the Cognitive Science of Religion

Richard Sosis

Introduction

In the summer of 2014, the cognitive science of religion (CSR) celebrated its twenty-fifth year at the fifth biennial conference of the International Association for the Cognitive Science of Religion (IACSR). The conference was held in Brno at Masaryk University, home of LEVYNA, one of CSR's premiere research labs. The conference was entitled "Religion Explained? The Cognitive Science of Religion after 25 Years." Yet, the question posed by the conference title was never seriously addressed, nor do I suspect there was any expectation that it should be. Nonetheless, a veritable army of young researchers, who presented the results of sophisticated and exciting new studies, indirectly answered the question. Religion explained? Not yet: we are just getting started, for the best is yet to come.

There is much to praise about what CSR has accomplished in twenty-five years and there is reason to celebrate. Here I aim to gauge that success by comparing CSR to two other young academic fields: human behavioral ecology and evolutionary psychology. Through these comparisons I seek to draw out lessons for CSR that I hope will help the field forge a successful future.

Assessing the cognitive science of religion

The marriage of religious studies and cognitive science

The obvious place to begin an assessment of any field is at its emergence, and there is no reason to diverge from that well-worn convention here. While the title of IACSR's biennial conference would suggest a precise founding date of CSR, such precision is hardly evident. I am not privy to the conversations of the conference organizers who determined the founding date of the field, but the organizers did reveal that there was substantial debate. And, indeed, there should have been. "Twenty-five years" was based on the publication of Lawson and McCauley's pathbreaking *Rethinking Religion* (1990). There are, however, seminal CSR papers that were published much earlier than 1990. What distinguishes these earlier works, although admittedly not all (e.g., Lawson 1967), is that they are anthropological.

For example, the founding date of CSR could be considered a full decade earlier with the 1980 publication of Stewart Guthrie's "A Cognitive Theory of Religion" in *Current Anthropology*. Or we could push the clock back even further and argue that Dan Sperber's *Rethinking Symbolism* (1974/1975) marks the founding of CSR. But Guthrie and Sperber would probably both argue that their works are built on a long tradition in anthropology that aims to understand the cognitive underpinnings of religious thought. For instance, prominent cognitive and psychological anthropologists Melford Spiro and Roy D'Andrade published a cross-cultural analysis of supernatural beliefs in *American Anthropologist* in the late 1950s. Their discussion seems to augur CSR's later interest in the role of cognition in the cultural transmission of religious ideas. In an illustrative passage, they write:

> Though analytically conceived as culturally-constituted fantasy, belief systems are not created de novo by the fantasy of each individual or of each generation; rather, they are transmitted as cognitive structures from one generation to the next as part of a group's cultural heritage. (Spiro and D'Andrade 1958: 456)

And of course the writings of Levy-Bruhl in the early twentieth century push the anthropological interest in religious cognition even further back (Levy-Bruhl [1910] 1926, [1922] 1923). Nonetheless, CSR is fundamentally a product of the so-called "cognitive revolution," beginning in the 1950s, so it rightfully does not draw upon such early works as those of Levy-Bruhl, or consider itself as the intellectual heir of this anthropological tradition.

If we accept the conference organizers' founding date and assume that CSR began in 1990, then we are not celebrating a twenty-fifth birthday of the field. For it is clear that the anthropological interest in religious cognition predates this time. Rather, CSR is celebrating an anniversary that marks the marriage of two prominent academic fields: religious studies and cognitive science. This is not to deny that anthropologists have played a critical role in the founding and development of CSR. Indeed, anthropologists Pascal Boyer, Scott Atran, Dan Sperber, Harvey Whitehouse, Stewart Guthrie, Armin Geertz, and others are considered leaders in the field; and frankly, CSR would be unrecognizable without their formative contributions. But to appreciate why CSR researchers are interested in the questions they pursue and why the field is institutionally structured as it is, it is necessary to recognize that CSR developed out of a very unlikely exogamous marriage that crossed not just disciplinary lines but also the science-humanities divide as well. By linking the founding of CSR with the publication of *Rethinking Religion*, the conference organizers were acknowledging that the primary direction and concerns of CSR have been defined by those trained in religious studies, even if much of the theoretical development and empirical research in CSR has been pursued by those outside religious studies.

Fundamental principles of the cognitive science of religion

To assess the health of an academic field, it is informative to examine how the field is upholding its core principles. In the case of CSR, this turns out to be more challenging than perhaps anticipated. Despite the fact that CSR has become a recognizable and firmly established subfield of religious studies, there have been surprisingly few attempts to articulate its core principles. Review articles of the field instead focus on delineating core topics pursued by CSR researchers, such as HADD (Hyperactive Agency Detection Device) and MCI (Minimally Counterintuitive) theory (e.g., Barrett 2000, 2007a Barrett and Lanman 2008; Boyer 2003). Incidentally, conference presentations by Robert McCauley, Armin Geertz, and others followed this approach precisely; core principles of CSR were never discussed until I raised the topic in the last lecture of the conference. It is an interesting state of affairs for a young field to offer polythetic, rather than essentialist, self-definitions. It suggests the lack of a common theoretical or methodological core, or possibly a field in flux. There is of course general agreement that CSR consists of research aimed at understanding the cognitive underpinnings of religious thought and behavior, but this is hardly a blueprint for an academic discipline.

Nonetheless, at risk of being accused of misrepresenting the literature, it must be acknowledged that there has been some limited discussion in the literature concerning fundamental principles of CSR. Consider a recent example by Dimitris Xygalatas (2014a: 345–6):

> CSR scholars by and large agree on a set of basic overarching assumptions. First of all, religion is not a *sui generis* domain of the human existence and therefore can and should be subject to explanatory scrutiny just like any other cultural expression. Second, a scientific study of religion must necessarily adopt a position of methodological naturalism; religious explanations of religious phenomena cannot be taken to have any explanatory value in themselves. In line with evolutionary psychology, it is accepted that cultural forms are subject to the biological constraints of the human brain and the universal mental capacities of the human species, as they have evolved through natural selection. In line with Cognitive Science, it is also accepted that the mind is neither a blank slate nor a general-purpose computational machine but comes pre-equipped with a host of specialized mechanisms, each with a specific function. Based on these premises, cognitive scientists of religion are interested in exploring the causal mechanisms that might account for the recurrent patterns of religious beliefs and practices found around the world.

While this summary is certainly not the final word on CSR's core principles, it does provide a helpful window into the field. There are three observations worth underscoring. First, these principles emphasize the point that CSR emerged out of religious studies. Indeed, these principles seem aimed precisely at distinguishing CSR from religious studies. For example, whether or not religion is sui generis is a long-standing debate in religious studies (Proudfoot 1985; Taves 2009), but it has received little attention from those outside of this discipline. Second, and following from the first, these core principles might be shared by CSR researchers, but they are shared for diverse reasons. Scientists interested in the study of religion would not pause to consider that religion might be beyond explanatory scrutiny. And methodological naturalism is such a basic assumption for scientists that it is rarely explicitly stated. In other words, for CSR researchers coming from the sciences, all of Xygalatas's assumptions will appear obvious and banal, but for those traditionally trained in religious studies, these principles are likely to appear revolutionary, or at least they did in the early years of CSR. Third, these principles are not distinctive to CSR, as those who claim to engage in the evolutionary psychology of religion also share them (e.g., Rossano 2010; Weeden and Kurzban 2013).

If Xygalatas's core premises of CSR are roughly correct, then the field is wide open to nearly anyone who wishes to study religion scientifically. This is intellectually exciting since it means that CSR can grow and adapt and will not be limited to particular methods or theories that are likely to be disproven or out of date any time soon. However, although this theoretical and methodological omnivory is exciting, it is institutionally challenging. Specifically, if CSR is unable to distinguish itself from other psychological and behavioral sciences studying religion, such as evolutionary psychology, what exactly is its raison d'etre? The original motivations for the development of the field, which distinguished CSR from religious studies, will soon become irrelevant. The marriage of religious studies and cognitive science has given birth to a young generation of researchers who, unlike their predecessors, are not trained in traditional religious studies programs and do not share the concerns and perspectives of the previous generation of CSR researchers.

CSR is indeed in the midst of a transition and I believe it is at a pivotal juncture in its development as a field. To gain perspective on its growth and potential trajectories, I compare CSR to two other young fields that I have been intimately connected with in my training, research, and teaching: human behavioral ecology and evolutionary psychology. Both of these fields emerged at roughly the same time as CSR and thus serve as useful comparators.

Comparing human behavioral ecology and the cognitive science of religion

Behavioral ecology is the application of the theory of natural selection to the study of behavioral adaptation and design in an ecological setting (Winterhalder and Smith 1992). Human behavioral ecology (HBE) extends the theoretical perspective and methodological tools of animal behavioral ecology (Krebs, Davies, and Parr 1993) to the study of human populations. Chroniclers of the field (Winterhalder and Smith 2000) situate its beginnings with optimal foraging studies, conducted in the 1970s, and notably collected in an edited volume published in the early 1980s (Smith and Winterhalder 1981). By the mid-1980s human behavioral ecologists began to extend their research beyond foraging behavior, applying optimality models to understand fertility decisions, mating tactics, childcare practices, life histories, and other aspects of human behavior.

There are various similarities between CSR and HBE, but two in particular are worth highlighting. First, similar to CSR, HBE emerged from two larger disciplines: biology and anthropology. Whereas CSR originally consisted primarily of religious studies scholars employing the theories and tools of the cognitive sciences, HBE was founded by anthropologists who borrowed the methods and models of evolutionary biologists and applied them to humans. In other words, both fields drew inspiration from a discipline that offered more rigorous methods and generalizable theories than were currently available in their home disciplines. Second, both CSR and HBE have benefited from the affections of larger academic fields. While archaeologists maintain a close relationship with behavioral ecology, using their models to interpret their data and advance theoretical development, theologians and philosophers of religion pay close attention to nearly every finding in CSR, scrutinizing them and reflecting on their significance for their own theories and theologies. The interests of archaeologists and theologians respectively should not be underestimated. CSR and HBE are both small fields; without the sustained and enthusiastic interest of larger disciplines, their survivorship as academic areas of study, especially in the early years, would have been much more precarious.

Despite these similarities there are considerable differences between the fields. First, whereas CSR researchers are trained and employed in various disciplines, including psychology, neuroscience, anthropology, religious studies, and history, HBE is primarily limited to one academic discipline: anthropology. HBE researchers collaborate with economists, biologists, psychologists, and others, but anthropology is the home discipline to the overwhelming majority of researchers who consider themselves human behavioral ecologists. Second, unlike CSR, HBE never developed its own academic organization. HBE researchers participated in the founding of the Human Behavior and Evolution Society in 1988, but they are just one subfield in this organization. Likewise, similar to CSR's section in the American Academy of Religion, HBE researchers were instrumental in creating the Evolutionary Anthropology Society (EAS) within the American Anthropological Association, but like the Human Behavior and Evolution Society, EAS does not consist exclusively of human behavioral ecologists. Third, HBE researchers never established their own journal, as CSR has, but rather publish in various evolutionary and anthropology journals. The publishing situation in HBE is somewhat similar to CSR before the founding of *Religion, Brain and Behavior*, and the *Journal for the Cognitive Science of Religion*. Specifically, *Human Nature* plays the role for HBE that the *Journal of Cognition and Culture* (*JCC*) played for CSR; *Human Nature* does not exclusively publish HBE research, but it does serve as an informal "home" for HBE researchers, similar to *JCC*'s status since its founding in 2000. Fourth, HBE researchers have not published any trade books. And the books they have published, with a few exceptions (e.g., Kramer 2005; Marlowe 2010), have been published primarily by Aldine de Gruyter, a respectable press but not of the stature or visibility of the university presses of Oxford, Cambridge, or Harvard. Fifth, one reason that human behavioral ecologists have not published trade books, in contrast to CSR researchers (e.g., Barrett 2012; Bering 2011; Boyer 2001; Slingerland 2014a), is because human behavioral ecologists employ research methods (e.g., testing of optimality models) and pursue research topics (e.g., foraging and life history theory) that lack mass appeal. In other words, HBE is understandably of little concern to lay people. The study of religion, in contrast, is of great interest to many religious, as well as nonreligious, laypeople. Sixth, HBE lacks the private financial support that is currently a significant engine in the growth of CSR. Specifically, HBE research is primarily funded by the Anthropology Section of the National Science Foundation (NSF), and the private agencies that are interested in HBE, such as the Leakey Foundation, are small. In contrast, the John Templeton Foundation (JTF) is directly or indirectly funding the majority of work in CSR and the amount of money that JTF has invested in CSR is remarkable. There are single projects being funded by JTF whose budget is larger than the entire annual NSF budget for anthropology. Quite simply, JTF has unquestionably transformed CSR and spurred its growth. While much has been made of JTF's possible religious and political influence on the field (e.g., Bains 2011; Martin and Wiebe 2014), JTF's greatest impact may be its preference for high-profile, newsworthy research.

Given the previous description of HBE, it is fair to ask whether HBE is a successful academic field. Of course, the answer to such a question would depend on the measuring instrument employed. If the number of headline-grabbing results or TED-talks were the assessment tool, then HBE would fall short. However, HBE researchers do not view themselves by these standards, and, importantly, neither do others. HBE has indeed been successful, but it is a quiet form of success. This is a field that has not sought publicity but has rather relied on the strength of its research to build its reputation. During the past fifteen years HBE research has been overrepresented (i.e., articles per capita) in *Current Anthropology*, which maintains the highest impact factor of anthropology's flagship journals. NSF also funds HBE research projects well out of proportion to what would be expected by the relative number of active HBE researchers. Moreover, anthropology departments have recognized the funding and publishing accomplishments of HBE, and despite inherent conflicts with the evolutionary and reductionist approaches of HBE, the job opportunities for HBE researchers continue to grow in an otherwise very difficult academic job market.

What can CSR learn from HBE? The message is simple: the key to success is high-quality research and not necessarily more journals, bigger organizations, media attention, or trade books. That said, in my estimation CSR cannot follow the path of HBE. First, CSR already has a thriving academic organization (IACSR) and successful specialized journals (*JCSR* and *RBB*). Moreover, CSR has already grabbed too much media attention to achieve "quiet" success. And given that JTF funds the majority of significant research programs in the field, CSR is likely to gain considerably more public attention in the coming years. Second, CSR's topic of study, religion, is inherently interesting to laypeople. Thus, even without the encouragement of JTF, CSR research will always have an attentive lay audience. Nonetheless, the importance of high-quality research is a message worth embracing, regardless of the path a field takes.

Comparing evolutionary psychology and the cognitive science of religion

Evolutionary psychology (EP) likely provides a better comparison with CSR than HBE, or at least they appear to have taken similar paths. The goal of EP is to uncover the selective pressures that have shaped human psychological mechanisms (Gaulin and McBurney 2001). The field itself views its roots in Darwin's own writings, but it was not until the publication of Donald Symons's *The Evolution of Human Sexuality* (1979) that the field genuinely began to take shape. The publication of *The Adapted Mind* (Barkow et al. 1992), however, launched EP into an exceptional growth pattern that, if not unique, is certainly rare among academic fields.

The similarities between EP and CSR are abundant. Here I focus specifically on similarities relating to the fast growth rate and high visibility of both fields. First, EP and CSR have their roots in the cognitive sciences and thus benefit from the sustained success of the cognitive sciences over the past several decades. Second, while CSR results often grab headlines, EP's research focusing on mate choice (e.g., Buss 1994) has resulted in extraordinary media attention. Sex is a topic, understandably, that interests nearly all humans, and consequently EP sells magazines and newspapers and attracts viewers and listeners in the visual and audio media. Third, while EP never developed its own independent centralized international organization, the Human Behavior and Evolution Society, which EP researchers dominate, essentially serves this role. It should be noted that despite their numerical prominence, evolutionary psychologists have equitably shared the power structure of HBEs with other fields, specifically HBE and cultural evolutionists. Rather than founding a centralized academic organization, EP has established various regional (e.g., NorthEastern Evolutionary Psychology Society, European Evolution and Human Behaviour Association) and topical (e.g., Feminist Evolutionary Psychology Society) organizations. Fourth, EP maintains several journals that exclusively publish evolutionary psychological research (e.g., *Evolutionary Psychology*, *Evolutionary Psychological Science*, and *Evolutionary Behavioral Sciences*). Fifth, EP has fostered an industry of trade books, some of which have become international best sellers (e.g., Buss 1994; Pinker 1997, 2002; Wright 1994).

The extraordinary growth of EP, however, has not been without costs. First, EP was stigmatized early in its development, with claims of shoddy research and the more persistent accusation of offering nothing more than "just-so-stories" (Buller 2005; Plotkin 2004). Many of these critiques were easily dismissed as genuine misunderstandings of EP research. And some critiques were based on fundamental disagreements about human behavioral sciences that were not particular to EP but were leveled against EP because concerns about determinism and ethical implications made EP an easy target (Buss 2005). But most objective observers would agree that EP research, especially in the early stages of its development, exhibited high variance in quality. As the field matured and became confident and stable enough to be self-critical, overall research quality improved, although EP's development of new journals may be outpacing the relative growth rate of its practitioners, which will inevitably put research quality at risk.

Second, the extraordinary success of EP has not only resulted in an increase in self-identified evolutionary psychologists but also transformed social psychology itself over the past several decades. It is now commonplace in psychology journals, possibly essential, to posit some evolutionary rationale for research findings. This trend, it should be emphasized, was never the goal or intention of EP. Evolutionary psychology offers a theoretical foundation that can generate, through informal inference, a priori hypotheses about psychological adaptations (Gaulin and McBurney 2001). Post hoc evolutionary interpretations of data—in other words, claiming that findings are consistent with what we would expect natural selection to produce—are a far cry from the methods envisioned by early proponents of EP. Ultimately, EP's success has meant that there are many psychologists who lack rigorous training in EP, yet apply evolutionary interpretations to their research. Again, whether fair or unfair, this situation has made EP an easy target for criticism.

There are several lessons that can be learned from the growth and development of EP. First, maintaining the high quality of research that was on display at the 5th biennial IACSR conference in Brno, and is generally manifest in CSR, will not be easy. With regard to EP's initial reputational problem, much of it was solved by an extraordinary husband and wife editorial team, Martin Daly and Margo Wilson. In 1997, when Daly and Wilson became editors of *Evolution and Human Behavior* (*EHB*), not only did they transform the journal but also their editorial work impacted the entire field. Daly and Wilson set the bar high and rejected papers from authors who had grown accustomed to regularly publishing in the previous incarnation of the journal, known as *Ethology and Sociobiology*. It did not take long for researchers to get the message: only the highest quality research would be published in *EHB*. Daly and Wilson's editorial decisions were undoubtedly difficult, but they ultimately placed the health of their field above long-term friendships and collegiate loyalties. Their efforts were ultimately appreciated, but it took such efforts to raise the bar in EP. I've described this history in some detail because CSR may find itself in similar circumstances, requiring someone to make comparable sacrifices. Moreover, it is worth recognizing that individuals, or in this case a couple, can have a lasting impact on a field.

Second, because EP is like CSR insofar as it is compelling to many outsiders, some researchers have been inclined to loosely draw upon CSR theories to interpret their data. The situation in CSR may be even more challenging than that faced by EP. Evolutionary psychologists may be concerned about how social psychologists use their theories for post hoc data interpretation, but social and evolutionary psychologists are at least part of the same family of disciplines, with both seeing themselves as pursuing human behavioral science. CSR, however, is a science that is attracting the wider attention of humanities scholars. These scholars recognize the power of CSR's theories and wish to apply them to their historical and ethnographic data. However, effectively using CSR's theoretical advances in the humanities will generally require some retraining (Slingerland 2008; Justin Barrett's former Cognition, Religion, and Theology project at Oxford University provides an excellent model).

Third, part of EP's success has been its ability to avoid splinter groups, especially in its earliest and obviously most fragile years of development. Splinter groups already seem to be sprouting in CSR, such as the emerging concentration on cognitive biblical studies (e.g., Luomanen et al. 2007), which maintains its own section (Mind, Society and Religion in the Biblical World) in the Society of Biblical Literature. Moreover, some of the evolutionary camps, such as D. S. Wilson's Evolutionary Religious Studies group, have not been fully integrated into CSR. The Human Behavior and Evolution Society might serve as a valuable model for CSR, as it represents the interests of diverse groups who share a common interest in employing evolutionary approaches to understanding the human condition. It would seem that IACSR could, and should, function similarly, developing into an umbrella organization that serves diverse groups interested in cognition and religion.

This suggestion leads to a fourth lesson that can be learned from EP: do not fear a name change. To keep evolutionary researchers, social psychologists, cultural anthropologists, and biblical studies scholars all in the same organization, CSR might need to consider an alternative name that broadly captures what all of these researchers have in common, which in truth, is probably not cognitive science. As noted, HBE went through this process with its flagship journal, changing its name from *Ethology and Sociobiology* to *Evolution and Human Behavior*, which reflected the reality that researchers in the organization no longer saw themselves as ethologists or sociobiologists, but rather as evolutionary anthropologists, psychologists, economists, political scientists, and so forth.

A fifth lesson that can be learned from EP is the importance of nurturing relationships with scholars who have a broad academic reach. *The Adapted Mind* (Barkow et al. 1992), as noted earlier, marked the initial ascent of EP, but its extraordinary growth was fostered by a number of highly popular trade books, such as *The Moral Animal* (Wright 1994), *The Evolution of Desire* (Buss 1994), and *The Origins of Virtue* (Ridley 1996). Stephen Pinker's conversion to EP and his subsequent writings (e.g., Pinker 1997, 2002) were also certainly not trivial. In contrast, CSR was marginalized, or notably absent, in several widely read and influential works on religion (e.g., Dawkins 2006; Dennett 2006; Wilson 2002). And most recently, CSR was summarily dismissed as irrelevant by the late, eminent sociologist Robert Bellah (2011), a point I return to later in this chapter.

EP and HBE offer very different models of how a field can develop. Lessons can be drawn from both of them, but most importantly, CSR needs to decide what its primary goals are. What kind of field does CSR want to be? What does CSR hope to accomplish? Once these questions are settled, the path that should be taken to reach its goals will be much clearer. Since IACSR has already been firmly established as the central institutional organization of the field, it would seem that it is the responsibility of its leaders to clearly articulate specific goals for the field and provide a compelling vision for achieving those goals.

Future directions and friendly advice

In this section I offer several recommendations concerning how CSR might develop. Whether or not my advice is heeded, if nothing else, I am hopeful that my comments can at least initiate several conversations that I believe are important for the health of CSR.

First, CSR needs to study religions. For the uninitiated, this recommendation might sound absurd, but CSR researchers, in fact, will recognize it as controversial. Many CSR researchers claim that religions do not really exist and therefore to study them, as religious studies scholars have always done, is misguided. The in vogue approach among CSR researchers is "fractionalization," as was discussed by both Whitehouse and Lanman at the IACSR conference. Fractionalization refers to the reductionist research strategy that seeks to understand elements of religion by breaking them down and isolating their features for scrutiny. For example, it is argued that to study ritual we need to isolate (in experimental labs) the various common components that constitute ritual, such as body movement, synchrony, singing, attire, and so forth. This is an approach that I have been sympathetic to in the past, and I continue to pursue research in this vein, but I have become increasingly concerned about its ability to explain religious experiences and lives. To be clear, fractionalization has its value and such an approach has already (and will continue to) advance our understanding of religion and the human condition. Admittedly, I am an unrepentant reductionist. But reductionism is just a first step in any analysis of complex phenomena. Ultimately, the pieces must be put back together, and therefore, drawing implications about religious systems from the results of analyses that focus on one corner of the system can be misleading. Even more problematic for fractionalization, however, is the strong possibility that religious systems exhibit emergent properties (Sosis 2012). If so, then religion cannot simply be broken down and studied in isolated fragments with the hope that by understanding the constituent parts we are gaining insight into the workings of the whole system (Purzycki et al. 2014; Sosis 2009; Sosis and Kiper 2014).

The fractionalization approach has been justified by arguing that religion is an illusion. As Whitehouse described at the IACSR meeting, "religion is like the big dipper." In other words, when we look at the Big Dipper, we believe we are seeing a pattern in which the sum of the parts constitute a meaningful and functional whole, but of course, classical constellations like the Big Dipper are arbitrary sets of stars. It is only human cognition that imposes a pattern on what is otherwise a capricious arrangement. Likewise, it is argued, religion looks like a "complex whole" only because human cognition detects such a pattern. But the analogy falls short. While the spatiotemporal relationships between stars in a constellation are independent, that is certainly not the case for religion. Indeed, one of the most interesting and pressing questions faced by CSR researchers concerns the interrelationships between the various elements, such as ritual, myth, and supernatural agent beliefs, that constitute the core of religions (Sosis 2009).

In *Religion in Human Evolution*, Robert Bellah (2011) recognized this problem within CSR, which is why he dismissed the field as irrelevant to his project. He singled out the two most widely read books in CSR, *Religion Explained* (Boyer 2001) and *In Gods We Trust* (Atran 2002), as "unhelpful." Bellah ignores CSR in his treatise because of its "lack of insight into religion as actually lived" (Bellah 2011: 629). Some may be inclined to write off Bellah as an outsider who failed to understand CSR's approach and achievements; nonetheless, the onus would still be on CSR to better articulate its value. I, rather, suspect Bellah understood CSR all too well. Indeed, many within CSR readily admit that they are not studying religions and thus Bellah was correct that his historical focus on the developmental trajectories of religions is an anathema to CSR.

Second, CSR may have emerged as a marriage between religious studies and cognitive science, but it did not remain a monogamous marriage for long. Anthropologists and philosophers were quickly seduced, and by the late 1990s the evolutionists had followed suit (Bulbulia et al. 2008). These were strange bedfellows, but CSR's religious studies scholars welcomed their new partners and overall it has been a fruitful polygamous arrangement. Nonetheless, the initial concerns that motivated disaffected religious studies scholars to break from their traditional discipline and develop CSR still linger in CSR discourse. It is time to recognize that CSR has extraordinary disciplinary diversity; the majority, or near majority, of the field no longer originates from religious studies. Rather than a subfield of religious studies, the next generation of CSR would be better conceived as a subfield within the behavioral sciences. As a subfield of the behavioral sciences, CSR will be in a position to reengage with religious studies in a healthier and more productive relationship as equals, rather than as parent and wayward teenager, which seems to characterize current relations.

Third, CSR has a responsibility to understand how it is impacting religious communities. It is clear that CSR is influencing the discourse in some religious communities, especially among religious intellectuals. But it remains unknown if media and other popular accounts of our work affect the beliefs and behaviors of those we seek to study. Admittedly, if it turns out that our research is genuinely impacting human lives, I am not sure what should, or could, be done with such information. Nevertheless, we cannot even begin to have such a conversation until we assess the possible impact of our research. Such a venture is not unusual: most scientists want to know how their work is influencing the populations—whether penguins or people—they are studying. And to be clear, I am not suggesting that if our work turns out to be impacting religious lives that we therefore alter our research to minimize effects. All I am saying is that a conversation about the impact of our research needs to occur, and such a conversation will only be useful if it is informed by accurate assessments.

CSR also has a responsibility to educate the public about religion. Indeed, we are in a unique position to help resolve the increasing frequency of secular and religious conflicts that plague our world. CSR cannot adjudicate such conflicts, but our research may be useful in diffusing such conflicts. One causal factor that pervades most conflicts between the religious and secular, as in most disputes, is a lack of understanding and appreciation for the beliefs and behaviors of the opposing side. Suggesting that observant Jews, for example, should light one less candle during Hanukkah, as proposed by Israeli conservationists ("Green Hanukkah Sparks Criticism," *United Press International*, December 5, 2007), indicates a complete lack of understanding and appreciation for the conviction of religious beliefs and practices. And disputes like this are commonplace (Sosis and Kiper n.d.). If CSR can offer anything in such disputes, it is the explanation of religious beliefs and behaviors in materialist terms. Of course, CSR theories and data are unlikely to be satisfying for many religious adherents, because they fail to capture the depth of meanings in their convictions. CSR research, however, may help religious adherents understand how even secular values can be sacralized by secularists and evoke strong commitments (Tetlock 2003). More importantly, CSR interpretations of religion are likely to provide satisfying materialist explanations for secularists. CSR should not aim to justify such beliefs and practices, but with an appreciation for why humans possess such strong convictions for religious beliefs, it can hopefully open up a fruitful dialogue to facilitate conflict resolution.

Conclusion

I conclude with a suggestion that CSR think deeply about its identity. To claim that CSR has an identity problem is a gross understatement. Consider the simple fact that

I am writing this chapter, not to mention that I serve on the executive committee of IACSR. My credentials as a cognitive scientist are simply nonexistent. Indeed, trained as a behavioral ecologist, I have largely been agnostic regarding cognitive mechanisms throughout my career (Sosis and Bulbulia 2011). It is a remarkable fact that I have been welcomed into the CSR community—no doubt, a testament to goodwill, but also surely evidence of CSR's lack of self-identity. My guess is that I, and those with similar backgrounds, have been welcomed into CSR because (1) we study religion scientifically as a natural phenomenon, and (2) we recognize that sociological, political, and economic models are insufficient for explaining religious thought and behavior. This makes for a stimulating, albeit amorphous, collection of scholars. Many of the researchers engaged in CSR are not actually doing cognitive science. In fact, at the IACSR conference in Brno, about half of the research papers presented could not be classified, even by the most liberal definitions, as cognitive science. It is not surprising, as previously discussed, that CSR defines itself by the research it pursues rather than a set of core principles.

The IACSR conference was not an aberration, but rather emblematic of changes brewing in CSR. Consider the recent research of IACSR's president, Joseph Bulbulia, and acting past president, Armin Geertz. In my mind they have published two of the most important, yet neglected, articles on religion in the past few years: "Charismatic Signaling" (Bulbulia 2010) and "Brain, Body, and Culture: A Bio-Cultural Theory of Religion" (Geertz 2010a). These papers, I suspect, have been relatively ignored because they offer theoretical models that do not fractionate religion, but rather point toward a holistic science of religion. Their models, therefore, are not easy to evaluate through experimental methods, although both Bulbulia and Geertz support their models by drawing upon many experimental results that are consistent with predictions from their respective models. Genuine evaluation of their models is likely to require the slow and labor-intensive methods of detailed ethnography. Ethnographic research requires more time and deeper commitment, and produces fewer headlines than experimental studies; it is therefore no surprise that their theories are not spurring evaluation and that even anthropologists are now frequently turning toward experimental methods. Ethnography, however, must complement experimental studies for CSR to remain a credible voice in the academic study of religion. The bottom line is that we cannot ignore the study of religions.

I highlight Bulbulia's and Geertz's articles because they offer, in my mind, a more compelling science of religion than the current dominant theoretical and methodological approaches in CSR. In some ways, these papers also characterize the direction in which CSR seems to be moving, although these papers are emblematic rather than causative of this movement. Self-identified CSR researchers are no longer just studying the cognitive mechanisms that underlie religious beliefs and behaviors. Neither Bulbulia nor Geertz, for example, focus on what is happening in the head, but rather their attention is drawn to the dynamic interaction between mind, body, and environment. Had Bellah been aware that CSR researchers were pursuing such approaches, I suspect he would have appreciated their efforts and given CSR the attention it deserves.

Moreover, it is clear that Bulbulia and Geertz are well aware that they are moving beyond the traditional CSR paradigm. Notably, neither draws on cognitive science to label their models. Bulbulia turns to sociology, specifically Weber's notion of religious charisma, and Geertz describes his model in anthropological terms as bio-cultural. Geertz was writing for a more traditional religious studies audience, and thus he was concerned that "[t]he term 'biocultural theory of religion' may strike some as too much hard science" (2010a: 313), although this is unlikely to trouble CSR researchers. I conclude with the vision for CSR of our acting past president, not only because he is better positioned to offer such a perspective, but because I endorse this vision and cannot say it any better myself:

> What I envision is a combination of theories from neurobiology, social psychology, anthropology, cognitive science, archaeology and comparative religion. I am convinced that nothing less can do the job and am therefore arguing that traditional disciplinary boundaries are in need of modification and change. . . . I hope that I have shown that the intimate interrelations between body, brain, culture and society, demand a "biocultural turn." (Geertz 2010a: 313)

Acknowledgments

I sincerely thank the Institute for the Advanced Study of Religion in Toronto, LEVYNA, and IACSR for graciously covering my expenses to IACSR's fifth biennial conference. I also thank the James Barnett Endowment for support, and Jordan Kiper and Benjamin Purzycki for very helpful comments on earlier drafts of this chapter.

14

Looking Back to Look Forward: From Shannon and Turing to Lawson and McCauley to …?

Justin E. Lane

Introduction

Although this volume is premised on the idea that the cognitive science of religion (CSR) is twenty-five years old and therefore represents an important milestone for reflection concerning our successes, failures, and future goals, to understand our future, however, we need to look more deeply into the past. I will present this chapter as a deliberate expansion—and violation—of the theme of this volume and argue that the interdisciplinary cognitive approach to culture and religion is best contextualized as an endeavor now almost seventy-five years old.

The Beginning

It has been argued in the past that the founding of CSR was with the publication of E. Thomas Lawson and Robert N. McCauley's seminal text, *Rethinking Religion* (Lawson and McCauley 1990). The field of religious studies acquired the name "Cognitive Science of Religion" a decade later (Lawson 2000). It was in this seminal paper that Lawson stated:

> It is difficult to miss the fact that the last fifty years has given birth to a revolution in the sciences. This revolution has not only transformed the way scientists theorize about the human mind but the means they have devised to test their theories. Cognitive science, which has emerged in the context of this revolution, has coordinated, distilled and extended the particular explanatory theories of human cognition provided by cognitive psychology, cognitive anthropology, linguistics, artificial intelligence (AI), philosophy, neuroscience and computer science. (2000: 338)

While CSR has focused many of its efforts in engaging with a few of these subdisciplines, such as psychology, anthropology, and—to a far lesser extent—neuroscience and philosophy, the rest of these concentrations (linguistics, artificial intelligence, and computer science) have been relatively untouched in CSR.

Naturally, it must be noted that the book *Rethinking Religion*, inspired by Chomsky's approach to syntactic structures of human language (1957), reshaped how we think about religion by offering a systematic framework for studying religion. This cognitive approach was embraced by anthropologists and psychologists for understanding religious ritual and began to generate discussion at conferences at Western Michigan University, University of Vermont, and Queen's University, Belfast; these were quickly followed by conferences at the University of Michigan Ann Arbor (Lawson, personal communication, May 12, 2015). The cognitive approach generated enough interest to form groups within larger academic bodies such as the American Academy of Religion (AAR; 2015) and the North American Association for the Study of Religion (NAASR; 2015) before forming its own academic body in 2006: the International Association for the Cognitive Science of Religion (IACSR; 2015).[1] These discussions generated a number of the early empirical studies of religious ritual (Barrett and Lawson 2001; Barrett 2002; Feeny et al. 2006; Malley and Barrett 2003). Around this time, the general theoretical assumptions of the cognitive approach were forming. This inspired many scholars of religion, interested in the comparative approaches of those like Mircea Eliade, to embrace the cross-cultural empirical framework provided by CSR's general theoretical orientation that the mind was an evolved organ and therefore a possible target for cross-cultural investigation (see Paden 2010 for a reflective discussion). This also inspired anthropologists such as Pascal Boyer (1992) and Harvey Whitehouse (1995) to reframe some of their work within a more cognitive paradigm (see Boyer 1994b; Whitehouse 1995, respectively). It was at this point that evolutionary theorizing was employed in order to reconstruct past narratives and situate the findings of these empirical studies within a broader context. This reframing also necessitated that we add the fields of (cognitive) archaeology and (cognitive) historiography[2] to the list of fields originally noted by Lawson mentioned previously.

CSR today largely fits within the previous narrative, and the impact of the past twenty-five years is, without a doubt, positive. The formative period that represents the foundation of what has been produced within the field and its books (such as the *Cognitive Science of Religion Series* with AltaMira Press; the *Religion, Cognition and Culture Series* with Routledge Press; and the *Scientific Studies of Religion: Inquiry and Explanation* series with Bloomsbury), papers, journals (such as the *Journal of Cognition and Culture* and the *Journal of Cognitive Science of Religion*), and academic conferences (hosted by the IACSR and its affiliates).

Regardless, it seems that CSR is currently at a crossroad. The direction of the field has many options—most of which are positive. However, my stance is that understanding cognitive approaches to the complex phenomena of religion, culture, and economics available prior to Lawson and McCauley (i.e., those approaches informing those who have, in turn, informed us) can provide a bearing for the future of the field; one which is far more promising than our past. It engages both present and past material and utilizes all of the sub-disciplines originally presented by Lawson (2000) in order to answer questions concerning the nature of human religiosity.

CSR's prehistory

The first time that an interdisciplinary group of people came together to discuss ways to understand topics such as religion, culture, and language was—debatably—the Haddon expedition. This expedition was a team of psychologists, linguists, musicians, and anthropologists who traveled, in 1898, to Torres Straits to study theories of memory and perception (Douglas 1980). This expedition was seminal at the time and has repercussions we can still feel today. For example, it inspired the work of Fredric Bartlett, who in 1932 published the monograph, *Remembering: A Study in Experimental and Social Psychology*, which outlined how individuals construct schemas that can change over time as we experience different events.[3] However, it wasn't until 1942 that a similar group came together with *information processing* as its primary focus at an invitation-only conference funded by the Josiah Macy Jr. Foundation (JMF). JMF was founded by an American businessman and philanthropist whose children used his wealth to fund research that they believed would better society (see macyfoundation.org). It seems that JMF's financial superiority was dominant in the field until corporate partnerships were formed with companies such as IBM.

This group of scientists, later called the Cybernetics Group, had founding members from anthropology (such as Margaret Mead and Gregory Bateson), neuroscience (Warren McCulloch), computer science, and informatics (such as Claude Shannon). Mead and Bateson's work on religion may be known to some CSR scholars who come from religious studies, anthropology, or sociology, as their work is considered by many to be seminal but may be foreign to those coming from cognitive science. Similarly, McCulloch may be foreign to the humanities, but his work is formative in computer science and psychology as the first formal model of neural networks and their integration with the brain. My last example, Claude Shannon, is virtually unknown in anthropology[4] and rarely brought up by cognitive scientists today. However, his work, to which I will return later in the context of Relevance theory (Sperber and Wilson 1995), was formative for cognitive science as he offered the first information processing approach to language (1948). These JMF meetings have effectively been forgotten by contemporary approaches to religion, despite a renewed interest in studying religion by understanding its information processing mechanisms.

During this period in the 1940s, scholars of religion such as Mead were directly engaged in theorizing about information processing. It is even reported that Mead broke a tooth during the first meeting out of sheer excitement (Gleick 2012). Ultimately, however, disciplinary cracks were revealed as Shannon proposed his theory of information transmission (Shannon 1948), a formal theory of how information is transmitted that is uninterested in semiotics. Purportedly, this made Mead and other people interested in "meaning" uncomfortable (Gleick 2012) although her attendance at these conference continued until at least 1953 (American Society for Cybernetics 2014).

After the end of World War II, information theory primarily influenced computer science with little impact on anthropology until the 1980s. The field of psychology was dominated by behaviorism, which continued to take a "blank slate" approach to the human mind. Both anthropology and psychology were vehemently engaged in showing how "culture" and the mind are constructed from environment alone. Ultimately, this blank slate view of the mind was subsumed by the information processing paradigm of cognitive science when Chomsky (1959) introduced an information processing approach to understanding language, exposing major flaws in the behaviorist views such as those of B. F. Skinner (1957). This shifted the understanding of the mind from a machine that copies information to one that has the ability to generate new information; it also resituated psychology within the cognitive framework that computer science—particularly its subfield of artificial intelligence—had been utilizing for a decade (Turing 1950).

This larger view of the mind was able to utilize the work of the other subfields because they now existed within a single paradigm; operating with enough similarity for theory to be built and for scientific understandings to be scaffolded on one another. This led to nearly unimaginable advances in the field of computer science. Questions about how we represent knowledge were presented by Minsky while working on artificial intelligence (AI; 1974) that were seemingly congruent with similar findings produced by psychologists (Collins and Quillian 1969; Quillian 1967). These advances in how information is processed and manipulated by both human and nonhuman structures—a question first posed in JMF meetings—were revolutionizing how humans themselves now thought. In parallel, Leon Festinger and a team of collaborators studied the effects of failed prophecy in a new religious movement, leading ultimately to his pioneering work *When Prophecy Fails* and to the theory of cognitive dissonance (Festinger et al. 2008).[5] This led to the first theoretical formulation—directly involving religious groups—that offered clear, falsifiable, and potentially replicable hypotheses about religious beliefs and behaviors.

A scientific approach to the study of religion was pushed further during the 1980s as sociology began to adopt an understanding of how human social interactions work to adopt—again from information processing—network analyses. For the first time since the 1940s, a quasi-cognitive approach to the topic of religion was reintroduced by Rodney Stark and W. S. Bainbridge (1996). This theory, which utilized axioms rooted in human psychology and posits more than 300 relationships derived from these statements, represented the first theory of religion explicitly to utilize cognitive science (Bainbridge 2007). However, this theory has been largely ignored by CSR despite the recent work of Bainbridge published in a CSR book series (2006). This theory produced many empirically viable studies of religion in the following decade. The research existed largely within sociological studies of demographics and the mechanisms that were not fully explored empirically for another two decades (see ibid.). Ultimately, it was the addition of linguistics and psychology to the "cognitive sciences" that set the stage for Lawson and McCauley. At this point it should be stressed that the study of religion and culture was not *incorporated* into the cognitive sciences nor did the cognitive sciences become *introduced* to the study of religion or culture. Rather, the work of Lawson and McCauley elegantly reintroduced the study of religion and culture to the cognitive sciences after a hiatus of more than four decades.

It is at this point that the introductory section of this chapter connects to the current narrative as well as to that of many of the other chapters of this book. Since many of this volume's authors were present for, or the authors of, the major findings in CSR,

I will not go into detail here. However, I will demonstrate how understanding this history can be fruitful for understanding religion within CSR now and how it can be a useful guide for our future as well. By taking the analysis of language as an example, I will show how the information processing approach has created a cohesive paradigm for studying one of the mind's most complex outputs and how this has direct relevance for CSR. I will conclude with a final section addressing how, if CSR moves forward within the *cognitive* sciences, it will likely experience greater success.

Cybernetics of language

Much like Lawson and McCauley's (1990) information processing approach disrupted the anthropological approach to ritual, Claude Shannon's information processing approach to language disrupted our understanding of language. After Shannon's presentation of an informatics of language, a number of anthropologists stopped participating in the Macy conferences (Gleick 2012); a rift that has generally continued since 1948. However, between the 1940s and now, the study of language has benefited greatly from information processing.

As language is the medium—either written or spoken—with which so much religious information is transmitted, it cannot be discounted. Shannon was able to elegantly show that by adding a twenty-seventh character to the Latin alphabet—the whitespace—much of the "noise" in the data disappears; this is because it can provide demarcations between coherent information sets. Like Sperber (1975), he addressed the crucial mechanisms by which language transmits information (Shannon 1948). His approach shows that controlling for and understanding this noise is crucial to being able to quantify the system itself (a point to which Sperber 1975 is owed a great debt by modern anthropology). Shannon's formalization was able to show that words appear with specific frequencies and that languages can be quantified by how efficiently they transmit information. While this may seem banal to some, his theory subsequently produced the tools needed to quantify both entropy and modern wireless communication (Gleick 2012) and is necessary for modern natural language processing and big data mining.

After parsing language to the level of a concept or word, Chomsky was later able to formulate the rules by which we create novel linguistic expressions by recombining rules using implicit understandings of how these bits of information interact. Subsequently, combining the psychological and computational approaches afforded Minsky the ability to formulate how concepts combine in structured ways to create mental schemas (Minsky 1974).

Learning from our past

The topics of the Macy conferences directly address many of those that CSR is only now addressing. For example, in 1948, the conference title was "Circular Causal and Feedback Mechanisms in Biological and Social Systems."[6] This topic of discussion is only recently found within CSR (for example, Xygalatas 2012 provides a discussion of this in regard to religion). We have to ask ourselves if the period from the 1950s to the 1990s was a "dark age" for the cognitive approach to religion and culture. If so, it seems that we are currently in our renaissance. However, we need to stop and ask ourselves about the earlier and current approaches that afforded so much success. Furthermore, we must also ask what about approaches to religion in between these two periods stifled the study of religion.

Concerning definitions

Scholars in the first generation of CSR were often drawn from the corner of academia interested in the secular study of religion. Many of them, such as Luther H. Martin (2008), have noted that one issue with the study of religion before CSR was that there was no consensus about a definition of religion (in many ways echoing Smith 2004). Indeed, scholars in the study of religion have noted for some time that religion is difficult, if not impossible, to define since it is not a natural kind. This has led most to adopt a stance that religion is an academic abstraction. CSR was able to move us past these shortcomings by taking a scientific approach where religion is operationally defined (most people adopting some version of the definition presented in Lawson and McCauley 1990). This definition attempts to include many relevant aspects of religion while excluding many irrelevant aspects. (It is interesting to note that Lawson and McCauley's definition is very similar to one offered by Spiro 1966, who was also looking for a more scientific definition.) Definitions are crucial to a scientific endeavor. As CSR moves forward it should be careful to define all variables at all times; a computer program cannot run without all variables being defined beforehand and neither can an appropriately formulated cognitive approach. Definitional insufficiencies can cause confusion and misinterpretation where they can often be avoided.

Concerning methodological and theoretical assumptions

Generally, the Lawson-McCauley definition of religion is accepted in CSR. This consensus has allowed scholars to build on one another's work. Crucial to this is that CSR scholars employ theories that are more or less compatible. The detailed formalization of theories (such as Lawson and McCauley 1990; Whitehouse 1995) has allowed scholars to work within common frameworks. This provides the ability for one to know to what extent the variables and phenomena are compatible between one theoretical approach and another. This specificity has allowed for precise revisions and critiques of theories (see McCauley and Lawson 2002; Whitehouse 2004). Most impressively, the cognitive approach has also allowed for a computational modeling of these theories (Braxton 2008; Whitehouse et al. 2012) and even some revisions based on these computer models (McCorkle and Lane 2012). What have resulted are generative research paradigms that are expanded through an iterative process that employs the scientific method.

Generally, CSR assumes biological evolution as its meta-theory; it is the theory within which almost all other CSR theories are situated (Liénard and Lawson 2008; Whitehouse 2004: Chapter 1). The attempt to create parsimony between evolutionary theory and CSR has caused considerable debate. Much of this debate has surrounded those who feel that functional definitions of religious groups allow us to treat groups or religions as units of inquiry (for example, see Wilson 2002) in contrast to those who look at the cognitive mechanisms possessed at the individual level as the appropriate level of selection (for examples, see Boyer 2001; Lawson 2000). This is not to say that there is no "function" to groups; indeed most people in CSR admit the evolutionary benefit of group living. Rather, the disagreement is whether or not groups are themselves viable targets for analysis. It seems that on the one hand, the individual is the clear and only plausible level of analysis for a psychological approach to the subject. Even earlier approaches to now defunct ideas such as the "collective unconsciousness" rejected the idea of a group mind (Bartlett [1932] 1995). However, a cognitive approach *can* incorporate a group approach to the study of religion. That is to say, while a cognitive *psychology* of religion cannot accommodate group-based analyses of religion, a cognitive *science* of religion can. This is because cognitive science as an information processing paradigm can study groups per se or groups as collections of individuals, whereas a cognitive psychology can only directly address the workings of the minds of individuals. If anything, a cognitive *psychology* of religion doesn't study even the individual as a unit; it studies intra-individual units (cognitive mechanisms). This is yet another example in which a clear and strict information processing approach, which necessitates clear definitions of variables, can serve to help CSR. Furthermore, employing new tools such as computational modeling in order to study both individual and group phenomena related to religion can overcome mereological issues (the micro/macro questions) and methodological limitations of working within a single area of cognitive science.

Currently, CSR stands with one road behind it and—at least—three choices ahead. To the far right we have the cognitive *psychology* of religion. This approach encompasses a similar one to that of the past by using carefully construed psychological experiments in order to better understand the psychological mechanisms that facilitate the production and maintenance of religious beliefs and behaviors. The strength of this approach is its largely reductionist methodological rigor, which allows researchers to distill religious phenomena to lower-level interactions. However, this approach's weakness is—almost paradoxically—its reductionism. Religion is a very "high-level" and abstract phenomenon (or collection of phenomena). While empirical psychology can test intra-individual psychological mechanisms and—to a lesser extent—interpersonal interactions, religion relies on many psychological mechanisms that only seem to instantiate themselves when observed at a social level (i.e., when multiple individuals interact). To reduce religion may explain how some of its aspects are caused at lower levels. But unless we also find a way to build religion back up, we may be missing a crucial aspect of our target of study (i.e., religion). This is not to argue for cultural causation—quite the contrary. While social psychology can empirically demonstrate many important relationships, its reliance on contextual variables to explain social phenomena gets us little further than the approaches, such as that of Durkheim before them. What is needed is a *generative* approach that allows for multiple interacting variables to generate higher-level phenomena due to complex interactions among their parts.

To the far left we have another choice, the cognitive study of *religions*. This approach posits that religions can be functional units, or can be viewed as traits, that can suffice as targets of analysis within an evolutionary framework. In this view, the religious group affords benefits and is therefore an adaptation; this seems to be backed up by the observation that religion is widespread—a clear opening, however, for the naturalistic fallacy. The religious group also facilitates certain ideas that are adaptive to their environment, they change and can die out—although this stance suffers from a lack of ability to define or quantify beliefs (see Kundt 2015 for a lengthy exposition on the subject). This approach to religion draws directly from cultural evolution (e.g., Boyd and Richerson 2005) and is increasing in its popularity. Its strength is that it seems elegant to describe how religions have changed over time. It employs many of the same mechanisms to explain how religion works as those at the level of ultimate explanations. Those who look to the evolution of religions look at groups of individuals, who admit that religion is a social phenomenon; this is true even given the definition of religion provided by Lawson and McCauley (1990). However, its ultimate explanations as to why religions have changed largely dissolve into historical narratives. The greatest shortcoming of the cultural evolution approach is that, unlike those of the cognitive psychology of religion, it cannot offer precise predictions. Outside of individual level cognition, they only provide elegant *post*-dictions of what has happened. This is largely because religion is defined by beliefs in *supernatural* agents and causally opaque actions that often do not make "rational" sense (see Talmot-Kaminski 2013). As such, religious beliefs cannot easily be measured by their "fitness," even in a utilitarian sense that is not reliant on some reproductive fitness measure. The supernatural aspect of religious beliefs can often only be said to be "good" or "bad" when supported by ideological, rather than empirical, frameworks—a trap long known to traditional scholars of religion.

Concerning interpretation/explanation

The above-mentioned strengths and shortcomings of CSR also reveal another crucial issue for its trajectory, past and present. Ultimately, this comes down to whether or not one is attempting to interpret or to explain religion. For many engaged in CSR, this is not so much an issue. Although CSR concentrates much of its efforts on experimental investigation, and an increasing amount of time at conferences and in papers that are targeted toward more explanatory endeavors, we cannot forget that many in the community utilize the findings for interpretive exercises aimed at understanding and making sense of religious phenomena. Indeed, the growing fields of "cognitive historiography" and "cognitive archaeology" are almost exclusively dedicated to interpreting such materials using the constraints and proclivities discerned by cognitive sciences operating in an explanatory mode. One could not—or should not—state that explanation has more inherent value than interpretation in the study of religion. However, we should be clear where these boundaries are drawn. Although experimental findings and explanations can provide statistically backed conclusions, these are often unavailable for use in any practical sense to make sense of religion at large. On the other hand, interpretations without an empirical framework can (and has in the past) run rampant with speculative conclusions. However, we can state that the cognitive *science* of religion is that which ultimately relies on the empirical testing of falsifiable and replicable studies that seek to uncover the mechanisms that underlie beliefs and behaviors related or referenced in relation to some supernatural agent. Concurrently, the cognitive *study* of religion is that which utilizes the framework of information processing (i.e., cognitive science) in order to better understand and describe beliefs and behaviors related or referenced in relation to some supernatural agent.

Given the depth and breadth of the field currently, if CSR is to move forward, it must adopt an explanatory framework that is amenable for use by those interested in interpretation. Those scholars of religion, such as historians and anthropologists who are interested in interpreting religion, can rely on the empirical results of scientific studies. They (interpretive scholars) can utilize their deep knowledge in combination with statistically backed trends and careful descriptions of mechanisms to better understand religion as it unfolds in the real world. Explanatory approaches aimed at reductionism have been the backbone of the generative and collaborative research programs in CSR that have provided the bulk of our progression over the past twenty-five years. These generative research programs provide a fruitful framework that can constrain and guide the interpretations of others.

The third option

At this point, we may note that in the broadest sense we have two positions with differing theoretical priorities: individual/group; evoked culture/cultural evolution; culture or religion as non-causal/culture or religion as a causal force; religion as non-sui generis/religion as sui generis; interpretation/explanation. At the root of the issues is a mereological problem, namely the relationship between the individual and the group. Those who study religion exclusively at the individual level do not, per se, have the theoretical specificity to adequately account for the social dynamics of religion and those who study religion at the group level do not have the theoretical specificity to adequately account for how individual mechanisms can aggregate to create something that can be a discernable "group" unit. Simply put, neither group seems to be able to account for how the individual and the group operate concurrently. As such, neither seems to be an ideal path for us moving forward.

There is, however, a third option before CSR: the middle path between our left and right options. This option returns to a strict information processing approach to the study of religion. However, it broadens the scope of CSR to incorporate multiple aspects of cognitive science in order to create an approach to religion as a social phenomenon. This approach seeks to formalize culture using techniques such as semantic networks (Carley and Kaufer 1993; Whitehouse 2013a). These networks can mathematically specify the conceptual schemas held by both individuals and groups (Lane 2015). These networks can be theoretically situated between agents who have predefined social (network) connections (Lane 2014). The behaviors of these agents can be described using psychologically realistic mechanisms, drawn from cognitive psychologists (Lane 2013; Sun and Hélie 2013). Such an approach can also leverage other techniques and findings from cognitive science, such as learning and pattern recognition through the use of artificial neural networks (Sørensen and Nielbo 2013b).

The power of this third approach lies in the fact that it is amenable to computational formalization. In such an approach, beliefs, behaviors, and potential social contacts are computationally traceable. Generally this approach is not utilized in the study of religion, including CSR (Braxton et al. 2012). Instead, we have chosen to formalize our theories either verbally or through the use of more formal representations. To date, only three theories of religion have undergone any computational testing: Lawson and McCauley's ritual competence theory (Braxton 2008), Stark and Bainbridge's theory of religion (Bainbridge 2006), and Whitehouse's divergent modes of religiosity (Whitehouse et al. 2012). All of these models are incomplete. However, only Whitehouse's divergent modes of religiosity model has undergone any extension since its publication (see McCorkle and Lane 2012).

Much of the research in CSR today is engaging in social psychology. That is to say, it is not interested in the *mechanisms* that produce religious beliefs and behaviors; rather, it interested in how religious beliefs and behaviors differ between social groups. The social psychological approach, dating back to the late-nineteenth-century/early-twentieth-century work of Wilhelm Wundt, is interested in how individuals interact with their social group or "society" (Hogg and Abrams 1988). This approach is compatible with the left option described previously. Indeed, Hogg and Abrams have even gone as far as to state that the mereological issue is "the 'master problem' of social psychology" (ibid.: 6). However, social psychology is not interested in information processing (i.e., cognitive science) as explanation. Instead, it utilizes a more Durkheimian notion whereby the social explains the social. Besides falling into traps long noted by religious scholars, such a reliance on the social group as an undefined "emergent" property is incompatible with a strict cognitive approach (Epstein 2006a; Lane 2013).

Social phenomena such as a social group or a religion or a culture are not "emergent properties" within a computational paradigm such as that envisioned by this third path. Higher level properties can of course be instantiated within a computational model. Many computationally defined systems have higher-level patterns with extreme randomness, even complete chaos—as defined mathematically not hyperbolically (Wolfram 2010). Some computational systems are fully predictive. However, if we are utilizing a cognitive approach that asks relevant questions, we can reconstruct the mechanisms of human sociality/religiosity and have the agents play out probable outcomes as we vary their computational environments. In this way, higher-level "group" processes may be observed, but they are in no way emergent; rather, they are predefined logically through complex interactive functions within the computer code (Epstein 2006b: chapter 1).

As an example, Joshua Epstein's cognitive architecture Agent Zero has modeled the process by which individual agents instantiate novel behaviors, which they are not programmed to do, simply by means of their associations with others in their social network (Epstein 2014)—a sort of cultural epidemiology of both information and affect. Naturally the *possibility* for the behavior is predefined, but not all agents have the same proclivity toward a behavior, nor are any agents told when to enact a target behavior. However, he was able to show how an agent without some predisposition would institute a behavior similar to one with a behavioral predisposition by means of information transfer between agents in variable environments in a social network. In this computational framework, the social group did not cause the behavior nor was the behavior emergent in any way. Instead, the target behavior resulted from the interactions of complex mechanisms and environments that were wholly defined within the computational system. It was neither nature (a genetic analog) nor nurture (a social analog) but both.

The next twenty-five years of CSR

The third modeling option is reliant on an approach that prioritizes a strict information processing approach very similar to that embraced originally by Margaret and George Mead, Gregory Bateson, Claude Shannon, and Alan Turing. By necessity, all of the interacting variables and mechanisms are logically defined. This is not to say that it is not complex. On the contrary, it can leverage contemporary breakthroughs in complex and dynamic systems theory for statistical testing. It can utilize increases in computational power and supercomputers in order to perform complex mathematical procedures. In these ways, it is simply pushing informatics as applied to religion into a modern computational world of online social networks, supercomputers, and data mining. However, it is different from the earlier work in that it denies that the social group is an emergent entity. In fact, it denies views of strong emergence entirely (for an extended discussion, see Epstein 2006b). Instead, it seeks to show how social phenomena can be "grown"—or generated—from their constituent parts and the complex interactions of agents. It goes against the right-hand option previously mentioned because it seeks also to construct what we have reduced and admits that a purely reductionist approach to religion is not enough to truly say something about religion as a human phenomenon. It also does not believe that "culture" or "religion" are causal variables. Instead they are artefacts of the recursive and iterative interactions of individuals that form a social environment that is in many ways inseparable from the biological environment. In this way it seeks to parsimoniously employ the evoked culture approach to religion[7] alongside those approaches that believe culture and religion can be defined as "socially transmitted information" (see Richerson and Christiansen 2013).

This approach can also leverage the massive amount of data stored online. Data mining algorithms targeting religious texts hosted online can be utilized to study massive corpora (see Lane 2015). It can also be utilized to study the massive amount of data that historians have amassed over the centuries (see Aiden and Michel 2014; Google Books 2015; Jackson 2010). This creates a framework that overcomes the presentism that results from relying only on modern experimental studies for empirical evidence. This third approach can facilitate results useful for interpretation because only through well-formulated complex computational models can we truly test potential counterfactual situations that may have arisen in the history of a religious group. This could provide even greater statistical parameters to those engaging in interpretation and open up the possibility for historical explanation without relying on ideas of "random mutations" that are neither random, nor mutations, and selection mechanisms without well-defined fitness parameters.

It is in the coming world of Big Data that this third way truly shines. There is more data hosted online than could reasonably ever be analyzed by human researchers. However, using data mining algorithms over massive samples of human data drawn from social networks, online blogs, and text archives, we can attempt to gather a representative sample of information drawn directly from social groups. This approach is truly Big as it utilizes information drawn from populations that are potentially numbered in the hundreds of millions (Lane 2013). If it is used to inform strict cognitive models such as those outlined here, we can then use big data to inform the study of religion in a cross-cultural and comparative framework that relies on the statistical certainty of our current empirical studies but expands them well beyond that imagined by early scholars of comparative religion.

So it seems that here we come full circle. The first proposal for a comparative and scientific study of religion (Müller 1882) focused on language. Now, it seems that we have used a scientific study of religion rooted in the study of language (Chomsky 1957) in order to study religion empirically. Now as we move forward, advanced algorithms aimed at allowing us to draw statistical inferences from unimaginably large corpora of data can inform computational models of religious behavior. It is likely that we will never fully understand the mind of *Homo religiosis*.[8] However, if we stay true to the cognitive science of religion in a way that has a keen eye for methodological and theoretical inconsistences as did our forebears, then we can create a viable science of religion; one that could potentially result in a "standard model" of human religiosity. However, we must keep in mind what we are studying (individuals/groups/both) and what we hope to do with those studies (explanation/interpretation). Ultimately, we must admit that it is information that we are studying. It is a truism that religion is nothing if not social. However, it is that "religious stuff" that we are interested in: the beliefs and behaviors appropriate to our definition of religion. Over the past twenty-five years, CSR has made great strides and is resulting in some basic computer models and databases mapping human social trends. However, over the next twenty-five years it can target larger-scale social networks with precise computational measurements using datasets of a league far larger than anything currently underway. And because of that, I for one, am extremely optimistic for a cognitive approach to religion.

15

A Neo-Victorian Cognitive Science of Religion

Steven Hrotic

Rediscovering the wheel

Several years ago, I was attempting to apply cognitive anthropology to a particular problem in music history related to the Catholic Church in the mid-1500s. For centuries, Western "art music"—including music sponsored by the Church and used during the more elaborate celebrations of the Mass—had become more and more complex. Unlike twentieth-century music, which typically employs a single melodic line (augmented by harmonies, for example), the music of the medieval Church employed several independent and simultaneous melodic "voices." The orthopraxy of "polyphony" was repeatedly challenged during the Counter-Reformation, however, with many arguing for a return to simpler music compositions. Polyphony survived; according to legend, Palestrina's *Missa Papae Marcellus* (ca. 1562) was partially responsible. My article attempted to understand the cognitive underpinnings of this purportedly theological challenge (Hrotic 2013).

There was an obvious complementary question I only cursorily addressed. Although polyphony survived the mid-sixteenth century, it did *not* last into the seventeenth. Why not? In my article, I wrote only that the "options offered by polyphony were essentially explored, and possibilities for further innovations under the same rules were exhausted," and that the accumulation of ever more complexity by innovative composers "strained to the breaking point both the physical resources needed to perform them and the cognitive resources required really to hear them" (20).

This is all that survived of a longer section, ultimately deleted. I had initially noted that complexity in medieval art music centered on *one kind* of complexity: the number and independence of melodic voices. I described how each successive period of art music was associated with the accumulation of another, characteristic domain of complexity, ending another hard constraint in conceptual difficulty. The Baroque period (ca. 1600–1750) explored (and ultimately exhausted) tonality and formalized structure (e.g., counterpoint); the Classical period (ca. 1750–1820) abandoned that domain of complexity and instead devised new ways to manipulate melodic phrases within a composition (changes in key, rhythm, etc.); the Romantic period then followed a parallel pattern.[1] Twentieth-century jazz developed in a similar fashion: for example, swing emphasized melodic improvisation; bebop emphasized *chordal* improvisation; hardbop then relaxed the rhythmic structure to allow imported elements of soul and avant-garde. In sum, each period "ratcheted" (sensu Tomasello 1999) complexity in a new domain.[2] As I was also reading about the history of astronomy at the time (especially models of Mars' retrograde motion), I had appropriated a term to describe this phenomenon as "epicycles of complexity."

I ultimately cut this section from my submitted version, not because I doubted the accuracy of my account—in fact, I still intend to explore the possibilities. However, it was a little disheartening to learn that my visions of "epicycles" were neither as new nor as original as I had thought: Alfred Kroeber (1944) had described a similar pattern decades earlier, which was later elaborated into an "epicyclical hypothesis" by Charles Gray (1958: esp. 17).[3]

My contention is that the history of anthropology is full of promising but abandoned beginnings, many of which involved biological and cognitive approaches to culture. This is to be expected, as we value most highly those academics who are brilliant and original, who make a name for themselves—and have the metrics (e.g., high H-indexes) to prove it: from my own field, for example, Tom Lawson, Robert McCauley, Pascal Boyer, and Harvey Whitehouse. There is somewhat less profit involved in *extending* these paths: we reward the trailblazers. As a result, not every promising idea is fully explored.

This volume is designed to celebrate one such beginning: Lawson and McCauley's 1990 book, *Rethinking Religion: Connecting Religion and Culture*. From my colleagues' accounts, I get the clear impression that we are celebrating not just a solid piece of scholarship but also the promise of a new field employing a "cognitive approach" to cultural behaviors. But to what extent was this approach unprecedented? And if earlier examples exist, why hadn't the field materialized long before 1990?

Anniversaries of cognitive anthropology

Depending on how one defines our field (and I must admit there is little consensus—which may be indicative of the problem), identifying the proper landmark as the beginning of our field is problematic. To start at the most specific level, our field involves several disciplines. As my approach to religion has an anthropological character, my revisionist, autobiographical memory places the beginnings slightly *after* Lawson and McCauley (1990), with Boyer's *The Naturalness of Religious Ideas: A Cognitive Theory of Religion* (1994b). On a more practical level, many of my peers were first introduced to cognitive approaches to religion via Whitehouse's *Arguments and Icons: Divergent Modes of Religiosity* (2000). But as all these scholars have acknowledged, Guthrie (1980) was an important step toward a new paradigm in the scientific study of religion. All of these are worthy candidates, but identifying *which* of these works published between 1980 and 2000 as most significant (or symbolic) is at least in part a matter of disciplinary perspective, of personal history, and of taste. Lawson and McCauley (1990) is an excellent compromise, as it appeared in the middle of this range, introduced a specific testability that distinguished our field, and, as described by Luther Martin, represented the "initial consolidation" of our field (pers. comm.).

If one takes an even slightly broader view, however, choosing *anything* published circa 1990 as our starting point is arguably off by decades or more. For example, if the key insight is that cognitive science can reveal patterns in the way we think that constrain cultural complexity, I would note that, at roughly the same date as Boyer's foundational work on the cognitive "naturalness" of religion (1994a, 1994b), D'Andrade (1995) described *decades* of precedents in cognitive anthropology. If we focus on the shaping of cultural information through cognitive transmission (as it evolves in our mental selection environments), we should choose Sperber (1975) and Dawkins (1976)—but even 1975 might be too conservative: what about Mead's *Continuities in Cultural Evolution* (1964)? Justin Lane (Chapter 14) connects her to a "cognitive paradigm [which] grew from informatics and cybernetics." Or, why not mark our anniversary from, for example, Miller's (1956) "The Magic Number Seven, Plus or Minus Two," with its critical insight into cognitive *constraints* on cultural forms, or, as Richard Sosis (Chapter 13) suggests, Spiro and D'Andrade (1958)? These are by no means the only candidates for antecedents to Lawson and McCauley; my point is merely that, like my epicycles, approaching religion (which for many academics was introduced via the humanities) from a cognitive perspective perhaps felt more novel to contemporary readers than it in fact was. The achievement of *Rethinking Religion* was a specific and well-reasoned argument—replicable observations resulting in a useful analytical tool—not in inventing a new approach to religion out of whole cloth.

As I mentioned previously, because of my anthropological stance, Boyer (1994b in particular) was my introduction to the cognitive science of religion, even though published several years after Lawson and McCauley (1990). It's ironic, then, that for me Boyer reads as rather Victorian—which I mean as a strong complement. Almost a century ago, Ernst Renan wrote, "I fear that the work of the twentieth century will consist in taking out of the waste paper basket a multitude of the excellent ideas which the nineteenth century has heedlessly thrown into it" (quoted in Modi 1922: ix). I have the opposite fear: that throughout the twentieth century we squandered much of our Victorian heritage and abandoned too many promising beginnings. These include, of course, early arguments for the application of scientific approaches to *all* human behavior, including religion. As a single but unexceptional example, Riale (1892) wrote:

> Now that we, as world peoples, are coming to sit under one umbrella in other things, we feel there must be some better understanding of the real relation of our religious beliefs than hitherto we have had. To make a long story short, the best are now firmly convinced that the hour is here when religious facts are to be brought under the influence of *scientific study*, if heart peace and soul harmony are ever to hold sway … [N]ow that we have brought the "no life," the plant life, the animal life, as well as the historic and economic life, into scientific harness, and have thought them all out in terms of evolution, we feel the next field for such study is the religious, the grandest and noblest of them all—the sphere of life which is the holy of holies, and of which we want to know the truth and the whole truth, that we may the better live. (243, emphasis original)

Recognition of the relevance of biological approaches to culture are just as old—similarities between cultures need not have only historical but also biological bases, based in our shared human nature. Folklorist Thomas Keightley wrote (before Darwin returned from the HMS *Beagle* expedition):[4]

> I [have] read a great quantity of poems, tales, romances, legends and traditions of various countries and in various languages. I here met such a number of coincidences where there could hardly have been any communication, that I became convinced that the original sameness of the human mind revealed itself as plainly in fiction as in the mechanical arts, or in manners and customs, civil or religious. (1834: 6)

Given my own interest in aesthetic as well as religious culture, I would offer J. W. Powell as an example of just how deep our history with these kinds of issues actually is, as some of his ideas were rediscovered decades later. In the first article of the first volume of the new series of *American Anthropologist*, Powell suggested teleological functions for aesthetic preferences like "sweet" rooted in the environments of our evolutionary past (1899: 6–7; cf. Barash 1979, Orians and Heerwagen 1992), emphasized individual variations on biological themes (Powell 1899: 6; cf. Carroll 2005), and even cultural evolution through exemplar choice:

> Growth is made usually by minute increments. Rarely indeed is there a sudden outgrowth, but the increments of development are all made by men with a genius for the activity.... A multitude is led by one, so that demotic evolution is dependent very largely for its initiative on the few which the many learn by imitation. (Powell 1899: 23; cf. Tomasello 1999; Kvadsheim's 1992 review of exemplar theory)

Victorians like Powell should be recognized as good exemplars for current scholars, not least because of their rich "evolutionism." The current volume celebrates twenty-five years of the cognitive science of religion; in this context, Powell and *his* predecessors suggest this anniversary may be off by a century.

Academic resistance

Despite elements of forward thinking, at other points Powell sometimes appears to us today as rather quaint, for example, in his naïve confidence in unidirectional social progress. ("In savagery men play for effects and control the causes, as they suppose, by necromantic figures... . In later stages of culture the sorcery to a greater or less extent is abandoned, and skill is recognized as the true cause" [1899: 10].) But to the generation *after* Powell, progressivism was anything but harmlessly quaint. Evans-Pritchard (1965: 53) called the belief in the superiority of Western cultures and academic perspectives "the basic assumption of all Victorian anthropologists"; as a result, this assumption necessarily became a prime target for their successors. Anthropology needed to move beyond the mistaken "impression that we, the civilized people, have advanced far beyond the stages on which other groups linger" (Boas 1938: 4). I would agree that this was necessary to move forward intellectually, to engage the racist ethnocentrism Boas fought against, and, in a separate and pragmatic vein, to allow the new generation to distinguish itself from the old. They were not successful in completely closing the door on these problems, however; the twin specters of social Darwinism and positivism lingered through the twentieth century, and I have no reason to believe that they do not continue into the twenty-first (Sahlins 1976; Lewontin 1993; see review in Laland and Brown 2002, chap. 3).

Note, however, that although early-twentieth-century anthropology distanced itself from evolutionism in service to the study of human culture, it did not "involve a systematic questioning of evolutionary assumptions" (Stocking 1987: 326), nor did it attempt to reverse all of its influences (e.g., an emphasis on "natural rather than supernatural origin and development of human capacities" [325]). Further, note that Boas responded in large part specifically to polygenic explanations of cultural differences (60), which does not describe twentieth-century (let alone current) evolutionism. Early-twentieth-century anthropology did not *disprove* evolutionary approaches per se, but rather excluded them en masse from anthropological discourse. When "sociobiology"—the claim of demonstrable relationships between biological selection and social behavior—gained intellectual ground in the 1970s (esp. Wilson 1975a), new objections (that sociobiology is constrictive, deterministic, speculative, politically insensitive) were raised, but subsequently struck down (Alcock 2001). Barkow (2005) engaged specific objections to evolutionism, reassuring his colleagues (yet again) that their fears were groundless: concerns were based on a misunderstanding of our goals and practices.

More fundamentally, empiricism itself (biologically sourced or otherwise) became suspect, leaving anthropology especially vulnerable to "external" political trends[5] (D'Andrade 2000) and to shifts toward postmodernism (Greenfield 2000). Slingerland (2008: 74–96) demonstrates that postmodernism was a necessary component of our collective journey toward understanding human social and cultural behavior, but argued it was not the destination; as Colin Robson argued, we should not "throw out the scientific baby with the positivist bath-water" (2002: 21). Nevertheless, despite the proven intellectual merit of our ideas, we must still be careful to situate any research with an element of sociobiology—or any of the parallel approaches from other disciplines—in the history of this conflict, even though its basic tenets are hardly controversial (e.g., Hrotic 2009: 120). After decades of supportive research, the "triumph of sociobiology" described by Alcock (2001) may not have been Pyrrhic, but has hardly been vindicating.

I believe we've made a strategic error in thinking that objections to our field are primarily intellectual and academic, when they are more firmly rooted in aesthetics and in the public sphere (Dennett 1995: 17–23). Like the apocryphal bishop's wife (O'Toole 2011), our opponents would *prefer* that we were not right; failing that, it would be best if our field does not become widely known. As a professor once wrote in response to one of my early undergraduate attempts: "You may be right, but I just don't buy it."

According to Stocking, it is somewhat ironic that the public sees the situation differently.

> From the point of view of the general reading public, the discipline [anthropology] never lost its evolutionary aura, and there have been important currents of

> popularizing anthropological literature which still sustain this image. . . . On the other hand, a great many of [American anthropologists] are equally disturbed by the potential effect on public attitudes and social policy of the sociobiologists' attempt to explain many aspects of human culture in terms of the evolution of "fitness." (1987: 327)

This sheds light on Steven Gould's and Richard Lewontin's decision to challenge the popularizer of sociobiology, their friend and colleague E. O. Wilson, not only in academic circles on intellectual grounds but also in a response to a *New York Times* review (Waddington 1975) of Wilson's book (1975a) with a review of their own focusing on the potential abuses of sociobiology (Allen et al. 1975)—an action Wilson described as "represent[ing] the kind of self-righteous vigilantism which not only produces falsehood but also unjustly hurts individuals and through that kind of intimidation diminishes the spirit of free inquiry and discussion crucial to the health of the intellectual community" (1975b).

> "Book burning, we are talking about book burning," Margaret Mead thundered during the annual meeting of the American Anthropological Association in November 1976. And a resolution to censure sociobiology was defeated—"but not by an impressive margin," recalls Edward O. Wilson. (Fisher 1994)

Given that my own respect for Wilson and his work biases my reactions, it is too easy for me to mistake Gould's and Lewontin's actions as unethical (especially given the occasional overreactions from the academy [Laland and Brown 2002: 65][6]), but the more important point is that, as their concerns related to (public) societal impacts, not just intellectual merit, perhaps a public response was justified. I can even partially agree with these concerns: not all popular accounts of sociobiology (by whatever name) are as careful as Waddington's review (e.g., Morris 1967; see Laland and Brown 2002: 63–4), as anyone who has seen too simplistic summaries of mate choice or parental investment in popular media can attest. Indeed, even fellow academics' accounts may be startlingly malformed.[7]

Nevertheless, I suspect that at least some of the concerns with Wilson (1975a) are based in matters of personal taste; individuals on *both* sides of the debate honestly prefer their side on *aesthetic* grounds, as well as intellectual. As an example, the 2005 court case *Kitzmiller et al. v. Dover Area School District* involved the appropriateness of teaching intelligent design in public schools. While the court was asked to evaluate intelligent design on intellectual merit, some religious individuals on *both* sides of the debate described the opposition with strikingly similar rhetoric: they claimed their perspective was more aesthetically pleasing (e.g., the beauty of connecting humans to God, the beauty of the continuity of humans with Nature), and questioned the moral character of the opposition and the wisdom of allowing these ideas to undermine education (McMaster et al. 2007).

Drawing again from my own experiences, on one hand, I see opponents of evolutionary approaches skeptical of the narratives we propose, calling them "Just So" stories after Kipling (following Gould 1978); on the other, many of us feel narratives are tools for generating testable hypotheses as well as pedagogical devices and who secretly enjoy them for their own sake. Cognitive anthropologist Jerry Barkow (pers. comm.) was once challenged to justify accounts that appeared to be "Just So" stories and responded forcefully—not to the suggestion that narratives shape how we interpret the data, but to the implications of triviality: "Not 'Just So' stories; heroic myths, maybe!"

In sum, it appears that most of the intellectual objections to evolutionary approaches have been met, and there is no reason to assume that any that remain will catastrophically undermine the field. However, it is important to recognize that objections have *never* been purely intellectual: they have been historical, popular, and aesthetic. We have been less successful answering the other kind of challenge: we have not yet demonstrated a positive societal impact of evolutionary approaches—and recall that in order to receive National Science Foundation grants, one must justify societal impacts as well as intellectual merit.; we have only partially communicated our work to the general public—as Boyer's work demonstrates, *minimally* counterintuitive ideas transmit most easily, and accurate evolutionary accounts of human behavior violate the untrained listener's expectations in many ways; we have failed to emphasize the narrative beauty of evolutionary accounts, tacitly accepting (ironically) that science and beauty occupy separate magisteria (sensu Gould 1997).

But these challenges can be met. Another problem is more subversive.

A nomadic field

My readers will likely have noticed the awkwardness with which I have attempted to avoid giving a label to "our field": is it evolutionism, sociobiology, cognitive anthropology, the cognitive science of religion (CSR), or so on?

If, as I believe, CSR (as represented by Lawson and McCauley) has a long history, it has been one of homelessness. Laland and Brown (2002) describe this history in a sequence of chapters on sociobiology, human behavioral ecology, evolutionary psychology, cultural evolution, and gene-culture coevolution: our center not only evolves but also continually shifts from academic department to department, always marginalized. (Compare to McCauley's discussion of "explanatory pluralism" [Chapter 2].) The results are the twin problems of polysemy and synonymy: the same words may have different meanings in different disciplines; the same idea may have different disciplinary labels (Tribble and Keene 2011: 2–14). Boyer (1994b), for example, relies heavily on the concept of "memory", but

> [t]his diversity is particularly apparent in the gulf between humanist understandings of memory and those of cognitive psychologists, so much so that we may ask whether "memory studies" can offer any true conceptual precision. To a psychologist, memory may look like an internal or individual process; to a cultural historian, it may look like a purely social phenomenon. The gap is reflected in the scholarly literature, in which it is relatively rare for psychologists to allude to the vast array of work on cultural and social memory, and for humanists to take

account of the enormous literature on cognitive psychology. (Tribble and Keene 2011: 9)

"Religion" as a conceptual category is even more problematic (see, e.g., Martin 2014b: 35–44; Smith 1998). Fifty years after the first dedicated departments (see Lawson 1967, e.g.), that challenge is exponentially more complex. Guthrie (citing *sympathetic* researchers McCutcheon 1997, Fitzgerald 1997, and Bloch 2008, among others) suggests that scholars of CSR define religion as fundamentally related to "the supernatural"—which, he argues, is itself a culture-bound and problematic term: "vague, often meaning little more than 'false'" (Chapter 5).

Academics (such as myself), then, who attempt to draw from both Boyer (2001)—the same Boyer wrote, "For anthropologists in particular, it is clear that the term 'religion' is of little use" (Boyer and Baumard, Chapter 6)—and Mary Carruthers' *The Book of Memory: A Study of Memory in Medieval Culture* (2008) face significant challenges.

So one word may refer to many different concepts; simultaneously, a single, focused concept may be given multiple labels by different groups of academics. Tribble and Keene (2011: 2–3), for example, demonstrate strong similarities between "extended mind" and "distributed cognition," before calling for "a historical cognitive science," which they call "cognitive ecology" (8; cf. *Journal of Cognitive Historiography*); Strauss and Quinn (1998: 47) note strong similarities between Dan Sperber's (1996) "schemas" (with which the CSR community is very familiar) and Pierre Bourdieu's "*habitus*" (e.g., in Bourdieu 1990).

This confusion of terms[8] makes the accumulation of progress and communication between even sympathetic scholars difficult (cf. Schjødt and Geertz, Chapter 4). For example, at the beginning of this essay I described my own attempt at cognitive historiography. At a cognitive science conference soon after, I presented a paper describing my views on polyphony. After twenty minutes of analyzing the music and theology of the Catholic Church, the very first question (asked by a well-established figure in "our field") was, "What does this have to do with religion?"

There are pragmatic problems more troublesome even than epistemological confusion. The academy is, for better or worse, built on disciplinary divisions: between departments, between professional organizations, and between academic journals. Collaborations even within a typical scientific field can be costly (Cummings and Kiesler 2007; Boudreau et al. 2014); "our field" combines aspects of biology, sociology, psychology, religion studies, anthropology, and more—without comfortably fitting within any. Note that anthropology itself is a multidisciplinary (and often uncomfortable) union between the humanities and "anthropological sciences," which produces its own personal and professional tensions (see, e.g., Leslie 2000). The scientific study of religion has had a similarly uncomfortable relationship with more generalized professional organizations for the academic study of religion (see, e.g., Martin and Wiebe 2016: 9–13, 17–23).

That is, even if one stipulates that we have a respectable field of study, we clearly do not have "a discipline." Disciplines may be simplistically defined as "data sets," but their functions within the academy are also in many senses *social*: disciplines have their own social networks and quasi-cultural schemas, along with assortative sociality patterns of identifying "in-group" members (Hrotic 2010a) and ethnocentric attitudes (Campbell 1969). Training in a discipline is not merely through explicit learning but also implicit indoctrination (see Merton 1972: 66–7). We trace our fictive-kin lineages, finding alliances through who studied with whom: the psychology department of my undergraduate alma mater has in the hallway a pedagogic family tree—and given the least opportunity, I claim Margaret Mead as my great-grandmother. (My mentor in anthropology was Jim Petersen, a student of Bill Haviland, who studied under Mead.) That is, disciplines are essentially "clans" within the Tribe of the Ivory Tower (Hrotic 2010a). Lacking a disciplinary coherence, we lack professional and institutional support—even successful individuals must often maintain a presence in a traditional field for pragmatic reasons, while justifying extra-disciplinary activities that may or may not contribute to tenure and promotion (see, e.g., Hrotic 2013: esp. 108–10)—and status building is a primary (and pragmatically necessary) activity within academia (Rhode 2006).

Without disciplinary associations, we have no niche within the academy; *with* them, our freedom to pursue studies unpopular with some disciplinary colleagues is curtailed.

> Most of what I want to do in the classroom grows out of decades of disciplinary and institutional involvement … [S]o long as faculty members receive disciplinary training and are appointed to disciplinary slots, and so long as colleges and universities are organized by departments that inherit disciplinary traditions, academic freedom will also embody unique discipline-specific expectations. (Nelson 2010: 3, 7)

Despite the potential restrictions, a disciplinary identity may be necessary. But where should we begin? Several years ago, I was exploring avenues of access to a University in Western Canada. The chair of anthropology expressed curiosity in my work, but explained that my approach was better suited to sociology in that particular university. The chair of sociology disagreed and suggested that cognitive approaches were better suited to the psychology department. The chair of psychology expressed confusion: "Aren't you an anthropologist?"[9]

Neo-Victorians

My goal with this essay is to wonder, Can we celebrate Lawson and McCauley's 1990 book as more significant than a particularly good book? Has it breathed new life into the multifaceted field with a hundred-year-old history variously named evolutionism, sociobiology, the cognitive science of religion, and so forth? I think the answer must be an unequivocal yes. Too much good work has followed, been inspired by, built on, and cited it to believe otherwise.

However, it is less certain that this impressive body of work is any closer to having a stable home in the academy. It is still most often the purvey of specialists from many disciplines building synergies with other disciplines (i.e., multidisciplinary efforts), perhaps integrating disciplinary studies (i.e., interdisciplinarity), or trying to engage directly the world outside academia (transdisciplinarity) (cf. Thompson Klein 2010). We alternate between trying to fill gaps between disciplines (i.e., interdisciplinarity, sensu Campbell 1969) or dispensing with disciplinary boundaries entirely (i.e., post-disciplinarity, sensu Book 2013). That is, we are trying everything *except* building a disciplinary niche.

We may argue—and I believe most within our field agree—that academic disciplines are not natural kinds; pursuing knowledge across disciplinary borders is essentially our most fundamental method. Pragmatically, however, this may well be politically unsustainable, as it ignores the fact that however non-disciplinary "knowledge" is, the academy is not.

> [T]he lines currently demarking one field of study from another are not natural but constructed by interested parties who have a stake in preserving the boundaries that sustain their claims to authority. The structure of the university and the curriculum is a political achievement that is always in the business of denying its origins in a repressive agenda. (Fish 1989: 15)

I suggested previously that we have yet to match (what I believe to be) the considerable intellectual merit of our field with complementary societal impacts—from Stanley Fish's perspective, our interdisciplinarity should grant us an *advantage* in this: "Disciplinary ghettos contain the force of our actions and render them ineffectual on the world's larger stage" (ibid.).

The anthropologists among us would presumably be able to shed some light on this process, perhaps through what Marshall Sahlins called "pseudoethnography" within academia (2009: 997), the first lesson of which is that the very nature of disciplinarity is in flux. "Jealous of their subject matters and defensive of their borders, the academic disciplines of a generation or two ago generally had little to do with one another … [But now t]he 'blurred genres' Clifford Geertz announced in 1980 are blurrier than ever" (997–8, in reference to Geertz 1980; Clark 1963). It is an unanswered question, however, if this "blurriness" in a period of financial crises within the academy merely encourages defensive xenophobia within disciplines.

In good structuralist *or* cognitive anthropological form, I have suggested that disciplines see themselves as linked "kin." If part of this essay concerns tracing the lineage of our field through the search for sacred ancestors (i.e., a search not just for our beginnings but for our origin), the reciprocal search—directly linked to the reason for marking the anniversary of Lawson and McCauley's seminal work—must be for our *descendants*. That is, *we lack undergraduates*: the sine qua non for multi-generational commitments, strong primary identities—currently impossible in a field populated by professionals with a primary affiliation with a traditional discipline.

In sum, we face a number of challenges related to our interdisciplinary status: we lack a common language (e.g., polysemy and synonymy) and institutional stability in the form of departmental support. *Organizational* stability (that is, professional networks to replace the support of individual institutions) would require that we recognize the "tribal" nature of academic disciplines (Becher and Trowler 2001)—which means that alongside developing epistemological clarity, we would benefit from a more coherent cultural identity (Hrotic 2010a).

But if we've failed to disciplinize our field, this is not entirely a bad thing: Godwin et al. (2013) argue that small-world networks tend to be the most creative and "that cross-campus but not within-campus or within-department connections optimize small-world properties" (136). Certainly the entrepreneurial spirit has motivated many of our most successful colleagues and has unexplored potential even within the university (see Brown 2015); Kaiserfeld (2013) argued that "hybrid" organizations have been historically more successful in connecting academic studies to real-world needs than traditional academic organizations.

This would require we give thought to "boundary maintenance," as William James did: perhaps we should strategically embrace our extra-disciplinary status. James fought *against* the departmental disciplining of philosophy, thinking that it is best for some avenues of academic exploration to remain unfettered to topical studies and healthiest for the academy to have some academics functioning as mortar between the departments. Philosophy was well suited to the task, as it was a fundamental part of *all* intellectual inquiry in every department. In essence, James fought against the disciplinary segregation of philosophy into philosophy—and ultimately lost (Bordogna 2008; cf. Frodeman and Rowland 2009).

In contrast, the academic study of religion was *necessarily* through "the emergence of new or expanded departments of religion" (Crary 1967: 35). In the first decade of secular, undergraduate departments of religion studies, Lawson acknowledged that those studying religion must be "conversant" with other disciplines (1967: 41).[10] However, he also argued specifically *against* perceiving religion as having a "depth dimension"—the kind of pervasive relevance James claimed for philosophy. "The logic of [the depth dimension] view is that there should be no department of religion at all, nor even any course in religion. Religion as the depth dimension of culture should qualify all inquiry in the university" (39). From Lawson's perspective, a depth dimension was unacceptable, as "it fails to distinguish between the presence and the study of religion in the university" (ibid). The academic study of religion—that is, a study distinct from pre-professional or theological studies of religion—had a "valid and defensible place *in the university curriculum*," that "[r]eligion can be conceived of as *an academic discipline* with its own methods, subject matter and propositions," that "the focus and purposes of the academic study of religion differs on the *undergraduate level*" (note that Lawson tacitly assumes it should not *exist* at the graduate level alone), and "[i]f the religion is an academic discipline, then the most adequate way to house such a discipline is in a *department of religion*" (43–4, emphases in original).

For better or worse, through our current interdisciplinary status—more, our *tradition* of being interdisciplinary, extra-departmental—we have what philosophy lost and what religion could not accept. The question we must ask, then, is, What shall we do with this?

The key, I think, lies in recognizing that there is a reason I tend to focus on *Religion Explained* (2001), rather than the careful but more opaque scholarship of Boyer's earlier *The Naturalness of Religious Ideas* (1994b): *Religion Explained* is inherently more accessible, a tool for colonization as well as scholarship. If you value the work of

E. Thomas Lawson and Robert McCauley, do not use it only to inspire more scholarly work for your colleagues' consumption; *teach* it to your undergraduate students, along with the skills needed to navigate disciplinary boundaries (e.g., Nambisan 2015: 182–4). If you *must* write, give us a textbook that gives our field cohesion, that is inclusive of the many disciplinary flavors of our field, and that establishes our origin myths and sacred ancestors.

In short, we need to become neo-Victorians. Ask the big questions, *embrace* the heroic narratives. Do not hyperspecialize: adopt a "polygnostic" ideal. Build our identities, develop our culture, and colonize both within and outside academic institutions. Some may see this as an intellectual sin; if so, sin boldly.

Conclusion

On Keeping Cognitive Science of Religion Cognitive and Cultural

Justin L. Barrett

In the fall semester of 1990, while I was an undergraduate psychology major at Calvin College, I took two courses that sparked my interest in a cognitive approach to the study of religion: cognitive psychology and an introduction to systematic theology. I attended the courses just hours apart on Mondays, Wednesdays, and Fridays, but the two courses could not have been any different. The cognitive psychology course was taught by a young, cool, and funny Chinese-Canadian-America with an iconoclastic bent and a love for film and narratives. The systematic theology course was taught by an elderly, polio-crippled German-Dutch-American who taught us from the manuscript of his magnum opus with straight lecture that occasionally lapsed into the German of his childhood, particularly when quoting scripture. Nevertheless, the juxtaposition of these two very different courses worked a connection in my mind. I could see from the theology class the importance of ideas, thoughts, feelings, attitudes, commitments, and beliefs that underwrite the cultural traditions known as religions. As my classmates and I wrestled with the justification for certain beliefs and practices in the Reformed Protestant tradition, sometimes strongly disagreeing with each other and the professor, we could see how much religious cognition mattered to us, to how we chose to live, and with whom we chose to identify. Meanwhile, in my psychology course I was learning how scientists could study cognition: the various methodologies, how cognition was subdivided into its various sub-processes and domains, and different ways to model cognitive processes and the mind. Those two courses convinced me that I wanted to become an academic and that I would study religious thought cognitively and scientifically. And so, the year 1990 marked the publication of Tom Lawson and Bob McCauley's landmark work in what would become known as the cognitive science of religion (CSR), and independently it was also the year that it occurred to me that the study of cognition could make a contribution to the study of religion. Unlike Lawson and McCauley, at that point I did not know how the science of cognition could contribute to the study of religion.

A few years later, however, my graduate advisor Frank Keil put me in touch with Pascal Boyer, with whom I struck up a correspondence using international postal mail, and Keil sent my first graduate paper to his former Cornell colleague Ulric Neisser (who named the field of cognitive psychology), who passed the paper onto his Emory

University colleague McCauley, who shared it with his former undergraduate mentor Lawson, who reached out to me to meet them at the Society for the Scientific Study of Religion annual meeting in October 1995, where I also met Don Wiebe and Luther Martin for the first time. Lawson graciously invited me to a conference he hosted at Western Michigan University in February 1996 where I finally met Pascal Boyer in person and also Harvey Whitehouse and Brian Malley. By then e-mail had become a more efficient way to communicate and Boyer, McCauley, Lawson, Whitehouse, Malley, and I struck up an e-mail conversation that lasted many months, sharing our half-thoughts, aspirations for the "cognitive conspiracy," and concerns about our fledgling area of study.

One of the concerns that we discussed gave impetus to the naming of this area of study, the worry that cognition would become a mere interpretive lens or useful frame for talking about religious thought, experience, and action, but that the insights and tools of the cognitive sciences would not be put into action in any serious or systematic way by this movement still almost entirely comprised of humanities scholars with little or no formal training in any science area. So in 1999 when I received an invitation to write an introduction to the area for the journal *Trends in Cognitive Sciences*, Lawson and I discussed the opportunity and agreed that it was time to highlight the scientific character of the enterprise by calling it "cognitive science of religion," which I did in my January 2000 article (Barrett 2000) and he did in a paper later that same year (Lawson 2000). Our attempt was to fortify the cognitive *and* scientific commitments of our movement, commitments that we should not easily surrender.

CSR, as I see it, has today matured well in maintaining a scientific anchoring, but we seem to have a weakening relationship with the C in CSR. It also appears to me that our general embrace of the science has had an unfortunate by-product: losing track of the cultural unit of analysis. In this chapter, I argue that CSR should reserve a special place for considering how human minds work such that humans entertain and communicate certain types of thoughts that become so widespread and stable as to become cultural. That is, at once CSR should remember that individual minds matter but that the primary target of explanation of religion is not that of individual cognition but of group-level cultural expression.

Some core commitments

As highlighted by many of the contributions to this volume, the scientific character of the cognitive science of religion has matured and diversified, and many of the same things that were attractive about the cognitive approach twenty-five years ago are still its primary draw, or should be. I sketch a few of these attractive features in this chapter as part of trying to make the case that, in looking forward, CSR will do well to keep at least two interrelated core commitments: to stay cognitive and to stay cultural.

Explanation of cultural expression commonly regarded as "religious"

Scholars in CSR have not been shy about aiming to explain why it is that various forms of cultural expression that are often called "religious" arise and persist in populations. The focus has typically been shared expression in populations and not merely idiosyncratic behavior—hence, the term "cultural expression"—and on these forms of expression as the explananda (but not "religion" as a whole). Many other scientific approaches to the study of religion have concerned how religious attitudes, beliefs, identifications, and practices may impact other factors such as health, morality, voting patterns, and belligerence. In CSR, however, the aim is almost always to account for the religious attitudes, beliefs, identifications, and practices themselves. When consequences or effects of these are examined, it is typically for the sake of determining where they fit in a selectionist paradigm: the outcomes of religious expression may encourage or discourage their continued expression.

Avoiding definitional quagmires

CSR has managed to make progress in the study of cultural expression commonly called "religious" without trying to seek consensus concerning the referent of the term "religion." Many strategies have been used. Guthrie (Chapter 5) addresses the definitional problem by identifying the cultural manifestation of a particular psychological tendency as "religion." If someone else wants to say that "religion" is not essentially about systematic anthropomorphism or that such a definition is too broad, Guthrie's contribution to our understanding of how basic, pan-cultural psychological tendencies to find human-like design, purpose, and agency in the world around us and to systemize such tendencies remains. Don't call it "religion" if you don't want to. Anthropomorphism happens essentially everywhere and demands an explanation. Lawson and McCauley (1990; McCauley and Lawson 2002) use a similar tactic concerning "religious rituals." Even if you find their theory of "religious rituals" too broad or too narrow from some particular perspective, it seems to meaningfully and defensibly pick out an interesting form of cultural expression and makes steps to better understanding its origins and persistence.

A related strategy for avoiding getting bogged down in disagreement over "religion" is noted by Whitehouse (Chapter 3): set aside the idea that "religion" is a useful scientific construct and instead explore the particular forms of cultural expression of interest (Barrett 2017). What are their cognitive properties? What psychological mechanisms give rise to their expression? What evolutionary processes may account for their recurrence? Do they tend to co-occur with other forms of cultural expression that are commonly called "religious," and if so, why? We see such a strategy at play in Liénard and Boyer's work on ritualized behavior (2006), and Kelemen's research on teleofunctional reasoning about the natural world (e.g., Kelemen et al. 2013). In neither of these research programs do we see a strong concern for mapping the phenomena of interest onto what either the folk or scholars may mean by "religion." Instead, they seek to get an explanatory handle on some pervasive and interesting features of human thought and behavior.

Methodological naturalism

CSR scholars try to explain the cultural expression under consideration by appealing only to natural entities and processes. Such an explanatory strategy enables scholars from various metaphysical commitments to find common ground from which to work. Of course scholars will carry metaphysical commitments with them that influence the assumptions they bring to their science and the interpretations they take away from the science, but a mutual commitment to seeing how much we can explain in entirely naturalistic terms while doing the science facilitates a collective enterprise in which we can check each other's evidence independently of whether one researcher is a Buddhist, another is a Muslim, and a third is an atheist.

Moving beyond particularism. Even when CSR scholars are concerned with a set of local phenomena such as the rise and fall of a Melanesian splinter-group (Whitehouse 1995), Afro-Brazilian spirit possession (Cohen 2007), or Spanish fire-walking rituals (Konvalinka et al. 2011), they enrich their explanations by identifying causal mechanisms that are a more general part of human psychology and not locally specific psychological peculiarities. Consequently, comparisons may be drawn with other forms of cultural expression that is underwritten by similar causes. It is not that local particulars do not matter, but that it is easier to see which ones do matter when broader comparisons are made. To appreciate a figure, we need to discern the ground.

Interdisciplinarity

Though interdisciplinarity is not necessarily a virtue in and of itself, the sort of interdisciplinarity that has characterized CSR holds great explanatory promise because it successfully integrates perspectives and methods from within the humanities and across the humanities-sciences divide. Consequently, disciplinary blind spots are more easily spotted and a diversity of data collection techniques are used to build cumulative cases for particular claims that would be under-supported if they relied on only one type of data. The body of work concerning Harvey Whitehouse's Modes of Religiosity is a case in point (e.g., see Whitehouse and Laidlaw 2004; Whitehouse and Martin 2004a; Whitehouse and McCauley 2005a). The complex interaction of social and psychological dynamics that is alleged to give rise to two very different types of religious communities require empirical evidence on both psychological and social levels of analysis.

Even research programs of a much narrower character benefit from diverse empirical scrutiny. Consider the claim that humans are intuitive dualists and that this dualism gives rise to cultural expression in terms of beliefs in souls, spirits, and afterlives (Bloom 2004). Because the intuitiveness or naturalness of a cognitive tendency can essentially never be established conclusively via experimental techniques alone (i.e., you cannot raise babies isolated from cultural influences to compare with a control group of babies), building a case may require developmental studies (and earlier is better to minimize cultural influences), cross-cultural studies, and cross-species comparisons (if chimps and three-year-olds think in a similar way, it is unlikely to be the result of cultural particulars). Establishing group-level pervasiveness of dualistic thinking concerning the composition of humans would benefit from historical and ethnographic study (e.g. Slingerland 2013). All of these empirical investigations may be sharpened by the analytic tools that philosophers and (in some cases) theologians may bring to the discussion (e.g., Hodge 2011).

The reason that CSR is fundamentally interdisciplinary is not merely because cognitive science is interdisciplinary. CSR is interdisciplinary because its primary explanatory strategy has been to try to explain cultural expression—a group-level phenomenon—by appeal to the psychological dynamics of the people that comprise these groups. As Lane (Chapter 14) observes, an examination of individual religious cognition is not CSR but a cognitive psychology of religion, a point I return to later in this chapter. Because CSR is concerned with accounting for cultural expression, it will always have a place for historical and anthropological studies (see Petersen, Chapter 8), and any fears that experimental or other psychological research methods will eliminate these other disciplinary approaches are unnecessary (e.g., Ambasciano, Chapter 9).

Focus on cognition

The interdisciplinary ability to move beyond particularism and do genuine explanatory and comparative study of phenomena commonly deemed "religious" without getting bogged down in definitional quagmire has been due, in large part, to CSR's focus on cognition. Specifically, CSR at its genesis was concerned with how human minds work in ways that help us better understand the sort of thought and action that becomes broadly recurrent enough to recognize as cultural. Much like the study of linguistics before it, CSR looks "downstairs" to the cognitive systems—conscious or nonconscious—that make some modes of thinking more attractive to people's minds. By focusing on cognition, CSR has been able to spawn many fruitful lines of scholarship concerning various cross-culturally recurrent forms of thought without having to spend too much effort attempting to situate these forms of expression in particular religious traditions. These areas of thought and related actions include afterlife beliefs (Astuti and Harris 2008; Bering and Bjorklund 2004; Huang et al. 2013), god concepts (Guthrie 1993; Pyysiäinen 2009; Wigger et al. 2013), magical thought (Legare and Souza 2014; Sørensen 2005), ritualized behavior (Liénard and Boyer 2006), rituals that appeal to superhuman agents (Hornbeck et al. 2015; Lawson and McCauley 1990; Malley and Barrett 2003), and the relationship among bodies, minds, souls, and spirits (Cohen and Barrett 2008a, 2008b; Roazzi et al. 2013). Once we have a good explanatory grasp on some of these forms of expression that causally cohere from a cognitive perspective, then we can better understand why it is that some of these co-occur historically and in particular cultural settings but not others. Whitehouse (Chapter 3) encourages this strategy of fractionation, and it bypasses such squabbles as whether magic is religion, religions must have gods, or religious practices must have certain social consequences, and so on.

Note that this focus on cognition has the additional virtue of giving scientific attention to areas of human thought and action that otherwise are neglected or even ignored by cognitive science. Surely a complete study of human thought includes how humans think about magical and ritual causation and how they conceptualize nonhuman minds that may or may not have visible bodies, and numerous other modes of thought that have characterized the day-to-day cognitive lives of humans around the world and throughout history. Rather than attend to these very common forms of thought, the cognitive sciences have paid more attention to many relatively rare and difficult cognitive concerns such as disinterested problem-solving and scientific and mathematical reasoning.[1] To illustrate, in a very good introductory text to the scientific study of cognition published before CSR's growth in prominence (Reisberg 1997), the topical index includes entries such as principles of evidence and disconfirmation in scientific reasoning, categorical syllogisms and conditional statements in logical reasoning, and a smattering of specific tasks and problems that are rarely part of ordinary human thought such as the Tower of Hanoi problem, the Hobbits and Orcs problem, and playing chess. Afterlife, gods, souls, rituals, and magic are nowhere to be seen. Surely understanding religious or theological expertise and decision-making is more important to understanding human thought and behavior than understanding chess expertise and decision-making concerning Hobbits and Orcs. CSR has been an important supplement to the cognitive sciences more generally.

Cognition and culture drifting apart

The cleverness of the Sperberian approach to the study of culture that has had such a large influence on CSR is its ability to explain cultural-level phenomena by appeal to the psychology of individuals interacting with each other and their environments (Sperber 1996). Such an approach, however, may be inherently fragile because it requires attention to both individual and group-level dynamics. It appears much easier to treat groups/cultures as occupying an ontologically meaningful level and study them or to focus on individual thought and action, especially once empirical testing begins. The psychological disciplines know how to study individuals and the sociological disciplines have good experience with groups, but how to empirically study the cognitive-cultural Sperberian bridge?[2] It appears that in CSR's recent history, the infusion of scientists into its ranks has encouraged either a drift away from culture or a drift away from cognition.

As noted by Saler and Ziegler (Chapter 11), one big change over the past twenty-five years is that more scientists have become involved in CSR, primarily psychological scientists. The resulting empirical testing has added considerably to our understanding of the cognitive mechanisms that underpin religious thought and greatly broadened and fortified what Guthrie (Chapter 5) has called the intuitionist approach. For instance, Deborah Kelemen's work on teleofunctional reasoning has both vindicated and nuanced Guthrie's anthropomorphism theory (e.g., Kelemen 2004). Her work suggests that, much as Guthrie anticipated, humans naturally see design and purpose in their environment and readily attribute this design and purpose to intentional agents, and importantly, this natural tendency seems to only be overridden through training and effort (Kelemen et al. 2013). Research by other developmental psychologists, including Jesse Bering and Paul Bloom, has prompted CSR to consider other natural foundations of religious beliefs such as intuitive dualism (Bloom 2004) and the intuitiveness of minds continuing after biological death (Bering 2011).

The infusion of psychological research into CSR has also impacted counterintuitivist treatments such as Pascal Boyer's claim that inferentially rich but slightly counterintuitive concepts are great candidates for transmission and stabilizing into cultural forms (Boyer 1994b, 2001). Though some continue to express doubt concerning how important counterintuitiveness is for concepts to become cultural (e.g., see Porubanova and Shaver, Chapter 10), empirical research across many cultural contexts has partially vindicated Boyer's account (Barrett 2016; McCauley, Chapter 2). Importantly, however, this psychological research has also suggested an important qualification on the original theory. It appears that minimally counterintuitive concepts are more memorable for adolescents and young adults than comparable intuitive concepts (e.g., controlling for familiarity, emotional valence, imagery, etc.), but the opposite may be true for older adults (e.g., see Gregory and Greenway 2016a and b). Familiarity with concepts may be more important for older adults. It may be, then, that younger people play an important role in acquiring and spreading the sorts of counterintuitive concepts that populate religious traditions, but older people may serve as a conservative force for checking unbridled innovation. More research in this area awaits as a direct result of the psychological investigation of one of the area's touchstone theories. Clearly CSR is scientifically fruitful (McCauley, Chapter 2).

As valuable as the infusion of genuinely scientific research into CSR has been, I fear that the enthusiasm with which many of us have embraced the psychological and neuroscientific turn in CSR has sometimes led to a neglect of theoretical precision and nuance. Specifically, because the methods of the neuro- and psychological sciences focus principally on individuals, we have occasionally forgotten that the primary explanatory units of analysis in CSR are not individuals but collectives: groups and cultures.

Focus on cultural-level, group expression began to grow fuzzy as soon as the first psychologist arrived on the scene: me. My entry research into CSR (e.g., Barrett and Keil 1996; Barrett and Van Orman 1996) affirmed Guthrie's general notion that anthropomorphism is rampant in theistic thought and, to a large extent, intuitive. Nevertheless, this theologically incorrect anthropomorphism of gods escaped the attention of many people that I studied in the United States and India (Barrett 1998). That is, they still held markedly counterintuitive concepts that they could successfully talk about when in a reflective mode. What accounts for the spread of such apparently counterintuitive concepts? They appear too counterintuitive to fit Boyer's proposed cognitive optimum (Boyer 1994b). Perhaps some of the non-anthropomorphic, allegedly counterintuitive features of gods are more intuitive than they first seem (at least at some early point in human cognitive development). This wondering drove me to begin theoretically and empirically exploring an intuitionist approach to god concepts even as Boyer's counterintuitionist approach was gaining empirical traction (Barrett and Nyhof 2001; Boyer and Ramble 2001). I began emphasizing the role of agency detection, for instance, and early developing intuitive foundations (such as what Kelemen's research was uncovering). My thinking was that perhaps some of the fancier theological concepts are not so counterintuitive after all. Perhaps given the right experiences, including exposure while conceptual systems were still developing during early childhood, that some theological ideas would be easier to understand and conceptually less burdensome. Hence, they would still fall within or near Boyer's proposed optimum. I still find such a proposition plausible for some (but not all) religious ideas.

At points through this attention to intuitions that underwrite religious thought, however, I began to lose sight of the genius of CSR: that it helps account for why *groups* of individuals converge on similar ways of thinking that we call "religious." Even the intuitionist approaches within CSR are not setting out to demonstrate that, barring developmental abnormalities, *every individual* is an intuitive theist. Rather, due to the natural intuitions that most individuals share, *plus* their interaction in social groups—including the sharing of experiences, ideas, and imagination—religious thought and action are likely to arise in nearly every typical cultural group. Individuals will catch religious ideas from others because of cognitive susceptibilities (including in-group conformity), but these susceptibilities will rarely be powerful enough to self-generate ideas that will become widely shared metaphysical commitments concerning gods, spirits, souls, afterlives, rituals, or other sorts of things typically considered religious.

Some forms of religious expression may be intuitive or natural in a way analogous to language: normally developing cognitive equipment plus normal social interaction and just a bit of cultural input to specify particulars may be enough. It may be that thinking dualistically about minds and bodies is natural in this way. Other forms of religious expression may be more analogous to dance: natural enough to occur in most all human groups but not necessarily so intuitive as to be expressed by all normally developing individuals within those groups at all points in the lifespan. Many if not most children will spontaneously move to music in a way that resembles dance, and this early proto-dance will typically be shaped through imitation and instruction to resemble a recognizable cultural form of dance. The result is that dance is a common feature of most cultures. Dance is natural, in this sense. It does not follow that everyone dances. Which forms of expression deemed "religious" are more like language or more like dance is an empirical question, but importantly both language and dance are forms of cultural expression that only arise as a consequence of group living and social dynamics.

I belabor the point about the difference between individual and group/collective phenomena because it carries important implications for motivating research. For instance, many CSR scholars have emphasized the importance of the human theory of mind system or mentalizing ability for representing god concepts (e.g., Barrett 2004a; Bering 2011; Boyer 2001; McCauley and Lawson 2002; Pyysiäinen 2009). It would not be unreasonable to assume, then, that weaker than normal mentalizing ability would predict weaker than normal commitment to the existence of gods. Research investigating this connection has been mixed (e.g., Jack et al. 2016; Norenzayan et al. 2012). This reasonable assumption, however, is not precisely what a CSR, focused on group-level phenomena, would necessarily predict. It may be that almost all humans have sufficient mentalizing ability to think about gods' minds, and those that reject belief in gods (or fail to form them in the first place) more often do so for reasons unrelated to their mentalizing ability. Consequently, we would not necessarily see a relationship between *individual* differences in mentalizing and *individual* differences in god beliefs.

Likewise, concerning the role of an agency detection system, if no members of a group ever detected the action of a postulated god through the working of an agency detection device (ADD) inside individuals' heads, then it seems less likely that such beliefs would persist. In fact, people do report seeing spirits, ghosts, and other gods and evidence of their actions, presumably due to the activity of their agency detection devices. Nevertheless, it does not follow that the general sensitivity of a specific individual's ADD will predict that individual's degree of religious commitment. A person may have very strong commitment to the reality of, say, the ancestors because some kind of persisting agency after death strikes them as intuitive (e.g., perhaps through simulation constraint challenges or intuitive dualism; see Bering 2011) and because they have been told by trusted family members about their encounters with the ancestors, some of which were perceived in a non-inferential sense by virtue of others' ADDs but some of which were inferred through a peculiar string of events (e.g., after having made an unusually large offering to the ancestors, a person received a remarkable windfall). Another person could have a truly hypersensitive ADD and be constantly detecting signs of agency but be raised in a home or cultural environment that provides a strong nonreligious narrative for explaining those experiences (e.g., "Schizophrenia runs in our family and so we often see agency where there isn't any. Just ignore it the best you can."). Such a person may measure very high on tests of agency detection sensitivity but very low on conventional religious beliefs or theism. What is needed for ADD to play a role in explaining the naturalness of religious thought on a group level is not identical to the role it would play on an individual level.

Returning to dance. Dance may be a common form of cultural expression in part because humans have a natural propensity toward rhythmic bodily movements. We may be tempted to predict, then, that individual variability in rhythm would be related to individual variability in the tendency to dance. Suppose studies are conducted in the United States and Canada and no such relationship is found. Would we conclude that rhythm does not matter to dance? Hopefully we would not be so hasty. It could be that in these cultural contexts, whether one dances or not is more strongly influenced by age, sex, ethnic heritage, religious tradition, and whether or not one participates in other physical and social pursuits such as sports instead of dance. It could be that many people inclined to dance who have weak rhythm end up doing dance forms that de-emphasize rhythm (e.g., ballet and modern), whereas many people disinclined to dance who have rhythm may end up in the band instead of on the dance floor. The point is that CSR's group-level theories do not always and easily translate into individual-level empirical testing, and null findings concerning individual psychology (whether concerning mentalizing, agency detection, or memory for counterintuitive concepts) do not necessarily threaten the proposed group-level theories.[3]

Keeping the cognition-culture connection strong not only requires attention to the fact that it is shared, distributed, group-level phenomena that are the ultimate explananda of CSR but also requires a central place for cognition: the mental states, representations, and processes that drive human action. CSR has benefited both the study of religion and the study of cognition by maintaining a strong focus on human minds, but these contributions could be lost by inattention. As noted by McCauley and Guthrie (Chapter 2 and Chapter 5 respectively), much of the research now flying under the banner of CSR is not particularly cognitive. It makes little or no reference to human psychology, let alone cognitive mechanisms as part of the explanatory framework. For instance, observing that those groups that have gods of war have outcompeted those that did not confers, thereby, a selective advantage for the cultural concept of gods of war is not a cognitive explanation (Johnson 2008). It is evolutionary, scientific, and interesting, but not cognitive, and yet, I suspect that many observers of the field (including those who identify with it) would place such evolutionary studies of religion in the CSR camp. Similarly, merely observing neural correlates of religious thought, practice, or experience does not make the project or explanation in question cognitive. It is clear that cognitive approaches may benefit from evolutionary perspectives and neuroscientific research methods (among others), but conflating these levels of explanation is not good conceptual hygiene.

My observation of conceptual differences should not be taken to mean that cognitive, evolutionary, and neuroscientific approaches should be sealed off from each other. These approaches have considerable ability to inform each other, and the boundaries are porous. For instance, a project that begins primarily as a multilevel adaptationist project with only glancing appeal to cognitive systems, such as Sosis's work on religious in-group cooperation (e.g., Sosis and Ruffle 2003), can be made more cognitively informed by more detailed exploration of the cognitive dynamics that make some kinds of signals of commitment more convincing than others. Neuroscience could help identify the biological dynamics that turn recognition of convincing displays (e.g., through coordinated religious rituals) into emotionally bonding events that boost trust and cooperation among group members. The various scientific approaches to the study of religion and the various methodologies that they bring are mutually beneficial, but probably more so when we recognize and appreciate their distinctiveness.

The cognitive science of religion has been a productive scientific research program that has shaken up (in mostly good ways) the study of religion as McCauley illustrates (Chapter 2). CSR has, for the most part, not forgotten what made it special in the 1990s, and so my encouragement is primarily for CSR to keep up the good work. Please, CSR, stay cognitive and stay cultural.

Notes

Introduction: Religion Explained? The Cognitive Science of Religion after Twenty-Five Years

1 This meeting of the IACSR was hosted by the Laboratory for the Experimental Research of Religion (LEVYNA) of Masaryk University, the Department for the Study of Religions at Masaryk University, and the Czech Association for the Study of Religions; it was co-sponsored by the Institute for the Advanced Study of Religion, Toronto. Martin and Wiebe were responsible for suggesting the theme of this conference.

2 Sponsored by the Culture and Cognition Program, University of Michigan, March 28–31, 2002.

3 Although we recognize the difficulty in securing funding for scientific research generally, we have criticized the corrosive effect of accepting funding from religiously oriented agencies in support of a new approach to the study of religion that is attempting to establish itself as scientific (Wiebe 2009).

2 Twenty-Five Years In: Landmark Empirical Findings in the Cognitive Science of Religion

1 Oddly, many interpretivists do not object when psychoanalytically oriented thinkers make similar claims on the basis of far *less* evidence arising from carefully designed tests.

2 However pointed the criticisms, the controversies (Norenzayan 2014; Stausberg 2014b) that have swirled around the synthetic proposals of Ara Norenzayan's *Big Gods* (2013) illustrate the sort of constructive exchanges that the interaction of CSR and religious studies can yield. None of the issues are settled. All of the auditors and, I suspect, all of the participants have benefited.

3 That, however, is not to imply that the scientific community is unconcerned with the moral and political implications of its work. Organizations such as the Union of Concerned Scientists and the Center for Science in the Public Interest have played a valuable and constructive role in enhancing public understanding of scientific issues and in contributing to thoughtful public policies.

4 Postmodernists' worries about the status of texts were not a mere coincidence.

5 During that time, CSR has arisen as a recognized contributor in religious studies with formal representation in some of the world's largest professional societies (International Association for the History of Religions, American Academy of Religion), a professional society of its own (International Association for the Cognitive Science of Religion—IACSR), and specialized journals (*Journal for the Cognitive Science of Religion*; *Journal of Cognition and Culture*; *Religion, Brain & Behavior*; *Journal of Cognitive*

Historiography). At the same time, it has also become a notable subfield within cognitive science with the IACSR regularly meeting periodically with and receiving formal recognition and support from the Cognitive Science Society, with papers and posters in the field accepted at that society's annual meetings, and with relevant articles by eminent mainstream cognitive scientists appearing in its flagship journal, *Cognitive Science* (Astuti and Harris 2008; Banerjee et al. 2013; Legare and Souza 2014).

6 Exactly how direct the role of such theory of mind capacities and of promiscuous teleology are in any such intuitive theism is a point of controversy (Lindeman et al. 2015).

7 This includes the radically counterintuitive representations in which the sciences traffic (McCauley 2011).

8 For an extended, systematic treatment of the issues at stake, see Barrett 2008.

9 Referring to these as "maximally counterintuitive" representations is an unhelpful convention (e.g., Norenzayan et al. 2006). Obviously, the number of violations exceeds one, but this hardly makes them *maximally* counterintuitive.

10 Arguably, it is probably 2Es too few. In addition to the importance of attending to cognition as *evolved*, it seems worthwhile to recognize how richly textured cognition is *emotionally* as well (Thagard 2006).

11 Henrich's observations illustrate lessons of explanatory pluralism: science does not traffic in comprehensive explanations; the vindication of an explanatory theory simply occasions a new set of questions about its un-explicated details and its further implications.

12 I am deeply grateful to Luther Martin and Don Wiebe for their aid, encouragement, patience, and support not only with regard to the production of this paper but also on myriad other fronts over the years.

8 The Indispensability of Cognitive Science for a Genuine History of Religion

1 Valéry 1919, 2: "Nous autres, civilisations, nous savons maitenant que nous sommes mortelles. Nous avions entendu parler de mondes disparus tout entiers, d'empires coulés à pic avous tous leurs hommes et tous leur engins; descendus au fond inexplorable des siècles avec leur dieux et leur lois, leur académies et leur sciences pures et applicquées, avec leur grammaires, leurs dictionaires, leurs classicques, leurs romantiques et leur symbolistes, leur critiques et les critiques de leur critiques. Nous savions bien que toute la terre apperente est faite de cendres, que la cendre signifie quelque chose. Nous apercevions à transvers l'épassieur de l'histoire, les fantômes d'immenses navires qui furent chargés de richesse et d'esprit. Nous ne pouvions par les compter. Mais ces naufrages, après tout, n'étaient pas notre affaire. *Élam, Ninive, Babylone* étaient de beaux noms vagues, et la ruine totale de ces mondes avait aussi peu de signification pour nous que leur existence même. Mais *France, Angleterre, Russie* … ce seraient aussi de beaux noms. *Lusitania* aussi est un beau nom. Et nous voyons maintenant quel'abîme de l'histoire est assez grand pour tout le monde. Nous sentons qu'une civilisationf a la même fragilité qu'une vie."

2 By universe of contact I designate—in accordance with the terminology of the German linguist and philosopher Karl Bühler—the physical space in which the sign, the object of the sign, and the user of the sign are all physically present.

3 See also Petersen 2011, 2013c, and the forthcoming essays.
4 This has been masterfully shown by Boyer (2001), who points to four basic domains likely to trigger religious representations among people. They are (1) ritual and ritualization; (2) moral questions; (3) matters pertaining to contingency; and (4) experiences relating to death and transmission of corpses.
5 See my 2016 essay in which I instantiate a dialogue between Frazer and Wittgenstein with focus on Wittgenstein's acidic remark that, were Frazer to imagine an ancient priest, he would think of him in terms of a contemporary English parson. See also the snappish comment: "Frazer is viel mehr savage als die meisten seiner savages denn diese werden nicht so weit vom Verständnis einer geistigen Angelegenheit entfernt sein wie ein Engländer des 20ten Jarhhunderts. Seine Erklärungen der primitiven Gebräuche sind viel roher als der Sinn dieser Gebräuche selbst" (*Kringel-Buch* 140. See also *KB* 127 and *Philosophical Occasions* 124 and 130).
6 This is based on personal correspondence with Bob Bellah for which I am most grateful.

9 Exiting the *Motel of the Mysteries*? How Historiographical Floccinaucinihilipilification Is Affecting CSR 2.0

1 It would be tempting to label this expanded research environment as "CSR 2.1" or even "3.0," but despite the disciplinary love for acronyms, I think that for the time being such an essentialized taxonomy would do very little in terms of avoiding confusion.
2 Last accessed: June 25, 2016.
3 That is, more or less widespread characters shared by different *taxa* but not inherited from recent common ancestry.

10 Minimal Counterintuitiveness Revisited, Again: The Role of Emotional Valence in Memory for Conceptual Incongruity

1 The concept list is included in the Appendix at the end of this chapter.

12 The Effects of Relative Stable Feedback Loops: Cognitive Science and Historical Explanations

1 E.g., Rydving (2008) on the cognitive science of religion in general (see Sørensen 2008 for an answer), and Levine (1998) on cognitive approaches to ritual (see McCauley and Lawson 1998 for reply).
2 See Geertz (2008) for a thorough discussion of how *not* to do cognitive studies of religion.
3 Both have, ironically, a legacy in the French structuralism of Lévi-Strauss. Whereas postmodernist scholars followed the deconstructive and poststructuralist route of understanding human behavior as eternally shifting patterns of difference, power, and discourse, cognitivists took Lévi-Strauss' reference to basic cognitive structures and neuroscience as their guiding principles.

4 E.g., Asad (1993) for a genealogical critique of "religion" as a scholarly concept.
5 E.g., Otto (2011) on magic as a tainted concept better abandoned in scholarly research.
6 Boyer 1994b, Sørensen 2007a. Lévi-Strauss preempted this reductive move at the end of totemism when he argued that treating religion as a proper scientific object would entail that it lose its specialness and become just another aspect of human conceptualization (Lévi-Strauss 1963).
7 Otherwise see Sørensen 2005 etc. for helpful reviews.
8 Cognitive science is not dependent on evolutionary theorizing—in fact a certain resistance has been aired toward this "biological turn" as it, at least implicitly, questions the functionalist "dogma" of multiple instantiability of cognitive processing (i.e., the notion that cognitive processing is independent of its material medium). However, from its early beginning, cognitive science of religion has been deeply influenced by evolutionary theory and by the evolutionary turn in the cognitive sciences.
9 See Boyer (1990, 1994b, 2001); Atran (1993); and Guthrie (1980, 1995).
10 On domain specificity, see Boyer and Barrett (2005). For a more recent discussion in light of epigenetics, see Karmiloff-Smith (2015).
11 Two examples of such hypothesis-driven studies are the experimental investigation of the memory-enhancing effects of minimally counterintuitive representations (e.g., Barrett and Nyhof 2001; Boyer and Ramble 2001; Harmon-Vukić and Slone 2009; Upal et al. 2007) and the investigation of the effects on action representation and semantic processing of ritualized behavior (e.g., Legare and Souza 2012; Nielbo and Sørensen 2011, 2013; Nielbo et al. 2013; Sørensen 2007b).
12 Even though structural-functionalist theory poses a number of different societal types (e.g., chiefdom, horticulturalist, hunter-gatherer), these are considered as stable types and transitions between these stable types as, relatively short, periods of nonequilibrium.
13 This development has, of course, had a significant impact on methods as experiments, and statistical inference has gained ground (Sørensen and Nielbo 2013a).
14 This argument is strongly inspired by arguments in developmental systems theory (DST); see Griesemer et al. (2005).
15 This is, of course, a very simplified description and I do not wish to imply that historiography is the simple description of these factors, nor that this is all that is to historiography.
16 This point has been highlighted by Berger (1967).
17 In fact, as argued by, for instance, Rosa (2003), in modern Western societies, technological development undergoes a process of acceleration. Previously, the process was characterized by being relatively gradual, even if subject to certain bursts of radical changes (e.g., gunpowder, printing press, steam engine). In modernity, humans must expect to experience several of such major technological changes within a lifetime that fundamentally alter human social interaction. Its effect on cognitive processing as well as on sexual selection is still rather unclear. However, it is likely that the frantic speed of development has added fuel to the erroneous idea that history has no causal impact on human cognition.
18 The discussion of how to isolate relevant systems in order to model causal effects is not merely a result of the level of description and the endemic difficulty of precisely defining and modeling inter-organismic systems (groups, cultures, languages, etc.). This discussion is highly pertinent in biology as well; see Griesemer et al. (2005).
19 See Holland and Quinn (1987); Strauss and Quinn (1997); and d'Andrade (1995).

20 An example of the direct influence of technological evolution on conceptual structures is how conceptualization of the human mind is seen in terms of contemporary technology, from Freudian depictions of the mind as a steam engine, with instinctual sexual drives adding pressure partly suppressed by the superego, to modern-day computer analogies seeing minds as run by algorithms, and divided into distinct functional "programs" controlled by a central executive system.
21 This view of cognitive processing is particularly prevalent in models building on predictive coding; see Clark (2013) for a thorough discussion.
22 For an example of such studies, see Moretti (2013).

14 Looking back to Look Forward: From Shannon and Turing to Lawson and McCauley to …?

1 It may be of interest to note that NAASR was founded in 1985 as a protest by E. Thomas Lawson, Luther H. Martin, and Donald Wiebe against the theological biases inherent in the AAR (Martin and Wiebe 2004). IACSR was founded in 2006, largely with leadership by Lawson, Martin, and Wiebe, in reaction to the rampant postmodernism (i.e., antiscientific) approach embraced by NAASR. One wonders if, in time, a new organization might not be formed in protest to theoretical disagreements within the IACSR.
2 It can be noted that this can be taken as different from "cognitive history." Whereas a cognitive history would be engaging in the study of new primary materials in order to construct some interpretation of the materials using theories developed and tested by cognitive science, the field as it currently stands is a cognitive *historiography*, a field that is utilizing cognitive science in order to study both how historians study history (similar to the "meta-critiques" of religious studies scholars from which the discipline comes) and larger bodies of historical corpora. The field is generally too young and is currently almost entirely populated by more senior scholars; however, this may change as the field grows and even publishes its own journal: *Journal of Cognitive Historiography.*
3 It is not hard to see how this idea inspired later cognitive scientists such as Minsky (1974) in formulating his concept of frames and is necessary for some of the current theories in CSR, such as the theory of Divergent Modes of Religiosity (Whitehouse 2004); these are discussed later.
4 Shannon is virtually unknown even though his approach is the direct inspiration for Sperber and Wilson's theory of relevance and communication (1995) and is the basis of modern electronic communication systems.
5 Paralleling these developments in psychology, the study of religion had a minor schism of its own. As individuals began to attempt a comparative study of religion, a rift opened between the theological and secular approaches in academic bodies supporting the study of religion. NAASR, for example, aimed to foster theoretically informed, secular studies of religion and began publishing the journal *Method and Theory in the Study of Religion* in order to facilitate dialogue in this new—largely atheological—approach to religion. I believe that it was the intent of its founders to initiate a "science" of religion. This idea, first expressed by F. Max Müller (1882), was to be a comparative, secular science, similar to philology. This point is returned to later in reference to the IACSR.

6 A full listing of their historical meeting minutes can be found on the Web site for the American Society for Cybernetics (2014)
7 A view of evoked culture was first proposed by Barkow, Cosmides, and Tooby (1992) and has been reformulated as the Evolution Without Cultural Evolution (EWCE) by Kundt (2015).
8 Furthermore, it should be noted that if we believe that *Homo religiosis* is anything other than Homo sapiens, then we have already shown our science to be incompatible with scientific naturalism and we are falling into the earlier traps of attributing sui generis qualities to "religion."

15 A Neo-Victorian Cognitive Science of Religion

1 As an example, "[Beethoven's last piano] sonata, Op. 111, is the most mysterious. It contains only two movements, a disconcertingly unclassical structure that has prompted much speculation. In Thomas Mann's *Doctor Faustus*, one of the characters gives a lecture entitled 'Why Did Beethoven Write No Third Movement in Op. 111?'—and the answer, in a nutshell, was that the second movement had effectively nailed the sonata form into its coffin" (Staines 2010: 61).
2 Note that in each period it took less time to exhaust the associated form of complexity. I would hypothesize that this is the result of improved technological and social tools for transmitting compositions, thus increasing the rate of innovation.
3 Gray (1958) was concerned primarily with the rise and fall of civilizations, and *his* epicycle was named not because it reminded him of astronomy, but because he found it "strikingly analogous to epicyclical trains (interacting geared mechanisms, like the wheels in clocks), as the term is used in modern engineering." Gray was inspired by Kroeber's *Configurations of Culture Growth* (1944), in particular his description of relatively small-scale cultural patterns—*including music* (ch. 9)—each of which was "a system [that] bears in itself specific potentialities, and also specific limitations, which enable it, even compel it, barring catastrophe from outside, to realize or fulfill itself to the terminus of these potentialities, but not to go beyond them ... there is no further growth until a thoroughly new set of impulses has arisen under new circumstances" (90). Both men later commented on each other's work (e.g., Kroeber 1958; Gray 1966), to the extent that epicyclical patterns in cultural development were briefly called "the Kroeber-Gray Hypothesis" (see Mead 1964: 392).
4 I find it interesting how often Darwin is cited as the beginning—or rather the *origin* —of our field, despite how rigorously he initially avoided discussions of humans and human culture. Only occasionally do we attempt to nominate earlier ancestors:

> Thomas Hobbes was the first sociobiologist, two hundred years before Darwin. As the opening words of his [1651] masterpiece make clear, he saw the creation of the state as fundamentally a matter of one artifact's making another, a sort of group-survival vehicle, "intended" for the "protection and defence" of its occupants ... [He saw,] clearly and confidently, the fundamental Darwinian task: he saw there *had* to be a story to be told about how the state first came to be created, and how it brought something altogether new on the face of the Earth: morality. (Dennett 1995: 453–4)

As another example, see how Martin and Wiebe (Introduction) and Guthrie (Chapter 5) treat David Hume's 1757 *The Natural History of Religions*.

5 I would argue that the shifts to the political left D'Andrade describes as external are also inherently *internal*, given anthropologists' strong identification with our discipline, as well as our blue-collar mentality and focus on marginalized cultures. For example, contrast the difficulties and ultimate failures of the U.S. military's attempt to engage anthropologists in the Human Terrain Systems program in Afghanistan (Jaschik 2015; Price 2008) versus our patriotic participation in World War II by Boas's student Ruth Benedict and many others (Hrotic 2010b; Price 2008: 173–7).

6 For example, the Seville Statement on Violence stated in part that "[i]t is scientifically incorrect to say that war or any other violent behavior is genetically programmed into our human nature"—a response to an exaggerated account of Lorenz (1963). As Laland and Brown (2002: 60–2) note, Lorenz never *quite* made such a bald statement, but from Lorenz, and ethology in general, as well anthropology, psychology, and history, it would not be controversial to say that an aggression, sometimes involving violence, is indeed natural for our species, as for very many others.

7 According to Dennett (1995: 391–2), among the arguments at an MIT conference against adaptationist explanations were: "Color vision has no function; we could tell red from green apples using intensity cues"; "Language can't be useful; it's led to war"; and "Hasn't that stuff about insect wings refuted Darwin?"

8 I am arguing that interdisciplinarity makes us vulnerable to conceptual difficulties such as polysemy; it is perhaps significant that "interdisciplinarity' is one such polysemous term (Klein 2010).

9 While I am, of course, biased, it would seem reasonable to promote Anthropology as the natural "home" of our field, and not only due to its explicit subject matter (i.e., "culture"). First, it is itself inclusive of multiple theory-and-methods sets, and also, unlike other similarly heterogeneous disciplines, has a tradition of robust debate regarding its boundaries. The second-most likely candidate, Religion, may be a less pragmatic choice, as it lacks the disciplinary identity (as former Chair of Religion at the University of Vermont once remarked, no one describes themselves as a "religionist," and even the use of the word is vanishingly rare [e.g., Lawson 1967: 40]), and—due to a pervasively theological character (Martin and Wiebe 2016)—may have fewer opportunities for synergy.

10 Compare Lawson's call to be "conversant," which I understand to be situational and piecemeal, to more recent calls to value and to develop more generalized approaches to "interactive expertise" (e.g., Engeström 1992).

Conclusion: On keeping Cognitive Science of Religion Cognitive and Cultural

1 I thank Bob McCauley for this observation.

2 I do not mean to suggest its impossibility, only its difficulty. Some of the strongest examples of successful bridging are from anthropologists who have cross-trained in psychology and/or who have closely collaborated with psychologists (e.g., Astuti and Harris 2008; Atran 1990; Boyer 2001; Hirschfeld 1996).

3 It may be tempting to think that it is the building blocks (Taves 2009) or fractionating approach to the study of culture that leads to a neglect of group-level dynamics. I certainly think that those of us of the fractionation ilk should take seriously Sosis's

(Chapter 13) encouragement for CSR to invest time studying the lived complexes of beliefs, identities, and practices that are called "religions," even if we reject the idea that "religion" and "religions" are meaningful scientific concepts (Barrett 2017). Looking at how the pieces go back together in the communities of real people is an important check on our theories and will undoubtedly inspire new research. I suspect, however, that it is the methodological tools of testing individual-level mechanisms—of the sort psychologists and neuroscientists possess—that lead to a neglect of group-level explanations rather than the theoretical perspective of the fractionaters. After all, fractionation was part of CSR from the beginning.

References

Aiden, E., and Michel, J.-B. (2014). *Uncharted: Big-Data as a Lens on Human Culture*. New York: Riverhead Books.

Alcock, John. (2001). *The Triumph of Sociobiology*. New York: Oxford University Press.

Alexander, R. D. (2013). Religion, Evolution, and the Quest for Global Harmony. In K. Summers and B. Crespi (Eds.), *Human Social Evolution: The Foundational Works of Richard D. Alexander* (384–425). New York: Oxford University Press.

Ali, S. S., Lifshitz, M., and Raz, A. (2014). Empirical Neuroenchantment: From Reading Minds to Thinking Critically. *Frontiers in Human Neuroscience*, *8*(357). doi: http://dx.doi.org/10.3389/fnhum.2014.00357.

Allen, E., Beckwith, B., Beckwith, J., Chorover, S., Culver, D., Duncan, M., Gould, S., Hubbard, R., Inouye, H., Leeds, A., Lewontin, R., Madansky, C., Miller, L., Pyeritz, R., Rosenthal, M., and Schreier, H. (1975). Against "Sociobiology." *New York Times Review of Books* (November 11). Available on-line: http://www.nybooks.com/articles/1975/11/13/against-sociobiology/ (accessed April 20, 2016).

Alvarez, W. (1997). *"T. rex" and the Crater of Doom*. Princeton: Princeton University Press.

Ambasciano, L. (2014). *Sciamanesimo senza sciamanesimo. Le radici intellettuali del modello sciamanico di Mircea Eliade: evoluzionismo, psicoanalisi, te(le)ologia* ["Shamanism without Shamanism. The Intellectual Roots of Mircea Eliade's Shamanic Paradigm: Evolutionism, Psychoanalysis, Theo-teleology"]. Rome: Nuova Cultura.

Ambasciano, L. (2016a). Mind the (Unbridgeable) Gaps: A Cautionary Tale about Pseudoscientific Distortions and Scientific Misconceptions in the Study of Religion. *Method & Theory in the Study of Religion*, *28*(2): 141–225.

Ambasciano, L. (2016b). Achilles' Historiographical Heel, or the Infelicitous Predominance of Experimental Presentism in Ara Norenzayan's *Big Gods*. *Studi e Materiali di Storia delle Religioni*, *82*(2): 1045–68.

American Academy of Religion. (2015). Cognitive Science of Religion Group | PAPERS. *American Academy of Religion*. Available on-line: https://papers.aarweb.org/content/cognitive-science-religion-group (accessed June 14, 2015).

American Anthropological Association. (2011). AAA LongRange Plan. Available on-line: http://www.aaanet.org/about/Governance/Long_range_plan.cfm (accessed March 11, 2015).

American Society for Cybernetics. (2014). *History of Cybernetics*. Available on-line: http://www.asc-cybernetics.org/foundations/history2.htm (accessed March 11, 2015).

Anderson, A. K. (2005). Affective Influences on the Attentional Dynamics Supporting Awareness. *Journal of Experimental Psychology: General*, *134*(2): 258–81.

Anderson, A. K., and Phelps, E. A. (2001). Lesions of the Human Amygdala Impair Enhanced Perception of Emotionally Salient Events. *Nature*, *411*(6835): 305–9.

Andersen, M., Schjødt, U., Nielbo, K. L., and Sørensen, J. (2014). Mystical Experience in the Lab. *Method & Theory in the Study of Religion*, *26*(3): 217–45.

Anttonen, V. (2002). Identifying the Generative Mechanisms of Religion. In I. Pyysiänen and V. Anttonen (Eds.), *Current Approaches in the Cognitive Science of Religion* (14–37). London: Continuum.

Armitage, D. (2012). What's the Big Idea? Intellectual History and the Longue Durée. *History of European Ideas*, *38*(4): 493–507.
Arnheim, R. (1969). *Visual Thinking*. Berkeley: University of California Press.
Arnheim, R. (1974). *Art and Visual Perception: The New Version*. Berkeley: University of California Press.
Asad, T. (1993). *Genealogies of Religion: Discipline and Reasons of Power in Christianity and Islam*. Baltimore, MD: Johns Hopkins University Press.
Astuti, R., and Harris, P. L. (2008). Understanding Mortality and the Life of the Ancestors in Rural Madagascar. *Cognitive Science*, *32*(4): 713–40.
Atkinson, Q. D., and Whitehouse, H. (2010). The Cultural Morphospace of Ritual Form: Examining Modes of Religiosity Cross-Culturally. *Evolution and Human Behavior*, *32*(1): 50–62.
Atran. S. (1990). *Cognitive Foundations of Natural History: Towards an Anthropology of Science*. Cambridge: Cambridge University Press.
Atran, S. (1993). Whither Ethnoscience. In P. Boyer (Ed.), *Cognitive Aspects of Religious Symbolism* (48–70). Berkeley: University of California Press.
Atran, S. (2002). *In Gods We Trust: The Evolutionary Landscape of Religion*. New York: Oxford University Press.
Bacon, F. ([1620] 1960). *The New Organon and Related Writings*. New York: Prentice Hall.
Bainbridge, W. S. (2006). *God from the Machine: Artificial Intelligence Models of Religious Cognition*. Lanham, MD: AltaMira Press.
Bainbridge, W. S. (2007). *Nanoconvergence: The Unity of Nanoscience, Biotechnology, Information Technology and Cognitive Science*. Upper Saddle River, NJ: Prentice Hall.
Bains, S. (2011). Questioning the Integrity of the John Templeton Foundation. *Evolutionary Psychology*, *9*(1): 92–115.
Banerjee, K., Haque, O., and Spelke, E. (2013). Melting Lizards and Crying Mailboxes: Children's Preferential Recall of Minimally Counterintuitive Concepts. *Cognitive Science*, *37*(7): 1251–89.
Banning, E. B. (2011). So Fair a House: Göbekli Tepe and the Identification of Temples in the Pre-Pottery Neolithic of the Near East. *Current Anthropology 52*(5): 619–40. Available on-line: http://www.jstor.org/stable/10.1086/661207 (accessed May 15, 2017).
Bar, M. (2009). The Proactive Brain: Memory for Predictions. *Philosophical Transactions of the Royal Society B*, *364*(1521): 1235–43.
Barash, D. (1979). Why Is Sugar Sweet? In *The Whisperings Within: Evolution and the Origin of Human Nature* (16–45). New York: Harper and Row.
Bargh, J. A. (2006). What Have We Been Priming All These Years? On the Development, Mechanisms, and Ecology of Nonconscious Social Behavior. *European Journal of Social Psychology*, *36*(2): 147–68.
Barkow, J. (Ed.) (2005). *Missing the Revolution: Darwinism for Social Scientists*. New York: Oxford University Press.
Barkow, J. H., Cosmides, L. E., and Tooby, J. E. (Eds.). (1992). *The Adapted Mind: Evolutionary Psychology and the Generation of Culture*. New York: Oxford University Press.
Barrett, J. L. (1998). Cognitive Constraints on Hindu Concepts of the Divine. *Journal for the Scientific Study of Religion*, *37*(4): 608–19.
Barrett, J. L. (2000). Exploring the Natural Foundations of Religion. *Trends in Cognitive Sciences*, *4*(1): 29–34.
Barrett, J. L. (2002). Smart Gods, Dumb Gods, and the Role of Social Cognition in Structuring Ritual Intuitions. *Journal of Cognition and Culture*, *2*(3): 183–93.

Barrett, J. L. (2004a). *Why Would Anyone Believe in God?* Walnut Creek, CA: AltaMira Press.

Barrett, J. L. (2004b). Bringing Data to Mind: Empirical Claims of Lawson and McCauley's Theory of Religious Ritual. In T. Light and B. C. Wilson (Eds.), *Religion as a Human Capacity: A Festschrift in Honor of E. Thomas Lawson* (265–88). Leiden: E. J. Brill.

Barrett, J. L. (2004c). The Naturalness of Religious Concepts. In P. Antes, A. W. Geertz, and R. R. Warne (Eds.), *New Approaches to the Study of Religion. Volume 2: Textual, Comparative, Sociological, and Cognitive Approaches* (401–18). Berlin: de Gruyter.

Barrett, J. L. (2007a). Cognitive Science of Religion: What Is It and Why Is It? *Religion Compass*, *1*(6): 768–86.

Barrett, J. L. (2007b). A Cognitive Typology of Religious Actions. *Journal of Cognition and Culture*, *7*(3): 201–11.

Barrett, J. L. (2008a). Coding and Quantifying Counterintuitiveness in Religious Concepts: Theoretical and Methodological Reflections. *Method & Theory in the Study of Religion*, *20*(4): 308–38.

Barrett, J. L. (2008b). Keeping "Science" in the Cognitive Science of Religion: Needs of the Field. In J. Bulbulia R. Sosis, E. Harris, R. Genet, C. Genet, and K. Wyman (Eds.), *The Evolution of Religion: Studies, Theories, and Critiques* (295–301). Santa Margarita, CA: Collins Foundation Press.

Barrett, J. L. (2009). Cognitive Science, Religion and Theology. In J. Schloss and M. J. Murray (Eds.), *The Believing Primate: Scientific, Philosophical, and Theological Reflections on the Origin of Religion* (76–100). Oxford: Oxford University Press.

Barrett, J. L. (2011). *Cognitive Science, Religion, and Theology: From Human Minds to Divine Minds*. West Conshohocken, PA: Templeton Press.

Barrett, J. L. (2012). *Born Believers: The Science of Children's Religious Belief*. New York: Simon and Schuster.

Barrett, J. L. (2016). The (Modest) Utility of MCI Theory. *Religion, Brain & Behavior* *6*(3): 249–51.

Barrett, J. L. (2017). Could We Advance the Science of Religion (Better) without the Concept "Religion"? *Religion, Brain & Behavior*. available on-line: http://dx.doi.org/10.1080/2153599X.2016.1249926 (accessed May 16, 2017).

Barrett, J. L., and Keil, F. C. (1996). Conceptualizing a Nonnatural Entity: Anthropomorphism in God Concepts. *Cognitive Psychology*, *31*(3): 219–47.

Barrett, J. L., and Lanman, J. A. (2008). The Science of Religious Beliefs. *Religion*, *38*(2): 109–24.

Barrett, J. L., and Lawson, E. T. (2001). Ritual Intuitions: Cognitive Contributions to Judgements of Ritual Efficacy. *Journal of Cognition and Culture*, *1*(2): 183–201.

Barrett, J. L., and Nyhof, M. A. (2001). Spreading Non-Natural Concepts: The Role of Intuitive Conceptual Structures in Memory and Transmission of Cultural Materials. *Journal of Cognition and Culture*, *1*(1): 69–100.

Barrett, J. L., and Trigg, R. (Eds.). (2014). *The Roots of Religion: Exploring the Cognitive Science of Religion*. Farnham, UK: Ashgate.

Barrett, J. L., and Van Orman, B. (1996). The Effects of Image Use in Worship on God Concepts. *Journal of Psychology and Christianity*, *15*(1): 38–45.

Barrett, J. L., Richert, R. A., and Driesenga, A. (2001). God's Beliefs versus Mother's: The Development of Non-Human Agent Concepts. *Child Development*, *72*(1): 50–65.

Barsalou, L. W. (1999). Perceptual Symbol Systems. *Behavioral and Brain Sciences*, *22*(4): 577–660.

Bartholow, B. D., Fabiani, M., Gratton, G., and Bettencourt, B. A. (2001). A Psychophysiological Analysis of Cognitive Processing of and Affective Responses to Social Expectancy Violations. *Psychological Science, 12*(3): 197–204.

Bartlett, F. C. ([1932] 1995). *Remembering: A Study in Experimental and Social Psychology*. Cambridge: Cambridge University Press.

Baumard, N., and Boyer, P. (2013). Explaining Moral Religions. *Trends in Cognitive Sciences, 17*(6): 172–80.

Baumard, N., and Chevallier, C. (2015). The Nature and Dynamics of World Religions: A Life-History Approach. *Proceedings of the Royal Society B: Biological Sciences, 282*(1818): 1–9.

Baumard, N., Hyafil, A., Morris, I., and Boyer, P. (2015). Increased Affluence Explains the Emergence of Ascetic Wisdoms and Moralizing Religions. *Current Biology, 25*(1): 10–15.

Baumeister, R. F., Vohs, K. D., and Funder, D. C. (2007). Psychology as the Science of Self-Reports and Finger Movements. Whatever Happened to Actual Behavior? *Perspectives on Psychological Science, 2*(4): 396–403.

Becher, T., and Trowler, P. R. (2001). *Academic Tribes and Territories: Intellectual Enquiry and the Culture of Disciplines* (2nd ed.). Buckingham, UK: Open University Press.

Bechtel, W., Abrahamsen, A., and Graham, G. (1998). The Life of Cognitive Science. In W. Bechtel and G. Graham (Eds.), *A Companion to Cognitive Science* (1–104). Oxford: Blackwell.

Bechtel, W., Abrahamsen, A., and Graham, G. (2001). Cognitive Science, History. In N. J. Smelser and P. B. Baltes (Eds.), *International Encyclopedia of the Social and Behavioral Sciences* (2154–8). Oxford: Elsevier Science.

Beck, L. W. (1949). The "Natural Science Ideal" in the Social Sciences. *Scientific Monthly, 68*(6): 386–94.

Beck, R. (2006). *The Religion of the Mithras Cult in the Roman Empire: Mysteries of the Unconquered Sun*. Oxford: Oxford University Press.

Beit-Hallahmi, B. (2002). Review of *Religion Explained: The Human Instincts that Fashion Gods, Spirits and Ancestors. Human Nature Review, 2*: 308–9.

Beit-Hallahmi, B. (2016). Prosociality and Religion: History and Experimentation. *Behavioral and Brain Sciences, 39*: e4. doi: 10.1017/S0140525X15000345.

Bellah, R. N. (1964). Religious Evolution. *American Sociological Review, 29*(3): 358–74.

Bellah, R. N. (2005). What Is Axial about the Axial Age? *Archives of European Sociology, 46*(1): 69–89.

Bellah, R. N. (2011). *Religion in Human Evolution: From the Paleolithic to the Axial Age*. Cambridge, MA: Belknap Press of Harvard University Press.

Bentley, A., and Ormerod, P. (2012). Agents, Intelligence, and Social Atoms. In E. Slingerland and M. Collard (Eds.), *Creating Consilience: Integrating the Sciences and the Humanities* (205–22). New York: Oxford University Press.

Berger, P. L. (1967). *The Sacred Canopy: Elements of a Sociological Theory of Religion*. New York: Doubleday.

Bering, J. M. (2002). Intuitive Perceptions of Dead Agents' Minds: The Natural Foundations of Afterlife Beliefs as Phenomenological Boundary. *Journal of Cognition and Culture, 2*(4): 263–308.

Bering, J. M. (2006). The Folk Psychology of Souls. *Behavioral and Brain Sciences, 29*(5): 453–62.

Bering, J. M. (2011). *The Belief Instinct: The Psychology of Souls, Destiny, and the Meaning of Life*. New York: W. W. Norton.

Bering, J. M., and Bjorklund, D. F. (2004). The Natural Emergence of Reasoning about the Afterlife as a Developmental Regularity. *Developmental Psychology, 40*(2): 217–33.
Bering, J. M., Blasi, C. H., and Bjorklund, D. F. (2005). The Development of "Afterlife" Beliefs in Religiously and Secularly Schooled Children. *British Journal of Developmental Psychology, 23*(4): 587–607.
Bettencourt, B. A., Dill, K. E., Greathouse, S., Charlton, K., and Mulholland, A. (1997). Predicting Evaluations of Ingroup and Outgroup Members: The Role of Category-Based Expectancy Violation. *Journal of Experimental Social Psychology, 33*(3): 244–75.
Bickle, J. (2003). *Philosophy and Neuroscience: A Ruthlessly Reductive Account.* Dordrecht: Kluwer Academic Publishers.
Blake, C. (1955). Can History Be Objective? *Mind, 64*(253): 61–78.
Blanke, O., Pozeg, P., Hara, M., Heydrich, L., Serino, A., Yamamoto, A., and Arzy, S. (2014). Neurological and Robot-Controlled Induction of an Apparition. *Current Biology, 24*(22): 2681–6.
Bloch, M. (2008). Why Religion Is Nothing Special but Is Central. *Philosophical Transactions of the Royal Society of London, Series B Biological Sciences, 363*(1499): 2055–61.
Bloom, P. (2004). *Descartes' Baby: How Child Development Explains What Makes Us Human.* London: William Heinemann.
Blume, M. (2009). The Reproductive Benefits of Religious Affiliation. In E. Voland and W. Schiefenhovel (Eds.), *The Biological Evolution of Religious Mind and Behavior* (117–27). New York: Springer Frontiers Collection.
Boas. F. (1938). *The Mind of Primitive Man* (rev. ed.). New York: Macmillan.
Book, L. (2013). Creativity in the Liberal Arts. In L. Book and D. P. Phillips (Eds.), *Creativity and Entrepreneurship: Changing Currents in Education and Public Life* (3–17). Cheltenham, UK: Edward Elgar.
Bordogna, F. (2008). *William James at the Boundaries: Philosophy, Science, and the Geography of Knowledge.* Chicago, IL: University of Chicago Press.
Boudreau, K., Brady, T., Ganguli, I., Gaule, P., Guinan, E., Hollenberg, T., and Lakhani, K. R. (2014). A Field Experiment on Search Costs and the Formation of Scientific Collaborations. Available on-line: papers.ssrn.com/sol3/papers.cfm?abstract_id=2486068 (accessed May 2, 2016).
Boudry, M., and De Smedt, J. (2011). In Mysterious Ways: On Petitionary Prayer and Subtle Forms of Supernatural Causation. *Religion, 41*(3): 449–69.
Bourdieu, P. (1990). *Homo Academicus,* trans. P. Collier. Stanford, CA: Stanford University Press.
Bower, G. H. (2000). A Brief History of Memory Research. In E. Tulving and F. I. M. Craik (Eds.), *The Oxford Handbook of Memory* (3–32). New York: Oxford University Press.
Boyd, R., and Richerson, P. J. (1985). *Culture and the Evolutionary Process.* Chicago, IL: University of Chicago Press.
Boyd, R., and Richerson, P. J. (2005). *The Origin and Evolution of Cultures.* New York: Oxford University Press.
Boyer, P. (1990). *Tradition as Truth and Communication: A Cognitive Description of Traditional Discourse.* Cambridge: Cambridge University Press.
Boyer, P. (1992). Explaining Religious Ideas: Elements of a Cognitive Approach. *Numen, 39*(1): 27–57. Available on-line: http://www.jstor.org/stable/3270074 (accessed).
Boyer, P. (1993). Cognitive Aspects of Religious Symbolism. In P. Boyer (Ed.), *Cognitive Aspects of Religious Symbolism* (4–47). Cambridge: Cambridge University Press.

Boyer, P. (1994a). Cognitive Constraints on Cultural Representations: Natural Ontologies and Religious Ideas. In L. A. Hirschfeld and S. Gellman (Eds.), *Mapping the Mind: Domain-Specificity in Cognition and Culture* (391–411). Cambridge: Cambridge University Press.

Boyer, P. (1994b). *Naturalness of Religious Ideas: A Cognitive Theory of Religion*. Berkeley: University of California Press.

Boyer, P. (1996a). Religion as an Impure Subject: A Note on Cognitive Order in Religious Representation in Response to Brian Malley. *Method & Theory in the Study of Religion, 8*(2): 201–13.

Boyer, P. (1996b). What Makes Anthropomorphism Natural: Intuitive Ontology and Cultural Representations. *Journal of the Royal Anthropological Institute, 2*(1): 83–97.

Boyer, P. (2001). *Religion Explained: The Evolutionary Origins of Religious Thought*. New York: Basic Books.

Boyer, P. (2003). Religious Thought and Behaviour as By-Products of Brain Function. *Trends in Cognitive Sciences, 7*(3): 119–24.

Boyer, P. (2005). A Reductionistic Model of Distinct Modes of Religious Transmission. In H. Whitehouse and R. N. McCauley (Eds.), *Mind and Religion: Psychological and Cognitive Foundations of Religion* (3–30). Walnut Creek, CA: Altamira Press.

Boyer, P. (n.d.). *Ten Problems in Search of a Research Programme: Towards Integrated Naturalistic Explanations of Human Culture* (version 3.1). St. Louis, MO. Available online: www.pascalboyer.net/articles/TenProblems3.1.pdf (accessed May 15, 2007).

Boyer, P., and Barrett, H. C. (2005). Intuitive Ontologies and Domain Specificity. In D. M. Buss (Ed.), *Handbook of Evolutionary Psychology* (96–118). Hoboken, NJ: John Wiley & Sons.

Boyer, P., and Baumard, N. (2016a). The Diversity of Religious Systems across History: An Evolutionary Cognitive Approach. In J. Liddle and T. Shackelford (Eds.), *The Oxford Handbook of Evolutionary Psychology and Religion*. Oxford: Oxford University Press. doi: 10.1093/oxfordhb/9780199397747.013.5.

Boyer, P., and Baumard, N. (2016b). Projecting WEIRD Features on Ancient Religions. *Behavioral and Brain Sciences, 39*(17): 272–280. doi: 10.1017/S0140525X15000369.

Boyer, P., and Liénard, P. (2006). Why Ritualized Behavior? Precaution Systems and Action Parsing in Developmental, Pathological and Cultural Rituals. *Behavioral and Brain Sciences, 29*(6): 595–650.

Boyer, P., and Ramble, C. (2001). Cognitive Templates for Religious Concepts: Cross-Cultural Evidence for Recall of Counter-Intuitive Representations. *Cognitive Science, 25*(4): 535–64.

Boyer, P., and Walker, S. (2000). Intuitive Ontology and Cultural Input in the Acquisition of Religious Concepts. In K. S. Rosengren, C. N. Johnson, and P. L. Harris (Eds.), *Imagining the Impossible: The Development of Magical, Scientific, and Religious Thinking in Contemporary Society* (130–56). Cambridge: Cambridge University Press.

Braudel, F. ([1958] 1980). History and the Social Sciences: The *Longue Durée*. In *On History* (25–54). Chicago, IL: University of Chicago Press.

Braxton, D. M. (2008). Modeling the McCauley-Lawson Theory of Ritual Forms (unpublished).

Braxton, D. M., Upal, A., and Nielbo, K. L. (2012). Computing Religion: A New Tool in the Multilevel Analysis of Religion. *Method & Theory in the Study of Religion, 24*(3): 267–90.

Bremmer, J. (1994). *Greek Religion*. Cambridge: Cambridge University Press.

Broesch, J., Barrett, H. C., and Henrich, J. (2014). Adaptive Content Biases in Learning about Animals across the Life Course. *Human Nature, 25*(2): 181–99.

Brown, P. (1988). *The Body and Society: Men, Women, and Sexual Renunciation in Early Christianity*. New York: Columbia University Press.

Brown, R. W. (2015). Challenges in Faculty Entrepreneurship in the Sciences: Becoming an Entrepreneur but Staying at the University. In S. Nambisan (Ed.), *Embracing Entrepreneurship Across Disciplines* (31–50). Cheltenham, UK: Edward Elgar.

Brown, R., and Kulik, J. (1977). Flashbulb Memories. *Cognition, 5*(1): 73–99.

Buhrmester, M. D., Fraser, W. T., Lanman, J. A., Whitehouse, H., and Swann, W. B., Jr. (2015). When Terror Hits Home: Identity Fused Americans Who Saw Boston Bombing Victims as "Family" Provided Aid. *Self & Identity, 14*(3): 253–70.

Bulbulia, J. (2004). The Cognitive and Evolutionary Psychology of Religion. *Biology and Philosophy, 18*(5): 655–86.

Bulbulia, J. (2006). Nature's Medicine: Religiosity as an Adaptation for Health and Cooperation. In P. MacNamara (Ed.), *Where Man and God Meet: The New Sciences of Religion and Brain* (87–121). Westwood, CT: Greenwood.

Bulbulia, J. (2010). Charismatic Signalling. *Journal for the Study of Religion, Nature and Culture, 3*(4): 518–51.

Bulbulia, J. (2013a). Bayes and the Evolution of Religious Belief. In J. P. Moreland, C. V. Meister, and K. A. Sweis (Eds.), *Debating Christian Theism* (223–41). Oxford: Oxford University Press.

Bulbulia, J. (2013b). The Arts Transform the Cognitive Science of Religion. *Journal for the Cognitive Science of Religion, 1*(2): 141–60.

Bulbulia, J., and Slingerland, E. (2012). Religious Studies as a Life Science. *Numen: International Review for the History of Religions, 59* (5–6): 564–613.

Bulbulia, J., and Sosis, R. (2011). Signalling Theory and the Evolution of Religious Cooperation. *Religion, 41*(3): 363–88.

Bulbulia, J., Sosis, R., Harris, E. Genet, C. Genet, R., and Wyman, K. (Eds.). (2008). *The Evolution of Religion: Studies, Theories, and Critiques*. Santa Margarita, CA: Collins Foundation Press.

Bulbulia, J., Geertz, A. W., Atkinson, Q. D., Cohen, E., Evans, N., François, P., Gintis, H., Gray, R. D., Henrich, J., Jordon, F. M., Norenzayan, A., Richerson, P. J., Slingerland, E., Turchin, P., Whitehouse, H., Widlok, T., and Wilson, D. S. (2013a). The Cultural Evolution of Religion. In P. J. Richerson and M. H. Christiansen (Eds.), *Cultural Evolution: Society, Technology, Language, and Religion* (381–404). Cambridge, MA: MIT Press.

Bulbulia, J., Greenhill, S., and Gray, R. (2013b). First Shots Fired for the Phylogenetic Revolution in Religious Studies: A Commentary on David Sloan Wilson. *Cliodynamics: Journal of Theoretical and Mathematical History, 4*(1): 128–33.

Bulbulia, J., Wilson, M. S., and Sibley, C. G. (2014). Thin and Thinner: Hypothesis-Driven Research and the Study of Humans. *Numen: International Review for the History of Religions, 61*(2–3): 166–81.

Bulbulia, J., Spezio, M. L., Sosis, R., and Wildman, W. J. (2016) Standards for Publishing in Religion, Brain & Behavior. *Religion, Brain & Behavior, 6*(4): 275–7.

Buller, D. J. (2005). *Adapting Minds: Evolutionary Psychology and the Persistent Quest for Human Nature*. Cambridge: MIT Press.

Burke, P. (1986). Strengths and Weaknesses of the History of Mentalities. *History of European Ideas, 7*(5): 439–51.

Burkert, W. (1985). *Greek Religion*. Cambridge: Harvard University Press.

Burkert, W. (1996). *Creation of the Sacred: Tracks of Biology in Early Religions*. Cambridge, MA: Harvard University Press.

Buss, D. M. (1994). *The Evolution of Desire: Strategies of Human Mating*. New York: Basic Books.

Buss, D. M. (Ed.). (2005). *The Handbook of Evolutionary Psychology*. Hoboken, NJ: Wiley.

Calin-Jageman, R. J., and Caldwell, T. L. (2014). Replication of the Superstition and Performance Study by Damisch, Stoberock, and Mussweiler (2010). *Social Psychology*, *45*(3): 239–45.

Campbell, D. (1969). Ethnocentrism of Disciplines and the Fish-Scale Model of Omniscience. In M. Sherif and C. W. Sherif (Eds.), *Interdisciplinary Relationships in the Social Sciences* (328–48). Chicago, IL: Aldine Publishing Company.

Carey, S. (1985). *Conceptual Change in Childhood*. Cambridge, MA: Bradford/MIT Press.

Carey, S. (1995). On the Origin of Causal Understanding. In D. Premack, A. Premack, and D. Sperber (Eds.), *Causal Cognition: A Multidisciplinary Debate* (268–302). Oxford: Oxford University Press.

Carey, S., and Spelke, E. (1996). Science and Core Knowledge. *Philosophy of Science*, *63*(4): 515–53.

Carley, K. M., and Kaufer, D. S. (1993). Semantic Connectivity: An Approach for Analyzing Symbols in Semantic Networks. *Communication Theory*, *3*(3): 183–213.

Carroll, J. (2005). Literature and Evolutionary Psychology. In D. M. Buss (Ed.), *Handbook of Evolutionary Psychology* (931–52). Hoboken, NJ: John Wiley & Sons.

Carruthers, M. (2008). *The Book of Memory: A Study of Memory in Medieval Culture* (2nd ed.). Cambridge: Cambridge University Press.

Casler, K., and Kelemen, D. (2008). Developmental Continuity in Teleo-Functional Explanation: Reasoning about Nature among Romanian Romani Adults. *Journal of Cognition and Development*, *9*(3): 340–62.

Chambers, C. D., Feredoes, E., Muthukumaraswamy, S. D., and Etchells, P. J. (2014). Instead of "Playing the Game" It Is Time to Change the Rules: Registered Reports at *AIMS Neuroscience* and Beyond. *AIMS Neuroscience*, *1*(1): 4–17.

Chang, L. J., and Sanfey, A. G. (2009) Unforgettable Ultimatums? Expectation Violations Promote Enhanced Social Memory Following Economic Bargaining. *Frontiers in Behavioral Neuroscience*, *3*: 36. doi: 10.3389/neuro.08.036.2009.

Cherry, J. (1992). Animism in Thought and Language, PhD diss., University of California, Berkeley.

Chiao, J. Y. (2009). *Cultural Neuroscience: Cultural Influence on Brain Function*. London: Elsevier.

Chomsky, N. (1957). *Syntactic Structures* (1st ed.). Berlin: Mouton.

Chomsky, N. (1959). A Review of B. F. Skinner's *Verbal Behavior*. *Language*, *35*(1): 26–58.

Christian, D. (2010). The Return of Universal History. *History and Theory*, *49*(4): 6–27.

Christianson, S. A. (1992). Emotional Stress and Eyewitness Testimony: A Critical Review. *Psychological Bulletin*, *112*(2): 284–309.

Claidière, N., Scott-Phillips, T. C., and Sperber, D. (2014). How Darwinian Is Cultural Evolution? *Philosophical Transactions of the Royal Society of London B: Biological Sciences*, *369*(1642): 1–8.

Clark, A. (2013). Whatever Next? Predictive Brains, Situated Agents, and the Future of Cognitive Science. *Behavioral and Brain Sciences*, *36*(3): 181–204.

Clark, B. R. (1963). Faculty Culture. In T. F. Lunsford (Ed.), *The Study of Campus Cultures* (39–54). Boulder, CO: Western Interstate Commission for Higher Education.

Claus, P., and Marriott, J. (2012). *History: An Introduction to Theory, Method and Practice.* Harlow: Pearson.

Clavin, W. (2011). Can WISE Find the Hypothetical "Tyche"? *NASA, Mission News* (February 18). Available on-line: http://www.nasa.gov/mission_pages/WISE/news/wise20110218.html (accessed June 25, 2016).

Cleland, C. E., and Brindell, S. (2013). Science and the Messy, Uncontrollable World of Nature. In M. Pigliucci and M. Boudry (Eds.), *Philosophy of Pseudoscience: Reconsidering the Demarcation Problem* (183–202). Chicago: University of Chicago Press.

Cohen, E. (2007). *The Mind Possessed: The Cognition of Spirit Possession in an Afro-Brazilian Religious Tradition.* Oxford: Oxford University Press.

Cohen, E., and Barrett, Justin L. (2008a). Conceptualizing Spirit Possession: Ethnographic and Experimental Evidence. *Ethos*, *36*(2): 246–267.

Cohen, E., and Barrett, J. L. (2008b). When Minds Migrate: Conceptualizing Spirit Possession. *Journal of Cognition and Culture*, *8*(1): 23–48.

Cohen, E., Ejsmond-Frey, R., Knight, N., and Dunbar, R. I. M. (2010). Rowers' High: Behavioural Synchrony Is Correlated with Elevated Pain Thresholds. *Biology Letters*, *6*(1): 106–8.

Cohen, J. E. (1971). Mathematics as Metaphor: Review of R. Rosen. *Dynamic System Theory in Biology*. *Science*, *172*(3984): 674–5.

Cohen, J., Burdett, E., Knight, N., and Barrett, J. (2011). Cross-Cultural Similarities and Differences in Person-Body Reasoning: Experimental Evidence from the United Kingdom and Brazilian Amazon. *Cognitive Science*, *35*(7): 1282–304.

Cohn, N. (1975). *Europe's Inner Demons*. London: Chatto and Windus.

Colborne, N. (2016). "It Was Like that When I Came in": David Sloan Wilson's Use of the Category "Religion". *Method & Theory in the Study of Religion*, *28*(4–5): 512–29.

Collar, A. (2013). *Religious Networks in the Roman Empire: The Spread of New Ideas.* Cambridge: Cambridge University Press.

Collins, A. M., and Quillian, M. R. (1969). Retrieval Time from Semantic Memory. *Journal of Verbal*, *8*(2): 240–7.

Cosmides, L. (1989). The Logic of Social Exchange: Has Natural Selection Shaped How Humans Reason? Studies with the Wason Selection Task. *Cognition*, *31*(3): 187–276.

Crary, S. T. (1967). A Department of Religion in a Privately Supported Liberal Arts College. In M. D. McLean (Ed.), *Religious Studies at Public Universities* (35–7). Carbondale, IL: Central Publications.

Crick, F. (1990). *What Mad Pursuit: A Personal View of Scientific Discovery.* New York: Basic Books.

Csibra, G. (2008). Goal Attribution to Inanimate Agents by 6.5-Month-Old Infants. *Cognition*, *107*(2): 705–17.

Cullen, H., Kanai, R., Bahrami, B., and Rees, G. (2014). Individual Differences in Anthropomorphic Attributions and Human Brain Structure. *Social Cognitive and Affective Neuroscience*, *9*(9): 1276–80.

Cummins, D. D., and Cummins, R. C. (1999). Biological Preparedness and Evolutionary Explanation. *Cognition*, *73*(3): B37–53.

Cummins, R. (2000). How Does It Work? Versus What Are the Laws Two Conceptions of Psychological Explanation. In F. Keil and R. Wilson (Eds.), *Explanation and Cognition* (117–44). Cambridge: MIT Press.

Cummings, J. N., and Kiesler, S. (2007). Coordination Costs and Project Outcomes in Multi-University Collaborations. *Research Policy*, *36*(10): 1624–34.

Currie, T. E., Bogaard, A., Cesaretti, R., Edwards, N. R., François, P., Holden, P. B., Hoyer, D., Korotayev, A., Manning, J., Moreno, J. C., Oyebamiji, O. K., Petrie, C., Turchin, P., Whitehouse, H., and Williams, A. (2015). Agricultural Productivity in Past Societies: Toward an Empirically Informed Model for Testing Cultural Evolutionary Hypotheses. *Cliodynamics: Journal of Quantitative History and Cultural Evolution*, *6*(1): 24–56.

Czachesz, I. (2013). The Evolutionary Dynamics of Religious Systems: Laying the Foundation of a Network Model. In A. W. Geertz (Ed.), *Origins of Religion, Cognition and Culture* (98–120). Durham, UK: Acumen.

Czachesz, I., and Uro, R. (Eds.) (2013). *Mind, Morality and Magic: Cognitive Science Approaches in Biblical Studies*. Durham: Acumen.

D'Andrade, R. (1995). *The Development of Cognitive Anthropology*. Cambridge: Cambridge University Press.

D'Andrade, R. (2000). The Sad Story of Anthropology, 1950–1999. *Cross-Cultural Research, 34*(3): 219–32.

Damasio, A. (2000). *The Feeling of What Happens: Body, Emotion and the Making of Consciousness*. San Diego, CA: Harvest.

Danziger, K. (2008). *Marking the Mind: History of Memory*. Cambridge: Cambridge University Press.

Darwin, C. ([1871] 2004). *The Descent of Man, and Selection in Relation to Sex*. New York: Penguin Classics.

Dawes, G. W., and Maclaurin, J. (2013). What Is Religion? In G. Dawes and J. Maclaurin (Eds.), *The New Science of Religion* (11–25). New York: Routledge.

Dawkins, R. (1976). *The Selfish Gene*. New York: Oxford University Press.

Dawkins, R. (2006). *The God Delusion*. London: Bantum Press.

Day, M. (2010). The Educator Must Be Educated: The Study of Religion at the End of the Humanities. *Method & Theory in the Study of Religion, 22*(1): 1–8.

de Certeau, M. (1975). *L'écriture de l'histoire*. Paris: Gallimard.

de Spinoza, B. ([1670] 1951). *A Theologico-Political Treatise*, R. H. M. Elwes (Ed.). New York: Dover.

Deeley, Q., Walsh, E., Oakley, D. A., Bell, V., Koppel, C., Mehta, M. A., and Halligan, P. W. (2013). Using Hypnotic Suggestion to Model Loss of Control and Awareness of Movements: An Exploratory fMRI Study. *PloS one*, *8*(10). doi: http://dx.doi.org/10.1371/journal.pone.0078324.

Deeley, Q., Oakley, D. A., Walsh, E., Bell, V., Mehta, M. A., and Halligan, P. W. (2014). Modelling Psychiatric and Cultural Possession Phenomena with Suggestion and fMRI. *Cortex, 53:* 107–19.

Dehaene, S., and Mehler, J. (1992). Cross-Linguistic Regularities in the Frequency of Number Words. *Cognition, 43*(1): 1–29.

DeLancy, S. (1990). Cross-Linguistic Evidence for the Structure of the Agent Prototype. *Papers and Reports on Child Language Development, 29:* 141–47.

deMause, L. (Ed.). (1975). *The New Psychohistory*. New York: Psychohistory Press.

deMause, L. (2002). *The Emotional Life of Nations*. New York: Karnac.

Dennett, D. C. (1987). *The Intentional Stance*. Cambridge: MIT Press.

Dennett, D. C. (1995). *Darwin's Dangerous Idea: Evolution and the Meanings of Life*. New York: Simon & Schuster.

Dennett, D. C. (2006). *Breaking the Spell: Religion as a Natural Phenomenon*. New York: Viking.

Diamond, J., and Robinson, J. (2010a). *Natural Experiments of History*. Cambridge, MA: Belknap Press of Harvard University Press.

Diamond, J., and Robinson, J. (2010b). Prologue. In J. Diamond and J. Robinson (Eds.), *Natural Experiments of History* (1–14). Cambridge, MA, and London: Belknap Press of Harvard University Press.

Dietrich, O., and Notroff, J. (2015). A Sanctuary, or So Fair a House? In Defense of an Archaeology of Cult at Pre-Pottery Neolithic Göbleki Tepe. In N. Laneri (Ed.), *Defining the Sacred: Approaches to the Archaeology of Religion in the Near East* (75–89). Oxford: Oxbow Books.

Dodds, E. R. (1951). *The Greeks and the Irrational*. Berkeley and Los Angeles: University of California Press.

Dolan, R. J., and Vuilleumier, P. (2003). Amygdala Automaticity in Emotional Processing. *Annals of the New York Academy of Sciences, 985*(1): 348–55.

Donald, M. (1991). *Origins of the Modern Mind: Three Stages in the Evolution of Culture and Cognition*. Cambridge, MA: Harvard University Press.

Donald, M. (2001). *A Mind So Rare: The Evolution of Human Consciousness*. New York, London: W. W. Norton.

Donald, M. (2009). The Roots of Art and Religion in Ancient Material Culture. In C. Renfrew and I. Morley (Eds.), *Becoming Human: Innovation in Prehistoric Material and Spiritual Culture* (95–103). New York: Cambridge University Press.

Douglas, M. (1966). *Purity and Danger: An Analysis of Concepts of Pollution and Taboo*. New York: Praeger.

Douglas, M. (1980). *Evans-Pritchard: His Life, Work, Writings, and Ideas*. Sussex: Harvester Press Limited.

Douglas, M., and Perry, E. (1985). Anthropology and Comparative Religion. *Theology Today, 41*(4): 410–27.

Doyen, S., Klein, O., Pichon, C.-L., and Cleeremans, A. (2012). Behavioral Priming: It's All in the Mind, but Whose Mind? *PLoS ONE, 7*(1). doi: 10.1371/journal.pone.0029081.

Durkheim, E. ([1915] 1965). *The Elementary Forms of Religious Life*. New York: Free Press.

Edelman, G. M. (1992). *Bright Air, Brilliant Fire: On the Matter of the Mind*. New York: Basic Books.

Eidinow, E., and Martin, L. H. (2014). Editor's Introduction. *Journal of Cognitive Historiography, 1*(1): 5–9.

Eisenstadt, S. (1986). *The Origins and Diversity of Axial Age Civilizations*. Buffalo, NY: SUNY Press.

Eldredge, N. (1999). *The Pattern of Evolution*. New York: W. H. Freeman.

Ellis, B. J. (2013). The Hypothalamic–Pituitary–Gonadal Axis: A Switch-Controlled, Condition-Sensitive System in the Regulation of Life History Strategies. *Hormones and Behavior, 64*(2): 215–25.

Ellis, B. J., Figueredo, A. J., Brumbach, B. H., and Schlomer, G. L. (2009). Fundamental Dimensions of Environmental Risk: The Impact of Harsh Versus Unpredictable Environments on the Evolution and Development of Life History Strategies. *Human Nature, 20*(2): 204–68.

Emmons, N. A. (2016). Dead People and Living Spirits: Lessons from Developmental Psychology on What Is Intuitive. *Religion, Brain & Behavior, 6*(3): 251–4.

Engeström, Y. (1992). *Interactive Expertise: Studies in Distributed Working Intelligence*. Research Bulletin 83. Helsinki: University of Helsinki, Dept.

of Education. Available on-line: https://pdfs.semanticscholar.org/a244/be8cb71342c27da3c561954008a0c0c663c0.pdf (accessed May 15, 2017).

Epstein, J. M. (2006a). Agent-Based Computational Models and Generative Social Science. In J. M. Epstein (Ed.), *Generative Social Science* (1–43). Princeton, NJ: Princeton University Press.

Epstein, J. M. (Ed.). (2006b). *Generative Social Science: Studies in Agent-Based Computational Modeling (Princeton Studies in Complexity)*. Princeton, NJ: Princeton University Press.

Epstein, J. M. (2014). *Agent_Zero: Toward Neurocognitive Foundations for Generative Social Science*. Princeton, NJ: Princeton University Press.

Evans, E. M. (2000). Beyond Scopes: Why Creationism Is Here to Stay. In K. S. Rosengren, C. N. Johnson, and P. L. Harris (Eds.), *Imagining the Impossible: The Development of Magical, Scientific, and Religious Thinking in Contemporary Society* (305–33). Cambridge: Cambridge University Press.

Evans, E. M. (2001). Cognitive and Contextual Factors in the Emergence of Diverse Belief Systems: Creationism Versus Evolution. *Cognitive Psychology*, *42*(3): 217–66.

Evans-Pritchard, E. E. (1965). *Theories of Primitive Religion*. Oxford: Clarendon Press.

Farah, M. J., and Heberlein, A. S. (2007). Personhood and Neuroscience: Naturalizing or Nihilating? *American Journal of Bioethics (AJOB-Neuroscience)*, *7*(1): 37–48.

Feeny, C., Liénard, P., and Sørensen, J. (2006). Agent and Instrument in Judgements of Ritual Efficacy. *Journal of Cognition and Culture*, *6*(3–4): 463–82.

Festinger, L., Riecken, H. W., and Schachter, S. (2008). *When Prophecy Fails* (2nd ed.). London: Pinter and Martin.

Fischer, D. H. (1971). *Historians' Fallacies: Towards a Logic of Historical Thought*. New York: Harper Perennial.

Fish, S. (1989). Being Interdisciplinary Is so very Hard to Do. *Profession*, *89*: 239–49.

Fisher, H. (1994). "Wilson," They Said, "Your [*sic*] All Wet!" *New York Times* (October 16). Available on-line: https://www.nytimes.com/books/98/12/06/specials/wilson-naturalist.html (accessed April 20, 2016).

Fitzgerald, T. (1997). A Critique of "Religion" as a Cross-Cultural Category. *Method & Theory in the Study of Religion*, *9*(2): 91–110.

Fitzgerald, T. (2000). *The Ideology of Religious Studies*. New York: Oxford University Press.

Foster, K., and Kokko, H. (2009). The Evolution of Superstitious and Superstition-Like Behavior. *Proceedings of the Royal Society B*, *276*(1654): 1–10.

Foucault, M. (1966). *Les Mots et les Choses*. Paris: Gailimard.

Frank, R. H. (1988). *Passions within Reason: The Strategic Role of Emotions*. New York: W. W. Norton.

Freud, S. ([1908] 2014). *On the Sexual Theories of Children* (Amazon: Kindle ed.). White Press.

Freud, S. ([1927] 1975). *The Future of an Illusion*. New York: Norton.

Frith, C. D. (2007). *Making Up the Mind: How the Brain Creates Our Mental World*. Oxford: Blackwell Publishing.

Frodeman, R., and Rowland, J. (2009). De-Disciplining the Humanities. *Alif: Journal of Comparative Poetics*, *29*: 62–72.

Galen, L. (2015). Atheism, Wellbeing, and the Wager: Why Not Believing in God (with Others) Is Good for You. *Science, Religion and Culture*, *2*(3): 54–69.

Gardner, H. (1985). *The Mind's New Science: A History of the Cognitive Revolution. With a New Epilogue by the Author: Cognitive Science after 1984*. New York: Basic Books.

Gaulin, S. J., and McBurney, D. H. (2001). *Psychology: An Evolutionary Approach*. New York: Prentice Hall.
Geertz, A. W. (2008). How *Not* to Do the Cognitive Science of Religion Today. *Method & Theory in the Study of Religion, 20*(1): 7–21.
Geertz, A. W. (2009). When Cognitive Scientists Become Religious, Science Is in Trouble: On Neurotheology from a Philosophy of Science Perspective. *Religion, 39*(4): 319–24.
Geertz, A. W. (2010a). Brain, Body and Culture: A Biocultural Theory of Religion. *Method & Theory in the Study of Religion, 22*(4): 304–21.
Geertz, A. W. (2010b). Too Much Mind and Not Enough Brain, Body and Culture: On What More Needs to Be Done in the Cognitive Study of Religion. *Historia Religionum, 2*: 21–37.
Geertz, A. W. (2013). Whence Religion? How the Brain Constructs the World and What This Might Tell Us about the Origins of Religion, Cognition and Culture. In A. W. Geertz (Ed.), *Origins of Religion, Cognition and Culture* (17–70). Durham, UK: Acumen.
Geertz, A. W. (2014). Long-lost Brothers: On the Co-histories and Interactions between the Comparative Science of Religion and the Anthropology of Religion. *Numen: International Review for the History of Religions, 61*(2–3): 255–80.
Geertz, A. W. (2016). Conceptions of Religion in the Cognitive Science of Religion. In P. Antes, A. W. Geertz, and M. Rothstein (Eds.), *Contemporary Views on Comparative Religion in Celebration of Tim Jensen's 65th Birthday* (127–39). Sheffield: Equinox.
Geertz, A. W., and Markússon, G. I. (2010). Religion Is Natural, Atheism Is Not: On Why Everybody Is Both Right and Wrong. *Religion, 40*(3): 152–65.
Geertz, C. (1973). *The Interpretation of Cultures*. New York: Basic Books.
Geertz, C. (1980). Blurred Genres: The Reconfiguration of Social Thought. *American Scholar, 49*(2): 165–79.
Gell, A. (1998). *Art and Agency: An Anthropological Theory*. Oxford: Oxford University Press.
Gergely, G., Nadasy, Z., Csibra, G., and Biro, S. (1995). Taking the Intentional Stance at 12 Months of Age. *Cognition, 56*(2): 165–93.
Gervais, W. M., and Henrich, J. (2010). The Zeus Problem: Why Representational Content Biases Cannot Explain Faith in Gods. *Journal of Cognition and Culture, 10*(3): 383–9.
Gibson, J. (1986). *The Ecological Approach to Visual Perception*. Hillsdale, NJ: Erlbaum.
Gigerenzer, G. (1997). The Modularity of Social Intelligence. In A. Whiten and R. Byrne (Eds.), *Machiavellian Intelligence, Vol. 2: Extensions and Evaluations* (264–88). Cambridge: Cambridge University Press.
Gleick, J. (2012). *The Information: A History, a Theory, a Flood*. New York: Vintage Books.
Godwin, D. W., Wiggins, W., Hayasaka, S., Laurienti, P., and Stapleton-Kotloski, J. (2013). Overcoming Obstacles to Creativity in Geographically Fragmented Environments. In L. Book and D. P. Phillips (Eds.), *Creativity and Entrepreneurship: Changing Currents in Education and Public Life* (127–40). Cheltenham, UK: Edward Elgar.
Goldin, P. (2002). *The Culture of Sex in Ancient China*. Honolulu: University of Hawaii Press.
Gombrich, E. (1960). *Art and Illusion: A Study in the Psychology of Pictorial Representation*. London: Phaidon.
Gombrich, E. (1969). *Art and Illusion: A Study in the Psychology of Pictorial Representation* (2nd ed.). Princeton, NJ: Princeton University Press.

Gombrich, E. (1973). Illusion and Art. In R. H. Gregory and E. H. Gombrich (Eds.), *Illusion in Nature and Art* (193–243). London: Duckworth.
Gomes, C. M., and McCullough, M. E. (2015). The Effects of Implicit Religious Primes on Dictator Game Allocations: A Preregistered Replication Experiment. *Journal of Experimental Psychology: General, 144*(6): e94–104.
Gonce, L. O., Upal, M. A., Slone, D. J., and Tweney, R. D. (2006). Role of Context in the Recall of Counterintuitive Concepts. *Journal of Cognition and Culture, 6*(3–4): 521–47.
Google Books. (2015). Google Books History. Available on-line: http://www.google.co.uk/googlebooks/about/history.html (accessed June 21, 2015).
Gopnik, A., Meltzoff, A., and Kuhl, P. (2000). *The Scientist in the Crib*. New York: Harper Collins.
Gottlieb, G. (2007). Probabilistic Epigenesis. *Developmental Science, 10*(1): 1–11.
Gould, S. J. (1978). Sociobiology: The Art of Storytelling. *New Scientist, 80*(1129): 530–3.
Gould, S. J. (1986). Evolution and the Triumph of Homology, or Why History Matters. *American Scientist, 74*(1): 60–9.
Gould, S. J. (1989). *Wonderful Life: The Burgess Shale and the Nature of History*. New York: W. W. Norton.
Gould, S. J. (1997). Nonoverlapping Magisteria. *Natural History, 106*: 16–22.
Gould, S. J. (2000). Second-Guessing the Future. In *The Lying Stones of Marrakech* (201–16). New York: Harmony Books.
Graham, J., Haidt, J., Koleva, S., Motyl, M., Iyer, R., Wojcik, S., and Ditto, P. H. (2013). Moral Foundations Theory: The Pragmatic Validity of Moral Pluralism. *Advances in Experimental Social Psychology, 47*: 55–130.
Granet, M. (1934). *La Pensée Chinoise*. Paris: Albin Michel.
Gray, C. E. (1958). An Analysis of Graeco-Roman Development: The Epicyclic Evolution of Graeco-Roman Civilization. *American Anthropologist, 60*(1): 13–31.
Gray, C. E. (1966). A Measurement of Creativity in Western Civilization. *American Anthropologist, 68*(6): 1394–417.
Greco, J. (1998). Perception as Interpretation. *Proceedings of the American Catholic Philosophical Association, 72*: 229–37.
Green, C. D. (1992). Of Immortal Mythological Beasts: Operationism in Psychology. *Theory & Psychology, 2*(3): 291–320.
Greenfield, P. M. (2000). What Psychology Can Do for Anthropology, or Why Anthropology Took Postmodernism on the Chin. *American Anthropologist, New Series, 102*(3): 564–76.
Gregory, J. P., and Greenway, T. S. (2016a). Is There a Window of Opportunity for Religiosity? Children and Adolescents Preferentially Recall Religious-Type Cultural Representations, but Older Adults Do Not. *Religion, Brain & Behavior*. doi: 10.1080/2153599X.2016.1196234.
Gregory, J. P., and Greenway, T. S. (2016b). The Mnemonic of Intuitive Ontology Violation Is Not the Distinctiveness Effect: Evidence from a Broad Age Spectrum of Persons in the UK and China during a Free-Recall Task. *Journal of Cognition and Culture, 17*(3–4): 169–97.
Gregory, R. L. (1997). *Eye and Brain: The Psychology of Seeing* (5th ed.). Princeton, NJ: Princeton University Press.
Greig, J. Y. T. (1932). *The Letters of David Hume*, 2 vols. Oxford: Oxford University Press.

Griesemer, J., Haber, M. H., Yamashita, G., and Gannett, L. (2005). Critical Notice: Cycles of Contingency-Developmental Systems and Evolution. *Biology and Philosophy*, *20*(2–3): 517–44.

Griffith, A. B. (2014). Dead Religion, Live Minds: Memory and Recall of the Mithraic Bull-Slaying Scene. *Journal of Cognitive Historiography, 1*(1): 72–89.

Gross, M. B., and Gilles, M. (2012). *The Last Myth: What the Rise of Apocalyptic Thinking Tells Us about America*. Amherst, NY: Prometheus Books.

Guldi, J., and Armitage, R. (2014). *The History Manifesto*. Cambridge: Cambridge University Press.

Guthrie, S. E. (1980). A Cognitive Theory of Religion [and comments and reply]. *Current Anthropology*, *21*(2): 181–203.

Guthrie, S. E. (1993). *Faces in the Clouds: A New Theory of Religion*. New York: Oxford University Press.

Guthrie, S. E. (2002). Animal Animism. In I. Pyysiäinen and V. Anttonen (Eds.), *Current Approaches in the Cognitive Science of Religion* (38–67). London: Continuum.

Guthrie, S. E. (2007). Anthropological Theories of Religion. In M. Martin (Ed.), *The Cambridge Companion to Atheism* (283–99). New York: Cambridge University Press.

Guthrie, S. E. (2016a). Commentary on Purzycki and Willard's Critique. *Religion, Brain & Behavior*, *6*(3): 254–6.

Guthrie, S. E. (2016b). Religion as Anthropomorphism: A Cognitive Theory. In J. Liddle and T. Shackelford (Eds.), *The Oxford Handbook of Evolutionary Psychology and Religion*. Oxford: Oxford University Press. doi: 10.1093/oxfordhb/9780199397747.013.6.

Haggett, P. (2001). *Geography: A Global Synthesis*. Harlow: Pearson.

Hallowell, I. (1960). Ojibwa Ontology, Behavior, and Worldview. In S. Diamond (Ed.), *Culture in History: Essays in Honor of Paul Radin* (19–52). New York: Columbia University Press.

Hamann, S. B. (2001). Cognitive and Neural Mechanisms of Emotional Memory. *Trends in Cognitive Science*, *5*(9): 394–400.

Hamann, S. B., Cahill, L., and Squire, L.R. (1997). Emotional Perception and Memory in Amnesia. *Neuropsychology*, *11*(1): 104–13.

Han, S., and Northoff, G. (2008). Culture-Sensitive Neural Substrates of Human Cognition: A Transcultural Neuroimaging Approach. *Nature Reviews Neuroscience*, *9*(8): 646–54.

Harmon-Vukić, M. E., and Slone, D. J. (2009). The Effect of Integration on Recall of Counterintuitive Stories. *Journal of Cognition and Culture*, *9*(1): 57–68.

Harmon-Vukić, M. E., Upal, M. A., and Sheehan, K. J. (2012). Understanding the Memory Advantage of Counterintuitive Concepts. *Religion, Brain & Behavior*, *2*(2): 121–39.

Harris, P. L., and Gimenéz, M. (2005). Children's Acceptance of Conflicting Testimony: The Case of Death. *Journal of Cognition and Culture*, *5*(1): 143–64.

Harrison, J., David, M., and McKay, R. T. (2013). Do Religious and Moral Concepts Influence the Ability to Delay Gratification? A Priming Study. *Journal of Articles in Support of the Null Hypothesis*, *10*(1): 25–40.

Harvey, G. (Ed.). (2014). *The Handbook of Contemporary Animism*. New York: Routledge.

Harvey, V. A. (1995). *Feuerbach and the Interpretation of Religion*. Cambridge: Cambridge University Press.

Harvey, V. A. (1996). Projection: A Metaphor in Search of a Theory? In D. Z. Philips (Ed.), *Can Religion Be Explained Away?* (6–82). New York: Palgrave Macmillan.

Hatano, G., and Inagaki, K. (2002). *Young Children's Naïve Thinking about the Biological World*. New York: Psychology Press.

Heberlein, A. (2008). Animacy and Intention in the Brain: Neuroscience of Social Event Perception. In T. F. Shipley and J. M. Zacks (Eds.), *Understanding Events: From Perception to Action* (363–88). Oxford: Oxford University Press.

Heider, F., and Simmel, M. (1944). An Experimental Study of Apparent Behavior. *American Journal of Psychology*, *57*(2): 243–59.

Heintz, C. (2010). Cognitive History and Cultural Epidemiology. In L. H. Martin and J. Sørensen (Eds.), *Past Minds: Studies in Cognitive Historiography* (11–28). London: Equinox.

Henrich, J. (2009). The Evolution of Costly Displays, Cooperation and Religion: Credibility Enhancing Displays and Their Implications for Cultural Evolution. *Evolution and Human Behavior*, *30*(4): 244–60.

Henrich, J. (2016). *The Secret of Our Success: How Culture Is Driving Our Evolution, Domesticating Our Species, and Making Us Smarter*. Princeton, NJ: Princeton University Press.

Henrich, J., and Gil-White, F. J. (2001). The Evolution of Prestige: Freely Conferred Status as a Mechanism for Enhancing the Benefits of Cultural Transmission. *Evolution and Human Behavior*, *22*(3): 1–32.

Henrich, J., Heine, S. J., and Norenzayan, A. (2010). The Weirdest People in the World? *Behavioral and Brain Sciences*, *33*(2–3): 61–135.

Herrmann, P. A., Legare, C. H., Harris, P. L., and Whitehouse, H. (2013). Stick to the Script: The Effect of Witnessing Multiple Actors on Children's Imitation. *Cognition*, *129*(3): 536–43.

Hirschfeld, L. A. (1996). *Race in the Making: Cognition, Culture and the Child's Construction of Human Kinds*. Cambridge, MA: MIT Press.

Hirshman E. (1988). The Expectation-Violation Effect: Paradoxical Effects of Semantic Relatedness. *Journal of Memory and Language*, *27*(1): 40–58.

Hodge, K. M. (2011). On Imagining the Afterlife. *Journal of Cognition and Culture*, *11*(3–4): 367–89.

Hogan, J. A., and Bolhuis, J. J. (2009). Tinbergen's Four Questions and Contemporary Behavioral Biology. In J. J. Bolhuis and S. Verhulst (Eds.), *Tinbergen's Legacy: Function and Mechanism in Behavioral Biology* (25–34). Cambridge: Cambridge University Press.

Hogg, M., and Abrams, D. (1988). The Social Identity Approach: Context and Content. In D. Abrams and M. A. Hogg (Eds.), *Social Identifications: A Social Psychology of Intergroup Relations and Group Processes* (6–30). New York: Routledge.

Holland, D., and Quinn, N. (1987). *Cultural Models in Language and Thought*. Cambridge: Cambridge University Press.

Holtz, T. R., Jr. (1994). The Arctometatarsalian Pes, an Unusual Structure of the Metatarsus of Cretaceous *Theropoda* (*Dinosauria: Saurischia*). *Journal of Vertebrate Paleontology*, *14*(4): 480–519.

Holtz, T. R., Jr. (2001). *Arctometatarsalia* Revisited: The Problem of Homoplasy in Reconstructing Theropod Phylogeny. In J. Gauthier and L. F. Gall (Eds.), *New Perspectives on the Origin and Early Evolution of Birds: Proceedings of the International Symposium in Honor of John H. Ostrom* (99–122). New Haven, CT: Yale University Press.

Hornbeck, R. G., Bentley, B., and Barrett, J. L. (2015). Examining Special Patient Rituals in a Chinese Cultural Context: A Research Report. *Journal for Cognition and Culture*, *15*(5): 530–41.

Horton, R. (1967). African Traditional Thought and Western Science: Part 1. *Africa: Journal of the International African Institute*, *37*(2): 50–71.

Horton, R. (1993). *Patterns of Thought in Africa and the West: Essays on Magic, Religion and Science*. Cambridge: Cambridge University Press.

Houston, R. and Frith, U. (2000). *Autism in History: The Case of Hugh Blair of Borgue*. Oxford, Malden: Blackwell.

Hrotic, S. (2009). Academic Peer-Evaluation Tactics: An Evolutionary Approach, PhD diss., Belfast: Queen's University.

Hrotic, S. (2010a). Disciplinary Clans. In P. Pachis and D. Wiebe (Eds.), *Chasing Down Religion: In The Sights of History and the Cognitive Sciences: Essays in Honour of Luther H. Martin* (197–214). Thessaloniki, Greece: Vanias Publications.

Hrotic, S. (2010b). War Stories: Social and Military Science Narratives of the Human Terrain System. Presentation at the Society for the Social Studies of Science conference, August 25–29, Tokyo (unpublished).

Hrotic, S. (2013). A Cognitive Analysis of the Palestrina Myth. *Religion, Brain & Behavior*, *3*(1): 16–38.

Huang, J., Cheng, L., and Zhu, J. (2013). Intuitive Conceptions of Dead Persons' Mentality: A Cross-Cultural Study and More. *International Journal for the Psychology of Religion*, *23*(1): 29–41.

Hughes, A. W. (2012). *Abrahamic Religions: On the Uses and Abuses of History*. New York: Oxford University Press.

Huizinga, J. ([1919] 1924). *The Waning of the Middle Ages*, trans. F. Hopman. London: E. Arnold.

Hume, D. ([1757] 1957). *The Natural History of Religion*. Stanford: Stanford University Press.

Hutchinson, J. R., and Padian, K. (1997). s.v. Arctometatarsalia. In P. J. Currie and K. Padian (Eds.), *Encyclopedia of Dinosaurs* (24–7). San Diego: Academic Press.

Hutton, P. H. (1981). The History of Mentalities: The New Map of Cultural History. *History and Theory*, *20*(3): 237–59.

Inagaki, K., and Hatano, G. (1987). Young Children's Spontaneous Personification as Analogy. *Child Development*, *58*(4): 1013–20.

International Association for the Cognitive Science of Religion. (2015). About the IACSR. Available on-line: http://www.iacsr.com/ (accessed June 14, 2015).

Inzlicht, M., and Tullett, A. M. (2010). Reflecting on God: Religious Primes Can Reduce Neurophysiological Response to Errors. *Psychological Science*, *21*(8): 1184–90.

Irons, W. (2001). Religion as a Hard-to-Fake Sign of Commitment. In R. Nesse (Ed.), *Evolution and the Capacity for Commitment* (292–309). New York: Russell Sage Foundation.

Jack, A. I., Friedman, J. P., Boyatzis, R. E., and Taylor, S. N. (2016). Why Do You Believe in God? Relationships between Religious Belief, Analytic Thinking, Mentalizing and Moral Concern. *PLoS ONE*, *11*(3): e0149989. doi: 10.1371/journal.pone.0149989.

Jackson, J. (2010). Google: 129 Million Different Books Have Been Published. *PCWorld* (August 6). Available on-line: http://www.pcworld.com/article/202803/google_129_million_different_books_have_been_published.html.

James, W. (1902). *The Varieties of Religious Experience*. New York: Modern Library.

Jaschik, S. (2015). Embedded Conflicts. *Inside Higher-Ed* (July 7). Available on-line: https://www.insidehighered.com/news/2015/07/07/army-shuts-down-controversial-human-terrain-system-criticized-many-anthropologists (accessed May 1, 2016).

Jaspers, K. (1953). *The Origin and Goal of History*, trans. M. Bullock. New Haven: Yale University Press.

Jensen, J. S. (2003). *The Study of Religion in a New Key*. Aarhus: Aarhus University Press

Jensen, J. S. (2016). How Institutions Work in Shared Intentionality and "We-Mode" Social Cognition. *Topoi. An International Review of Philosophy*, *35*(1): 301–12.

Johnson, D. (2008). Gods of War: The Adaptive Logic of Religious Conflict. In J. Bulbulia, R. Sosis, C. Genet, R. Genet, E. Harris, and K. Wyman (Eds.), *The Evolution of Religion: Studies, Theories, and Critiques* (111–17). Santa Margarita, CA: Collins Foundation Press.

Johnson, D. (2016). *God is Watching You: How Fear of God Makes Us Human*. New York: Oxford University Press.

Johnson, M. (2006). Mind Incarnate: From Dewey to Damasio. *Daedalus*, *135*(3): 46–54.

Jolly, A. (1996). Primate Communication, Lies, and Ideas. In A. Lock and C. R. Peters (Eds.), *Handbook of Human Symbolic Evolution* (167–77). Oxford: Clarendon Press.

Jong, J. (2013). On Faith and the Fear of Fatality: A Review of Recent Research on Deities and Death. *Journal for the Cognitive Science of Religion*, *1*(2): 193–214.

Jong, J., Whitehouse, H., Kavanagh, C., and Lane, J. (2015). Shared Negative Experiences Lead to Identity Fusion via Personal Reflection. *PlosOne*. doi: 10.1371/journal.pone.0145611.

Jordan, F. M., van Schaik, C., François, P., Gintis, H., Haun, D. B. M., Hruschka, D. J., Janssen, M. A., Kitts, J. A., Lehmann, L., Mathew, S., Richerson, P. J., Turchin, P., and Wiessner, P. (2013). Cultural Evolution of the Structure of Human Groups. In P. J. Richerson and M. H. Christiansen (Eds.), *Cultural Evolution: Society, Technology, Language, and Religion* (87–116). Cambridge, MA: MIT Press.

Josephson, J. A. (2012). *The Invention of Religion in Japan*. Chicago, IL: University of Chicago Press.

Kaiserfeld, T. (2013). Why New Hybrid Organizations Formed: Historical Perspectives on Epistemic and Academic Drift. *Minerva*, *51*(2): 171–94.

Kaptchuk, T. J. (2002). The Placebo Effect in Alternative Medicine: Can the Performance of a Healing Ritual Have Clinical Significance? *Annals of Internal Medicine*, *136*(11): 817–25.

Karmiloff-Smith, A. (2015). An Alternative to Domain-general or Domain-specific Frameworks for Theorizing about Human Evolution and Ontogenesis. *AIMS Neuroscience*, *2*(2): 91–104.

Keightley, T. (1834). *Tales and Popular Fictions: Their Resemblance and Transmission from Country to Country*. London: Whittaker.

Kelemen, D. (1999a). Beliefs about Purpose: On the Origins of Teleological Thought. In M. C. Corballis and S. E. G. Lea (Eds.), *The Descent of Mind: Psychological Perspectives on Hominid Evolution* (278–94). Oxford: Oxford University Press.

Kelemen, D. (1999b). The Scope of Teleological Thinking in Pre-School Children. *Cognition*, *70*(3): 241–72.

Kelemen, D. (1999c). Why Are Rocks Pointy? Children's Preference for Teleological Explanations of the Natural World. *Developmental Psychology*, *35*(6): 1440–52.

Kelemen, D. (2004). Are Children "Intuitive Theists"? Reasoning about Purpose and Design in Nature. *Psychological Science*, *15*(5): 295–301.

Kelemen, D., and Rosset, E. (2009). The Human Function Compunction: Teleological Explanation in Adults. *Cognition*, *111*(1): 138–43.

Kelemen, D., Rottman, J., and Seston, R. (2013). Professional Physical Scientists Display Tenacious Teleological Tendencies: Purpose-Based Reasoning as a Cognitive Default. *Journal of Experimental Psychology, General, 142*(4): 1074–83.

Kensinger, E. A., Brierley, B., Medford, N., Growdon, J. H., and Corkin, S. (2002). Effects of Normal Aging and Alzheimer's Disease on Emotional Memory. *Emotion, 2*(2): 118–34.

Klass, M. (1995). *Ordered Universes: Approaches to the Anthropology of Religion*. Boulder, CO: Westview Press.

Klein, J. T. (2010). A Taxonomy of Interdisciplinarity. In R. Frodeman, J. T. Klein, and C. Mitcham (Eds.), *Oxford Handbook of Interdisciplinarity* (15–30). New York: Oxford University Press.

Koch, G. (2009). The Cognitive Origins of Soul Belief: Dualism, Moral Judgement, and Perceptions of Intentionality, PhD thesis, Aarhus University.

Konvalinka, I., Xygalatas, D., Bulbulia, J., Schjødt, U., Jegindø, E.-M., Wallot, S., Van Orden, G., and Roepstorff, A. (2011). Synchronized Arousal between Performers and Related Spectators in a Fire-Walking Ritual. *Proceedings of the National Academy of Sciences, 108*(20): 8514–19.

Kramer, Karen L. (2005). *Maya Children: Helpers at the Farm*. Cambridge, MA: Harvard University Press.

Krebs, John R., Nicholas B. Davies, and Jan Parr. (1993). *An Introduction to Behavioural Ecology*. Oxford: Blackwell Scientific Publications.

Kroeber, A. L. (1944). *Configurations of Culture Growth*. Berkeley: University of California Press.

Kroeber, A. L. (1958). Gray's Epicyclical Evolution. *American Anthropologist, 60*(1): 31–8.

Kuhn, T. (1970). *The Structure of Scientific Revolutions* (2nd ed.). Chicago, IL: University of Chicago Press.

Kundt, R. (2015). *Contemporary Evolutionary Theories of Culture and the Study of Religion*. London: Bloomsbury.

Kvadsheim, R. (1992). *The Intelligent Imitator: Towards an Exemplar Theory of Behavioral Choice*. Amsterdam: Elsevier Science Publishers.

LaBar, K. S., and Phelps, E. A. (1998). Arousal-Mediated Memory Consolidation: Role of the Medial Temporal Lobe in Humans. *Psychological Science, 9*(6): 490–3.

Lakatos, I. (1978). *The Methodology of Scientific Research Programmes: Volume 1: Philosophical Papers*. J. Worrall and G. Currie (Eds.). Cambridge: Cambridge University Press.

Lakoff, G., and Johnson, M. (1980). *Metaphors We Live By*. Chicago, IL: University of Chicago Press.

Lakoff, G., and Johnson, M. (1999). *Philosophy in the Flesh: The Embodied Mind and Its Challenge to Western Thought*. New York: Basic Books.

Laland, K. N., and Brown, G. R. (2002). *Sense and Nonsense*. Oxford: Oxford University Press.

Laland, K. N., Odling-Smee, J., and Myles, S. (2010). How Culture Shaped the Human Genome: Bringing Genetics and the Human Sciences Together. *Nature Reviews Genetics, 11*(2): 137–48.

Lane, J. E. (2013). Method, Theory, and Multi-Agent Artificial Intelligence: Creating Computer Models of Complex Social Interaction. *Journal for the Cognitive Science of Religion, 1*(2): 161–80.

Lane, J. E. (2014). Data "Mining" and Automated Analysis of Cultural Material: Comparing British and Singaporean Social Groups (unpublished).
Lane, J. E. (2015). Semantic Network Mapping of Religious Material. *Journal for Cognitive Processing, 16*(4): 333–41.
Lawson, E. T. (1967). A Rationale for a Department of Religion in a Public University. In M. D. McLean (Ed.), *Religious Studies at Public Universities* (38–44). Carbondale, IL: Central Publications.
Lawson, E. T. (2000). Towards a Cognitive Science of Religion. *Numen: International Review for the History of Religions, 47*(3): 338–49.
Lawson, E. T. (2004). The Wedding of Psychology, Ethnography, and History: Methodological Bigamy or Tripartite Free Love? In H. Whitehouse and L. H. Martin (Eds.), *Theorizing Religious Past: Archaeology, History, and Cognition* (1–6). Walnut Creek, CA: AltaMira Press.
Lawson, E. T., and McCauley, R. N. (1990). *Rethinking Religion, Connecting Cognition and Culture*. Cambridge: Cambridge University Press.
Lawson, E. T., and McCauley, R.N. (1993). Crisis of Conscience, Riddle of Identity: Making Space for a Cognitive Approach to Religious Phenomena. *Journal of the American Academy of Religion, 61*(2): 201–23.
Lazarus, R. S. (1994). *Emotion and Adaptation*. New York: Oxford University Press.
Leder, D. (1990). *The Absent Body*. Chicago, IL: University of Chicago Press.
LeDoux, J. E. (1992). Brain Mechanisms of Emotion and Emotional Learning. *Current Opinions in Neurobiology, 2*(2): 191–8.
LeDoux, J. E. (1998). *The Emotional Brain: The Mysterious Underpinnings of Emotional Life*. New York: Simon and Shuster.
Lefebvre, G. (1932). *La Grande Peur de 1789*. Paris: A. Colin.
Legare, C. H., and Souza, A. L. (2012). Evaluating Ritual Efficacy: Evidence from the Supernatural. *Cognition, 124*(1): 1–15.
Legare, C. H., and Souza, A. (2014). Searching for Control: Priming Randomness Increases the Evaluation of Ritual Efficacy. *Cognitive Science, 38*(1): 152–61.
Legare, C. H., Wen, N. J., Herrmann, P. A., and Whitehouse, H. (2015). Imitative Flexibility and the Development of Cultural Learning. *Cognition, 142*: 351–61.
Leslie, M. (2000). Divided They Stand. *Stanford Magazine* (January/February). Available on-line: https://alumni.stanford.edu/get/page/magazine/article/?article_id=40546 (accessed April 23, 2016).
Levine, M. P. (1998). A Cognitive Approach to Ritual: New Method or No Method at All? *Method & Theory in the Study of Religion, 10*(1): 30–60.
Lévi-Strauss, C. (1963). *Totemism*. Boston: Beacon Press.
Lévi-Strauss, C. (1966). *The Savage Mind*. Chicago, IL: University of Chicago Press.
Levy-Bruhl, L. ([1922] 1923). *Primitive Mentality*. New York: Macmillan.
Levy-Bruhl, L. ([1910] 1926). *How Natives Think*. London: Allen and Unwin.
Lewontin, R. C. (1993). *Biology as Ideology: The Doctrine of DNA*. New York: Harper Perennial.
Lewontin, R. C., and Levins, R. ([1999] 2007). Does Culture Evolve? In R. Lewontin and R. Levins (Eds.), *Biology under the Influence: Dialectical Essays on Ecology, Agriculture, and Health* (267–96). New York: Monthly Review Press.
Liénard, P., and Boyer, P. (2006). Whence Collective Rituals? A Cultural Selection Model of Ritualized Behavior. *American Anthropologist, 108*(4): 814–27.
Liénard, P., and Lawson, E. T. (2008). Evoked Culture, Ritualization and Religious Rituals. *Religion, 38*(2): 157–71.

Lienhardt, G. (1961) *Divinity and Experience: The Religion of the Dinka*. Oxford: Clarendon Press.

Lillard, A. S., and Skibbe, L. (2005). Theory of Mind: Conscious Attribution and Spontaneous Trait Inference. In R. R. Hassin, J. S. Uleman, and J. A. Bargh (Eds.), *The New Unconscious* (277–305). Oxford: Oxford University.

Lindeman, M., Svedholm-Häkkinen, A. M., and Lipsanen, J. (2015). Ontological Confusions but Not Mentalizing Abilities Predict Religious Belief, Paranormal Belief, and Belief in Supernatural Purpose. *Cognition, 134*: 63–76.

Lohmann, R. (2003). The Supernatural Is Everywhere: Defining Qualities of Religion in Melanesia and Beyond. *Anthropological Forum, 13*(2): 175–85.

Lorenz, K. (1963). *On Aggression*. London: Methuen.

Lotman, Y. M. (2005). On the Semiosphere. *Sign Systems Studies, 33*(1), 205–29.

Luckmann, T. (1976). Comment on: *Malinowski's Magic: The Riddle of the Empty Cell*, by Karl E. Rosengren. *Current Anthropology, 17*(4): 678–9.

Luomanen, P., Pyysiäinen, I., and Uro, R. (Eds.). (2007). *Explaining Christian Origins and Early Judaism: Contributions from Cognitive and Social Science*. New York: Brill.

Lyons, D. E., Young, A. G., and Keil, F. C. (2007). The Hidden Structure of Overimitation. *Proceedings of the National Academy of Sciences, 104*(50): 19751–6.

Macaulay, D. (1979) *Motel of the Mysteries*. Boston: Houghton Mifflin.

Madigan, S. A. (1969). Intraserial Repetition and Coding Processes in Free Recall. *Journal of Verbal Learning and Verbal Behavior, 8*(6): 828–35.

Malafouris, L. (2010). The Brain-Artefact Interface (BAI): A Challenge for Archaeology and Cultural Neuroscience. *Social Cognitive and Affective Neuroscience, 5*(2–3): 264–73.

Malinowski, B. (1948). *Magic, Science and Religion, and Other Essays*. Boston, MA: Beacon Press.

Malley, B., and Barrett, J. L. (2003). Can Ritual Form Be Predicted from Religious Belief? A Test of the Lawson-McCauley Hypotheses. *Journal of Ritual Studies, 17*(2): 1–14.

Mandler, J. M. (1984). *Stories, Scripts, and Scenes: Aspects of Schema Theory*. Hillsdale, NJ: Lawrence Erlbaum Associates.

Marlowe, F. (2010). *The Hadza: Hunter-Gatherers of Tanzania*. Berkeley: University of California Press.

Martin, L. H. (2003). Cognition, Society and Religion: A New Approach to the Study of Culture. *Culture and Religion: An Interdisciplinary Journal, 4*(2): 207–31.

Martin, L. H. (2004a). Performativity, Narrativity and Cognition: Demythologizing the Roman Cult of Mithras. In W. Braun (Ed.), *Rhetoric and Reality in Early Christianity* (187–217). Waterloo: Wilfrid Laurier University Press.

Martin, L. H. (2004b). Ritual Competence and Mithraic Ritual. In T. Light and B. Wilson (Eds.), *Religion as a Human Capacity: A Festschrift in Honor of E. Thomas Lawson* (245–63). Leiden: E. J. Brill.

Martin, L. H. (2005a). Towards a Cognitive History of Religion. In H. Whitehouse and L. H. Martin (Eds.), *Theorizing Religions Past: Archaeology, History, and Cognition* (7–14). Walnut Creek, CA: AltaMira Press.

Martin, L. H. (2005b). Religion and Cognition. In J. R. Hinnells (Ed.), *The Routledge Companion to the Study of Religion* (473–88). London: Routledge.

Martin, L. H. (2007). The Promise of Cognitive Science for the Historical Study of Religions, with Reference to the Study of Early Christianity. In P. Luomanen, I. Pyysiäinen, and R. Uro (Eds.), *Explaining Early Judaism and Christianity: Contributions from Cognitive and Social Sciences* (37–56). Leiden: E. J. Brill.

Martin, L. H. (2008). Can Religion Really Evolve? (And What Is It Anyway?). In *The Evolution of Religion: Studies, Theories, and Critiques* (349–55). Santa Margarita, CA: Collins Foundation Press.

Martin, L. H. (2014a). Great Expectations for Ara Norenzayan's *Big Gods*. *Religion, 44*(4): 628–37.

Martin, L. H. (2014b). *Deep History, Secular Theory: Historical and Scientific Studies of Religion*. Boston, MA: de Gruyter.

Martin, L. H. (2015). *The Mind of Mithraists: Historical and Cognitive Studies in the Roman Cult of Mithras*. London: Bloomsbury.

Martin, L. H., and Eidinow, E. (2014). Editors' Introduction. *Journal of Cognitive Historiography, 1*(1): 5–9.

Martin, L. H., and Pyysiäinen, I. (2013). Conclusion: Moving towards a New Science of Religion; Or, Have We Already Arrived? In D. Xygalatas and W. W. McCorkle Jr. (Eds.), *Mental Culture: Classical Social Theory and the Cognitive Science of Religion* (213–26). Durham, UK Acumen.

Martin, L. H., and Sørensen, J. (Eds.). (2011). *Past Minds: Studies in Cognitive Historiography*. London: Equinox.

Martin, L. H., and Whitehouse, H. (2004). *Theorizing Religious Past: Archaeology, History, and Cognition*. Walnut Creek, CA: AltaMira Press.

Martin, L. H., and Wiebe, D. (2004). *Establishing a Beachhead: NAASR, Twenty Years Later*. In Conversations and Controversies in the Scientific Study of Religion: Collaborative and Co-authored Essays by Luther H. Martin and Donald Wiebe (36–41). Brill. Leiden.

Martin, L. H., and Wiebe, D. (2012). Religious Studies as a Scientific Discipline: The Persistence of a Delusion. *Journal of the American Academy of Religion, 80*(3): 587–97.

Martin, L. H., and Wiebe, D. (2014). Pro- and Assortative-sociality in the Formation and Maintenance of Religious Groups. *Journal for the Cognitive Science of Religion, 2*(1): 1–12.

Martin, L. H., and Wiebe, D. (2016). *Conversations and Controversies in the Scientific Study of Religion: Collaborative and Co-authored Essays by Luther H. Martin and Donald Wiebe*. Leiden: Brill.

Mayr, E. (2004). *What Makes Biology Unique? Considerations on the Autonomy of a Scientific Discipline*. Cambridge: Cambridge University Press.

McCauley, R. N. (1986). Intertheoretic Relations and the Future of Psychology. *Philosophy of Science, 53*(2): 179–99.

McCauley, R. N. (1996). Explanatory Pluralism and the Coevolution of Theories in Science. In R. N. McCauley (Ed.), *The Churchlands and Their Critics* (17–47). Oxford: Blackwell Publishers.

McCauley, R. N. (2000). Overcoming Barriers to a Cognitive Psychology of Religion. In A. Geertz and R. McCutcheon (Eds.), *Perspectives on Method and Theory in the Study of Religion* (141–61). The Hague: Brill.

McCauley, R. N. (2001). Ritual, Memory, and Emotion: Comparing Two Cognitive Hypotheses. In J. Andresen (Ed.), *Religion in Mind: Cognitive Perspectives on Religious Belief, Ritual, and Experience* (115–40). Cambridge: Cambridge University Press.

McCauley, R. N. (2007). Reduction: Models of Cross-Scientific Relations and Their Implications for the Psychology-Neuroscience Interface. In P. Thagard (Ed.), *Handbook of the Philosophy of Psychology and Cognitive Science* (105–58). Amsterdam: Elsevier.

McCauley, R. N. (2009). Time Is of the Essence: Explanatory Pluralism and Accommodating Theories about Long Term Processes. *Philosophical Psychology*, *22*(5): 611–35.
McCauley, R. N. (2011). *Why Religion Is Natural and Science Is Not*. New York: Oxford University Press.
McCauley, R. N. (2013). Explanatory Pluralism and the Cognitive Science of Religion: Or Why Scholars in Religious Studies Should Stop Worrying about Reductionism. In D. Xygalatas and W. W. McCorkle Jr. (Eds.), *Mental Culture: Classical Social Theory and the Cognitive Science of Religion* (11–32). London: Acumen.
McCauley, R. N., and Lawson, E. T. (1996). Who Owns "Culture"? *Method & Theory in the Study of Religion*, *8*(2): 171–90.
McCauley, R. N., and Lawson, E. T. (1998). Interactionism and the Non-Obviousness of Scientific Theories: A Response to Michael P. Levine. *Method & Theory in the Study of Religion*, *10*(1): 61–77.
McCauley, R. N., and Lawson, E. T. (2002). *Bringing Ritual to Mind: Psychological Foundations of Cultural Forms*. Cambridge: Cambridge University Press.
McCorkle, W. W. (2014). Cognitive Historiography: Religion as an Artifact of Culture and Cognition. *Journal of Cognitive Historiography*, *1*(2): 140–5.
McCorkle, W. W., and Lane, J. E. (2012). Ancestors in the Simulation Machine: Measuring the Transmission and Oscillation of Religiosity in Computer Modeling. *Religion, Brain & Behavior*, *2*(3): 215–18.
McCutcheon, R. T. (1997). *Manufacturing Religion: The Discourse on Sui Generis Religion and the Politics of Nostalgia*. New York: Oxford University Press.
McCutcheon, R. T. ([2010] 2012). Will Your Cognitive Anchor Hold in the Storms of Culture? In W. E. Arnal and R. T. McCutcheon (Eds.), *The Sacred Is the Profane: The Political Nature of Religion* (91–101). New York: Oxford University Press.
McGuire, M. B. (1977). Testimony as a Commitment Mechanism in Catholic Pentecostal Prayer Groups. *Journal for the Scientific Study of Religion*, *16*(2): 165–8.
McKay, R. (2015). Editorial: Religion and Agency. *Journal for the Cognitive Science of Religion*, *2*(2): 93–6.
McKay, R., and Whitehouse, H. (2015). Religion and Morality. *Psychological Bulletin*, *141*(2): 447–73.
McMaster, J., Johnstone, G., and Tovell, V. (producers). (2007). *Judgment Day: Intelligent Design on Trial* [documentary]. United States: NOVA.
McNeill, W. H. (2001). Passing Strange: The Convergence of Evolutionary Science with Scientific History. *History and Theory*, *40*(1): 1–15.
Mead, M. (1964). *Continuities in Cultural Evolution*. New Haven, CT: Yale University Press.
Medawar, P. B. (1969). *Induction and Intuition in Scientific Thought*. London: Methuen.
Medawar, P. B., and Medawar, J. (1983). s.v. Historicism. In *Aristotle to Zoos: A Philosophical Dictionary of Biology* (143–4). Oxford: Oxford University Press.
Meltzoff, A. N. (2007). The "Like Me" Framework for Recognizing and Becoming an Intentional Agent. *Acta Psychologica*, *124*(1): 26–43.
Meltzoff, A. N., and Moore, M. K. (1977). Imitation of Facial and Manual Gestures by Human Neonates. *Science*, *198*(4312): 75–8.
Menary, R A. (2010). Introduction to the Special Issue on 4E Cognition. *Phenomenology and the Cognitive Sciences*, *9*(4): 459–63.
Merton, R. K. (1972). Institutional Imperatives of Science. In B. Barnes (Ed.), *Sociology of Science* (65–79). Baltimore, MD: Penguin Books.

Mesoudi, A. (2011). *Cultural Evolution: How Darwinian Theory Can Explain Human Culture and Synthesize the Social Sciences*. Chicago: University of Chicago Press.

Methys, C. D., Lomakina, E. I., Daunizeau, J., Iglesias, S., Brodersen, K. H., Friston, K. J., and Stephen, K. E. (2014). Uncertainty and the Hierarchical Gaussian Filter. *Frontiers in Human Neuroscience*, *8*(825): 1–24.

Miller, G. (1956). Magical Number Seven, Plus or Minus Two: Some Limits on Our Capacity for Processing Information. *Psychological Review*, *63*(2): 81–97.

Minsky, M. (1974). *A Framework for Representing Knowledge: The Psychology of Computer Vision*. Cambridge, MA. Available on-line: Mithen, S. (1998). Comment: The Origins of Anthropomorphic Thinking. *Journal of the Royal Anthropological Institute*, *4*(8): 129–32.
https://web.media.mit.edu/~minsky/papers/Frames/frames.html (accessed May 15, 2015)

Mithen, S. (1999). Symbolism and the Supernatural. In R. Dunbar, C. Knight, and C. Power (Eds.), *The Evolution of Culture: An Interdisciplinary View* (147–69). New Brunswick, NJ: Rutgers University Press.

Modi, J. J. (1922). *The Religious Ceremonies and Customs of the Parsees*. Bombay: British India Press. Available on-line: www.avesta.org/ritual/rcc1937.pdf

Moretti, F. (2013). *The Bourgeois: Between History and Literature*. London: Verso.

Morin, O. (2013). How Portraits Turned Their Eyes upon Us: Visual Preferences and Demographic Change in Cultural Evolution. *Evolution and Human Behavior*, *34*(3): 222–9.

Morin, O. (2016). *How Traditions Live and Die*. Oxford: Oxford University Press.

Morinis, A. (1985). The Ritual Experience: Pain and the Transformation of Consciousness in Ordeals of Initiation. *Ethos*, *13*(2): 150–74.

Morris, D. (1967). *The Naked Ape*. London: Vintage.

Müller, F. M. (1882). *Introduction to the Science of Religion* (new ed.). London: Longmans, Green.

Murray, M. J. (2009). The Evolution of Religion: New Adaptationist Accounts. In M. Y. Stewart (Ed.), *Science and Religion in Dialogue, Two Volume Set* (458–71). Hoboken, NJ: Wiley.

Musick, M. A., Traphagan, J. W., Koenig, H. G., and Larson, D. B. (2000). Spirituality in Physical Health and Aging. *Journal of Adult Development*, *7*(2): 73–86.

Nagel, T. (1974). What Is It Like to Be a Bat? *Philosophical Review*, *83*(4): 435–50.

Nambisan, S. (2015). Towards a Cross-Disciplinary Understanding of Entrepreneurship. In S. Nambisan (Ed.), *Embracing Entrepreneurship across Disciplines* (173–88). Cheltenham: Edward Elgar.

Näreaho, L. (2008). The Cognitive Science of Religion: Philosophical Observations. *Religious Studies*, *44*(1): 83–98.

Nelson, C. (2010). *No University Is an Island: Saving Academic Freedom*. New York: New York University Press.

Nettle, D. (2010). Dying Young and Living Fast: Variation in Life History across English Neighborhoods. *Behavioral Ecology*, *21*(2): 387–95.

Nettle, D., Grace, J. B., Choisy, M., Cornell, H. V., Guégan, J. F., and Hochberg, M. E. (2007). Cultural Diversity, Economic Development and Societal Instability. *PLoS One*, *2*(9). doi: http://dx.doi.org/10.1371/journal.pone.0000929.

Newberg, A. B. (2005). s.v. Neurotheology. In L. Jones (Ed.), *Encyclopedia of Religion* (2nd ed.; Vol. 10) (6492–5). Farmington Hills, MI: Macmillan Reference USA, an imprint of Thomson Gale.

Newberg, A. B. (2010). *Principles of Neurotheology*. Farnham-Burlington: Ashgate.

Nichols, S. (2002). On the Genealogy of Norms: A Case for the Role of Emotion in Cultural Evolution. *Philosophy of Science*, *69*(2): 234–55.
Nielbo, K. L., and Sørensen, J. (2011). Spontaneous Processing of Functional and Non-Functional Action Sequences. *Religion, Brain & Behavior*, *1*(1): 18–30.
Nielbo, K. L., and Sørensen, J. (2013). Prediction Error during Functional and Non-Functional Action Sequences: A Computational Exploration of Ritual and Ritualized Event Processing. *Journal of Cognition and Culture*, *13*(3–4): 347–65.
Nielbo, K. L., Schjødt, U., and Sørensen, J. (2013). Hierarchical Organization of Segmentation in Non-Functional Action Sequences. *Journal for the Cognitive Science of Religion*, *1*(1): 71–97.
Nietzsche, F. (1966). *Werke in Drei Bänden*, vol. 3. Munich: Carl Hanser.
Nietzsche, F. (1968). *The Will to Power*, W. Kaufmann (Ed.). New York: Vintage Books.
Nongbri, B. (2013). *Before Religion. A History of a Modern Concept*. New Haven, CT: Yale University Press.
Norenzayan, A. (2013). *Big Gods: How Religion Transformed Cooperation and Conflict*. Princeton, NJ: Princeton University Press.
Norenzayan, A. (2014). Big Questions about Big Gods: Response and Discussion. *Religion, Brain & Behavior* 5(4): 327–42.
Norenzayan, A., and Shariff, A. F. (2008). The Origin and Evolution of Religious Prosociality. *Science*, *322*(5898): 58–62.
Norenzayan, A., Atran, S., Faulkner, J., and Schaller, M. (2006). Memory and Mystery: The Cultural Selection of Minimally Counterintuitive Narratives. *Cognitive Science*, *30*(3): 531–53.
Norenzayan, A., Hansen, I., and Cady, J. (2008). An Angry Volcano? Reminders of Death and Anthropomorphizing Nature. *Social Cognition*, *26*(2): 190–7.
Norenzayan, A., Gervais W. M., and Trzesniewski, K. H. (2012). Mentalizing Deficits Constrain Belief in a Personal God. *PLoS ONE*, *7*: 1–8. doi: 10.1371/journal.pone.0036880.
Norenzayan, A., Shariff, A. F., Gervais, W. M., Willard, A. K., McNamara, R. A., Slingerland, E., and Henrich, J. (2016). The Cultural Evolution of Prosocial Religions. *Behavioral and Brain Sciences 39*: e1. doi: http://dx.doi.org/10.1017/S0140525X14001356.
North American Association for the Study of Religion. (2015). Available on-line: http://naasr.com/ (accessed June 14, 2015).
O'Toole, G. (2011). Darwinism: Let Us Hope It Is Not True, but If It Is, Let Us Pray It Does Not Become Widely Known. Available on-line: http://quoteinvestigator.com/2011/02/09/darwinism-hope-pray/ (accessed April 26, 2016).
Odling-Smee, F. J. (1995). Niche Construction, Genetic Evolution and Cultural Change. *Behavioural Processes*, *35*(1): 195–205.
Odling-Smee, F. J., Laland, K. N., and Feldman, M. W. (2003). *Niche Construction: The Neglected Process in Evolution*. Princeton: Princeton University Press.
Öhman A., Flykt, A., and Esteves, F. (2001). Emotion Drives Attention: Detecting the Snake in the Grass. *Journal of Experimental Psychology: General*, *130*(3): 466–78.
Orians, G. H., and Heerwagen, J. H. (1992). Evolved Responses to Landscapes. In J. Barkow, L. Cosmides, and J. Tooby (Eds.), *Adapted Mind: Evolutionary Psychology and the Evolution of Culture* (555–79). Oxford: Oxford University Press.
Oster, E. (2004). Witchcraft, Weather and Economic Growth in Renaissance Europe. *Journal of Economic Perspectives*, *18*(1): 215–28.

Otto, B.-C. (2011). *Magie. Rezeptions- und diskursgeschichtliche Analysen von der Antike bis zur Neuzeit*. Berlin: de Gruyter.

Pachis, P., and Martin, L. H. (Eds.). (2009). *Imagistic Traditions in the Graeco-Roman World*. Thessaloniki: Vanias.

Pachis, P., and Wiebe, D. (Eds.). (2010). *Chasing Down Religion: In the Sights of History and the Cognitive Sciences. Essays in Honor of Luther H. Martin*. Sheffield, UK: Equinox.

Paden, W. E. (2010). The History of Religions and Evolutionary Models: Some Reflections on Framing and Mediating Vocabulary. In P. Pachis and D. Wiebe (Eds.), *Chasing Down Religion: In the Sights of History and the Cognitive Sciences* (293–304). Sheffield, UK: Equinox.

Paldam, E., and Schjødt, U. (2016). Miracles and Pain Relief: Experienced Health Effects of Charismatic Prayer Healing in a Large Collection of Christian Testimonies. *Archive for the Psychology of Religion, 38*(2): 210–31.

Parker, R. (1998). Pleasing Things: Reciprocity in Greek Religion. In C. Gill, N. Postlethwaite, and R. Seaford (Eds.), *Reciprocity in Ancient Greece* (105–25). Oxford: Oxford Press Clarendon.

Parsons, K. M. (2001). *Drawing Out Leviathan: Dinosaurs and the Science Wars*. Bloomington: Indiana University Press.

Pearsall, J. (Ed.). (1999). *The Concise Oxford Dictionary* (10th ed.). Oxford: Oxford University Press.

Penner, H. (1989). *Impasse and Resolution: A Critique of the Study of Religion*, vol. 8 (Toronto Studies in Religion). New York: Peter Lang.

Penner, H. H., and Yonan, E. A. (1972). Is a Science of Religion Possible? *Journal of Religion, 52*(2): 107–33.

Pessoa, L., and Ungerleider, L. G. (2004). Neuroimaging Studies of Attention and the Processing of Emotion-Laden Stimuli. *Progress in Brain Research, 144*: 171–82.

Petersen, A. K. (2011). Rituals of Purification, Rituals of Initiation: Phenomenological, Taxonomical, and Cultural Evolutionary Reflections. In D. Hellholm, T. Vegge, Ø. Norderval, and C. Hellholm (Eds.), *Ablution, Initiation, and Baptism*, vol. 1 (3–40). Berlin: de Gruyter.

Petersen, A. K. (2013a). Justin Martyr in Search of the Self. In J. Rüpke and G. Woolf (Eds.), *Religious Dimensions of the Self in the Second Century CE* (104–29). Tübingen: Mohr-Siebeck.

Petersen, A. K. (2013b). Attaining Divine Perfection through Different Forms of Imitation. *Numen, 60*(1): 7–38.

Petersen, A. K. (2013c). Paul and Magic: Complementary or Incongruent Entities? Scholarly Use of the Concept of Magic, Its Definition, and Prevalence of Magic in Paul's World. In H. Jacobus, A. K. de Hemmer Gudme, and P. Guillaume (Eds.), *Studies on Magic and Divination in the Biblical World* (181–210). Piscataway, NJ: Gorgias Press.

Petersen, A. K. (2015). Zur Zeit Wird Hier das Raum: A Cultural Evolutionary Perspective on Paul and His Religion as Epitomised by His Letters. In G. van den Heever (Ed.), *Spatialising Practices*. Leiden: Brill.

Petersen, A. K. (2016). Wittgenstein's Remarks on Frazer: An Emotional Philosophical Puppet. In J. Rothaupt and L. Albinus (Eds.), *Wittgenstein's Remarks on Frazer: The Text and the Matter* (369–401). Berlin: de Gruyter.

Petersen, A. K. (2017a). 1 Maccabees from an Axial Age Perspective. In F. Avemarie, P. Bukovec, S. Krauter and M. Tilly (Eds.), *DieMakkabäer/The Maccabees. Festschrift for Professor Hermann Lichtenberger* (41–67). Tübingen: Mohr-Siebeck.

Petersen, A. K. (2017b). Geburtswunder der Maitresse des Mörders: Andreasakten nach Gregor von Tours, Liber de miraculis beati Andreae apostoli 25. In R. Zimmermann and I. Czaschez (Eds.), *Kompendium der frühchristlichen Wundererzählungen* (714–20). Gütersloh: Gütersloh Verlaghaus.

Petersen, A. K. (2017c). Die Austreibung der Dämonen aus dem Haushalt des Antiphanes: Andreasakten nach Gregor von Tours, Liber de miraculis beati Andreae apostoli 29. In R. Zimmermann and I. Czaschez (Eds.), *Kompendium der frühchristlichen Wundererzählungen* (721–31). Gütersloh: Gütersloh Verlaghaus.

Petersen, A. K. (2017d). Strafwunder durch Fluch aus Statue: Akten des Andreas und Matthias 29–32. In R. Zimmermann and I. Czaschez (Eds.), *Kompendium der frühchristlichen Wundererzählungen* (753–63). Gütersloh: Gütersloh Verlaghaus.

Petersen, A. K. (forthcoming). The Naturalness of Rhapsodomantics. In A. K. Petersen and J. Sørensen (Eds.), *Divination and Magic and Their Interactions*. Leiden: Brill.

Peterson, G. R. (2013). McCauley, the Maturationally Natural, and the Current Limits of the Cognitive Science of Religion. *Religion, Brain & Behavior*, *3*(2): 141–51.

Pigliucci, M. (2010). *Nonsense on Stilts: How to Tell Science from Bunk*. Chicago University of Chicago Press.

Pigliucci, M. (2015). The Debate about Funding of Basic Scientific Research. In *Scientia Salon: Philosophy, Science, and all interesting things in between* (June 18). Available online: https://scientiasalon.wordpress.com/2015/06/18/the-debate-about-funding-of-basic-scientific-research/ (accessed June 25, 2016).

Pigliucci, M., and Boudry, M. (Eds.). (2013). *Philosophy of Pseudoscience: Reconsidering the Demarcation Problem*. Chicago: University of Chicago Press.

Pinker, S. (1997). *How the Mind Works*. New York: Norton.

Pinker, S. (2002). *The Blank Slate: The Modern Denial of Human Nature*. New York: Penguin.

Pinker, S. (2011). *The Better Angels of Our Nature: Why Violence Has Declined*. New York: Viking.

Plotkin, H. (2004). *Evolutionary Thought in Psychology: A Brief History*. London: Blackwell.

Popper, K. (1934). *The Logic of Scientific Discovery*. New York: Harper and Row.

Popper, K. (1957). *The Poverty of Historicism*. London: Routledge & Kegan Paul.

Porubanova, M. (2016). MCI Theory: Can MCI Theory Alone Explain the Abundance of Religious Ideas? *Religion, Brain & Behavior*, *6*(3): 262–4.

Porubanova, M., Shaw, D., and Xygalatas, D. (2013). Minimal-Counterintuitiveness Revisited: Effects of Cultural and Ontological Violations on Concept Memorability. *Journal for the Cognitive Science of Religion*, *1*(2): 181–92.

Porubanova, M., Shaw, D.J., McKay, R., and Xygalatas, D. (2014). Memory for Expectation-Violating Concepts: The Effects of Agents and Cultural Familiarity. *PLoS ONE*, *9*(4): e90684.

Powell, J. W. (1899). Esthetology, or the Science of Activities Designed to Give Pleasure. *American Anthropologist, New Series*, *1*(1): 1–40.

Preus, S. J. (1987). *Explaining Religion: Criticism and Theory from Bodin to Freud*. New Haven: Yale University Press.

Price, D. (2008). *Anthropological Intelligence: The Deployment and Neglect of American Anthropology in the Second World War*. Durham, NC: Duke University Press.

Proudfoot, W. (1985). *Religious Experience*. Berkeley: University of California Press.

Purzycki, B. G. (2006). *Myth, Humor, and Ontological Templates: A Study of the Retention and Transmission of Mythical-Religious Ideas*, MA thesis, University of Nebraska-Lincoln.

Purzycki, B. G. (2010). Cognitive Architecture, Humor, and Counterintuitiveness: Retention and Recall of MCIs. *Journal of Cognitive and Culture, 10*(1): 189–204.
Purzycki, B. G. (2011). Humor as Violation and Deprecation: A Cognitive Anthropologist Account. *Journal of Cognition and Culture, 11*(1–2): 217–30.
Purzycki, B. G. (2013). Toward a Cognitive Ecology of Religious Concepts: An Example from the Tyva Republic. *Journal for the Cognitive Science of Religion, 1*(1): 99–120.
Purzycki, B. G., and Willard, A. K. (2016). MCI Theory: A Critical Discussion. *Religion, Brain & Behavior, 6*(3): 207–48.
Purzycki, B. G., Haque, O., and Sosis, R. (2014). Extending Evolutionary Accounts of Religion beyond the Mind: Religions as Adaptive Systems. In F. Watts and L. Turner (Eds.), *Evolution, Religion, and Cognitive Science: Critical and Constructive Essays* (74–91). New York: Oxford University Press.
Purzycki, B. G., Kiper, J., Shaver, J., Finkel, D. N., and Sosis, R. (2015). Religion. In R. A. Scott and S. M. Kosslyn (Gen. Eds.), N. Pinkerton (Man. Ed.), *Emerging Trends in the Social and Behavioral Sciences: An Interdisciplinary, Searchable, and Linkable Resource* (1–16). Wiley Online Library. doi: 10.1002/9781118900772.etrds0280.
Pyysiäinen, I. (2001). *How Religion Works: Towards a New Cognitive Science of Religion.* Leiden: Brill.
Pyysiäinen, I. (2009). *Supernatural Agents: Why We Believe in Souls, Gods, and Buddhas.* New York: Oxford University Press.
Pyysiäinen, I. (2012). Putting Cognition and Culture Back Together Again: Religion in Mind and Society. *Method & Theory in the Study of Religion, 24*(1): 29–50.
Pyysiäinen, I. (2013). Cognitive Science of Religion: State-of-the-Art. *Journal for the Cognitive Science of Religion, 1*(1): 5–28.
Pyysiäinen, I., and Uro, R. (Eds.) (2007). *Explaining Christian Origins and Early Judaism: Contributions from Cognitive and Social Science.* Leiden: Brill.
Quillian, M. R. (1967). Word Concepts: A Theory and Simulation of Some Basic Semantic Capabilities. *Behavioral Science, 12*(5): 410–30.
Quinlan, R. J., and Quinlan, M. B. (2007). Parenting and Cultures of Risk: A Comparative Analysis of Infidelity, Aggression, and Witchcraft. *American Anthropologist, 109*(1): 164–79.
Raihani, N. J., and Bshary, R. (2015). Why Humans Might Help Strangers. *Frontiers in Behavioral Neuroscience, 9*(39). doi: 10.3389/fnbeh.2015.00039.
Rappaport, R. A. (1999). *Ritual and Religion in the Making of Humanity.* Cambridge: Cambridge University Press.
Raup, D. M. ([1987] 1999). *The Nemesis Affair: A Story of the Death of Dinosaurs and the Ways of Science* (rev. and expand. ed). New York: W. W. Norton.
Raup, D. M., and Jablonski, D. (Eds.). (1986). *Patterns and Processes in the History of Life. Report of the Dahlem Workshop on Patterns and Processes in the History of Life, Berlin 1985, June 16–21.* Berlin,: Springer Verlag.
Raup, D. M., and Sepkoski, J. J., Jr. (1984). Periodicity of Extinctions in the Geologic Past. *Proceedings of the National Academy of the United States, 81*(3): 801–5.
Reisberg, D. (1997). *Cognition: Exploring the Science of the Mind.* New York: W. W. Norton.
Reisberg, D., and Heuer, F. (1992). *Remembering the Details of Emotional Events.* In E. Winograd and U. Neisser (Eds.), *Affect and Accuracy in Recall: Studies of "Flashbulb" Memories* (162–90). Cambridge: Cambridge University Press.
Renfrew, C. (2009). Introduction. Becoming Human: Changing Perspectives on the Emergence of Human Values. In C. Renfrew and I. Morley (Eds.),

Becoming Human: Innovation in Prehistoric Material and Spiritual Culture (1–9). Cambridge: Cambridge UniversityPress.
Rhode, D. L. (2006). *In Pursuit of Knowledge: Scholars, Status and Academic Culture.* Stanford, CA: Stanford Law & Politics.
Riale, F. N. (1892). The World's Religions at the World's Fair. *Arena, 6*: 243–9.
Richerson, P. J., and Boyd, R. (2006). *Not By Genes Alone: How Culture Transformed Human Evolution.* Chicago, IL: University of Chicago Press.
Richerson, P. J., and Christiansen, M. H. (Eds.). (2013). *Cultural Evolution: Society, Technology, Language, and Religion.* Cambridge, MA: MIT Press.
Richerson, P. J., and Newson, L. (2009). Is Religion Adaptive? Yes, No, Neutral. But Mostly We Don't Know. In J. Schloss and M. J. Murray (Eds.), *The Believing Primate: Scientific, Philosophical, and Theological Reflections on the Origin of Religion* (100–17). New York: Oxford University Press.
Ridley, M. (1996). *The Origins of Virtue.* New York: Penguin.
Ristau, C. A. (1998). Cognitive Ethology: The Minds of Children and Animals. In D. D. Cummins and C. Allen (Eds.), *The Evolution of Mind* (127–61). New York: Oxford University Press.
Rizzolatti, G., Fadiga, L., Fogassi, L., and Gallese, V. (2011). From Mirror Neurons to Imitation: Facts and Speculations. In A. N. Meltzoff and W. Prinz (Eds.), *The Imitative Mind: Development, Evolution, and Brain Bases* (247–66). Cambridge: Cambridge University Press.
Roazzi, M., Nyhof, M. A., and Johnson, C. N. (2013). Mind, Soul, and Spirit: Conceptions of Immaterial Identity in Different Cultures. *International Journal for the Psychology of Religion, 23*(1): 75–86.
Robson, Colin. (2002). *Real World Research: A Resource for Social Scientists and Practitioner-Researchers.* Oxford: Blackwell.
Roepstorff, A., Niewöhner, J., and Beck, S. (2010). Enculturing Brains through Patterned Practices. *Neural Networks, 23*(8–9): 1051–9.
Rosa, H. (2003). Social Acceleration: Ethical and Political Consequences of a Desynchronized High-Speed Society. *Constellations, 10*(1): 3–33.
Rosen, R. (1970). *Dynamic System Theory in Biology, Vol. 1, Stability Theory and Its Applications.* New York: Wiley Interscience Series on Biomedical Engineering.
Rossano, M. (2010). *Supernatural Selection: How Religion Evolved.* Oxford: Oxford University Press.
Rozin, P., Haidt, J., and McCauley, C. R. (1993) Disgust. In M. Lewis and J. M. Haviland (Eds.), *Handbook of Emotions* (69–73). New York: Guilford.
Rudwick, M. J. S. (2014). *Earth's Deep History: How It Was Discovered and Why It Matters.* Chicago: University of Chicago Press.
Rumelhart, D. E. (1980). *Schemata: The Building Blocks of Cognition.* In R. J. Spiro, B. C. Bruce, and W. F. Brewer (Eds.), *Theoretical Issues in Reading Comprehension* (33–58). Hillsdale, NJ: Lawrence Erlbaum.
Russell, Y. I., and Gobet, F. (2013). What Is Counterintuitive? Religious Cognition and Natural Expectation. *Review of Philosophy and Psychology, 4*(4): 715–49.
Russell, Y. I., Gobet, F., and Whitehouse, H. (2016). Mood, Expertise, Analogy, and Ritual: An Experiment Using the Five-Disc Tower of Hanoi. *Religion, Brain & Behavior,* 6(1): 67–87.
Rydving, H. (2008). A Western Folk Category in Mind? *Temenos, 44*(1): 73–99.
Sahlins, M. (1976). *Use and Abuse of Biology: An Anthropological Critique of Sociobiology.* Ann Arbor: University of Michigan Press.

Sahlins, M. (2009). The Conflicts of the Faculty. *Critical Inquiry, 35*(4): 997–1017.
Saler, B. (1993). *Conceptualizing Religion: Immanent Anthropologists, Transcendent Natives and Unbounded Categories*. Leiden: Brill.
Saler, B. (1977). Supernatural as a Western Category. *Ethos, 5*(1): 31–52.
Saler, B. (2008). Conceptualizing Religion: Some Recent Reflections. *Religion, 38*(3): 219–25.
Saler, B. (2009). Anthropomorphism and Animism: On Stewart E. Guthrie, *Faces in the Clouds*, 1993. In M. Vingberg (Ed.), *Contemporary Theories of Religion: A Critical Companion* (39–52). Oxon, UK: Routledge.
Saler, B. (n.d.). Observations on the Construction of the Supernatural in Euro-American Cultures (unpublished manuscript).
Schachter, D. L. (1995). Memory Distortion: History and Current Status. In D. L. Schacter (Ed.), *Memory Distortion: How Minds, Brains, and Societies Reconstruct the Past* (1291–4). Cambridge: MIT Press.
Schank, R. C., and Abelson, R. (1977). *Scripts, Plans, Goals, and Understanding*. Hillsdale, NJ: Earlbaum Associates.
Schilbach, L., Eickhoff, S. B., Rotarska-Jagiela, A., Fink, G. R., and Vogeley, K. (2008). Minds at Rest? Social Cognition as the Default Mode of Cognizing and Its Putative Relationship to the "Default System" of the Brain. *Consciousness and Cognition, 17*(2): 457–67.
Schilbrack, K. (2010). Religions: Are There Any? *Journal of the American Academy of Religion, 78*(4): 1112–38.
Schimmack, U. (2005). Attentional Interference Effects of Emotional Pictures: Threat, Negativity, or Arousal? *Emotion, 5*(1): 55–66.
Schjødt, U. (2007). Homeostasis and Religious Behaviour. *Journal of Cognition and Culture, 7*(3): 313–40.
Schjødt, U. (2009). The Religious Brain: A General Introduction to the Experimental Neuroscience of Religion. *Method & Theory in the Study of Religion 21*(3): 310–39.
Schjødt, U., Stødkilde-Jørgensen, H., Geertz, A. W., and Roepstorff, A. (2008). Rewarding Prayers. *Neuroscience Letters, 443*(3): 165–8.
Schjødt, U., Stødkilde-Jørgensen, H., Geertz, A. W., and Roepstorff, A. (2009). Highly Religious Participants Recruit Areas of Social Cognition in Personal Prayer. *Social Cognitive and Affective Neuroscience, 4*(2): 199–207.
Schjødt, U., Stødkilde-Jørgensen, H., Geertz, A. W., Lund, T. E., and Roepstorff, A. (2011). The Power of Charisma—Perceived Charisma Inhibits the Frontal Executive Network of Believers in Intercessory Prayer. *Social Cognitive and Affective Neuroscience, 6*(1): 119–27.
Schjødt, U., Sørensen, J., Laigaard Nielbo, K., Xygalatas, D., Mitkidis, P., and Bulbulia, J. (2013). Cognitive Resource Depletion in Religious Interactions. *Religion, Brain & Behavior, 3*(1): 39–55.
Schmidt, S. R. (1985). Encoding and Retrieval Processes in the Memory for Conceptually Distinctive Events. *Journal of Experimental Psychology: Learning, Memory, and Cognition, 11*(3): 565–78.
Scholl, B., and Tremoulet, P. (2000). Perceptual Causality and Animacy. *Trends in Cognitive Sciences, 4*(8): 299–309.
Schultz, W. T. (2005). *Handbook of Psychobiography*. New York: Oxford University Press.
Scott-Phillips, T. C. (2014). *Speaking Our Minds: Why Human Communication Is Different, and How Language Evolved to Make It Special*. New York: Palgrave Macmillan.

Sepkoski, J. J., Jr. (2001). s.v. Mass Extinctions, Concept of. In S. A. Levin (Ed.), *Encyclopedia of Biodiversity*, vol. 4 (97–110). San Diego: Academic Press.

Sepkoski, D., and Ruse, M. (Eds.). (2010). *The Paleobiological Revolution: Essays on the Growth of Modern Paleontology*. Chicago: University of Chicago Press.

Shannon, C. E. (1948). A Mathematical Theory of Communication. *Bell System Technical Journal, 27*(3): 379–423.

Shea, J. J. (2011). *Homo sapiens* Is as *Homo sapiens* Was: Behavioral Variability vs. "Behavioral Modernity" in Paleolithic Archaeology. *Current Anthropology, 52*(1): 1–35.

Shermer, M. ([2011] 2012). *The Believing Brain. From Spiritual Faiths to Political Convictions: How We Construct Beliefs and Reinforce Them as Truths.* London: Robinson.

Shermer, M., and Grobman, A. ([2001] 2009). *Denying History: Who Says the Holocaust Never Happened and Why Do They Say It? Updated and Expanded.* Berkeley: University of California Press.

Shryock, A., and Smail, D. L. (Eds.). (2011). *Deep History: The Architecture of Past and Present.* Berkeley: University of California Press.

Simmel, G. ([1905] 1977). *The Problems of the Philosophy of History: An Epistemological Essay*, trans. Guy Oakes (2nd ed.). New York: Free Press.

Sjöblom, T. (2005). "Going, Going, Gone"--Or How I Fail to Gain Hermeneutic Understanding. In R. Gothoni (Ed.), *How to Do Comparative Religion? Three Ways, Many Goals* (35–44). Berlin and New York: de Gruyter.

Skinner, B. F. (1957). *Verbal Behavior*. Acton, MA: Copley .

Skjoldli, J. (2015). The Backwash of Norenzayan's *Big Gods*: A Post-review Essay. *Numen: International Review for the History of Religions, 62*(5–6): 639–60.

Slaughter, V., Jaakkola, R., and Carey, S. (1999). Constructing a Coherent Theory: Children's Biological Understanding of Life and Death. In M. Siegal and C. C. Petersen (Eds.), *Children's Understanding of Biology and Health* (71–96). New York: Cambridge University Press.

Slingerland, E. (2008). *What Science Offers the Humanities: Integrating Body and Culture.* New York: Cambridge University Press.

Slingerland, E. (2013). Body and Mind in Early China: An Integrated Humanities–Science Approach. *Journal of the American Academy of Religion, 81*(1): 6–55.

Slingerland, E. (2014a). *Trying Not to Try: The Art and Science of Spontaneity.* New York: Crown.

Slingerland, E. (2014b). Toward a Second Wave of Consilience in the Cognitive Scientific Study of Religion. *Journal of Cognitive Historiography, 1*(1): 121–30.

Slone, D. J. (2004). *Theological Incorrectness: Why Religious People Believe What They Shouldn't.* New York: Oxford University Press.

Slone, D. J. (Ed.). (2006). *Religion and Cognition: A Reader.* London and Oakville, CT: Equinox.

Slone, D. J., Gonce, L. O., Upal, A., Tweney, R., and Edwards, K. (2007). Imagery Effects on Recall of Minimally Counterintuitive Concepts. *Journal of Cognition and Culture, 7*(3): 355–67.

Smail, D. L. (2008). *On Deep History and the Brain.* Berkeley University of California Press.

Smith, B. H. (2009). *Natural Reflections: Human Cognition at the Nexus of Science and Religion.* New Haven, CT: Yale University Press.

Smith, E. A., and Winterhalder, B. (Eds.). (1981). *Hunter-Gatherer Foraging Strategies: Ethnographic and Archaeological Analyses*. Chicago, IL: University of Chicago Press.

Smith, J. Z. (1982). *Imagining Religion: From Babylon to Jonestown*. Chicago, IL: University of Chicago Press.

Smith, J. Z. (1990). *Drudgery Divine: On the Comparison of Early Christianities and the Religions of Late Antiquity*. London: School of Oriental and African Studies-University of London.

Smith, J. Z. (1993). *Map Is Not Territory: Studies in the History of Religions*. Chicago, IL: University of Chicago Press.

Smith, J. Z. (1998). Religion, Religions, Religious. In M. C. Taylor (Ed.), *Critical Terms for Religious Studies* (269–84). Chicago: Chicago University Press.

Smith, J. Z. (2004). A Twice-Told Tale: The History of the History of Religions' History. In *Relating Religion: Essays in the Study of Religion* (362–74). Chicago: University of Chicago Press.

Snow, C. P. (1959). *The Two Cultures and the Scientific Revolution*. Cambridge: Cambridge University Press.

Sørensen, J. (2004). Religion, Evolution, and an Immunology of Cultural Systems. *Evolution and Cognition, 10*(1): 61–73.

Sørensen, J. (2005). Religion in Mind: A Review Article of the Cognitive Science of Religion. *Numen, 52*(4): 465–94.

Sørensen, J. (2007a). *A Cognitive Theory of Magic*. Plymouth: AltaMira.

Sørensen, J. (2007b). Acts that Work: A Cognitive Approach to Ritual Agency. *Method & Theory in the Study of Religion, 19*(3): 281–300.

Sørensen, J. (2008). Cognition and Religious Phenomena: A Response to Håkan Rydving. *Temenos, 44*(1): 111–22.

Sørensen, J. (2010). Past Minds: Present Historiography and Cognitive Science. In L. H. Martin and J. Sørensen (Eds.), *Past Minds: Studies in Cognitive Historiography* (179–96). London: Routledge.

Sørensen, J., and Nielbo, K. L. (2013a). The Experimental Study of Religion: or, There and Back Again. *Journal for the Cognitive Science of Religion, 1*(2): 215–32.

Sørensen, J., and Nielbo, K. L. (2013b). Prediction Error during Functional and Non-Functional Action Sequences: A Computational Exploration of Ritual and Ritualized Event Processing. *Journal of Cognition and Culture, 13*(3–4): 347–65.

Sosis, R. (2009). The Adaptationist-Byproduct Debate on the Evolution of Religion: Five Misunderstandings of the Adaptationist Program. *Journal of Cognition and Culture, 9*(3): 315–32.

Sosis, R. (2012). Religions as Complex Adaptive Systems. Center of Theological Inquiry Colloquium paper, Princeton, NJ (unpublished).

Sosis, R., and Bulbulia, J. (2011). The Behavioral Ecology of Religion: The Benefits and Costs of One Evolutionary Approach. *Religion, 41*(3): 341–62.

Sosis, R., and Kiper, J. (2014). Religion Is More than Belief: What Evolutionary Theories of Religion Tell Us about Religious Commitment. In M. Bergman and P. Kain (Eds.), *Challenges to Religion and Morality: Disagreements and Evolution* (256–76). New York: Oxford University Press.

Sosis, R., and Kiper, J. (n.d.). If Religion Is Natural, Does It Deserve Political Protection? The Cognitive and Evolutionary Sciences of Religion and Its Implications for Religious Freedom (unpublished).

Sosis, R., and Ruffle, B. J. (2003). Religious Ritual and Cooperation: Testing for a Relationship on Israeli Religious and Secular Kibbutzim. *Current Anthropology*, *44*(5), 713–22.

Sperber, Dan. (1974). *Symbolisme en General*. Paris: Hermann.

Sperber, D. (1975). *Rethinking Symbolism*, trans. A. E. Morton. Cambridge, UK: Cambridge University Press.

Sperber, D. (1985a). *On Anthropological Knowledge*. Cambridge: Cambridge University Press.

Sperber, D. (1985b). Anthropology and Psychology: Towards an Epidemiology of Representations. *Man*, *20*(1): 73–89.

Sperber, D. (1996). *Explaining Culture: A Naturalistic Approach*. Oxford: Blackwell.

Sperber, D., and Hirschfeld, L. A. (2004). The Cognitive Foundations of Cultural Stability and Diversity. *Trends in Cognitive Sciences*, *8*(1): 40–6.

Sperber, D., and Wilson, D. (1986). *Relevance: Communication and Cognition*. Cambridge, MA: Harvard University Press.

Sperber, D., and Wilson, D. (1995). *Relevance: Communication and Cognition* (2nd ed.). Oxford: Blackwell.

Spinney, L. (2012). Human Cycles: History as Science. *Nature*, *488*(7409): 24–6.

Spiro, M. (1966). Religion: Problems of Definition and Meaning. In M. Banton (Ed.), *Anthropological Approaches to the Study of Religion* (85–126). London: Tavistock.

Spiro, M. E., and D'Andrade, R. G. (1958). A Cross-Cultural Study of Some Supernatural Beliefs. *American Anthropologist*, *60*(3): 456–66.

Staines, J. (Ed.). (2010). *The Rough Guide to Classical Music*. London: Penguin Books.

Stangor, C., and McMillan, D. (1992). Memory for Expectancy-Congruent and Expectancy-Incongruent Information: A Review of the Social and Social Developmental Literatures. *Psychological Bulletin*, *111*(1): 42–61.

Stark, R. W., and Bainbridge, W. S. (1996). *A Theory of Religion*. New Brunswick, NJ: Rutgers University Press.

Stark, R. W., and Finke, R. (1992). *The Churching of America 1776–1992: Winners and Losers in Our Religious Economy*. New Brunswick, NJ: Rutgers University Press.

Starr, C. G. (1965). *A History of the Ancient World*. Oxford: Oxford University Press.

Stausberg, M. (2014a). *Big Gods* in Review: Introducing Ara Norenzayan and his Critics. *Religion*, *44*(4): 592–608.

Stausberg, M. (Ed.). (2014b). Review symposium on Ara Norenzayan: *Big Gods: How Religion Transformed Cooperation and Conflict*. *Religion*, *44*(4): 592–683.

Stocking, G. (1987). *Victorian Anthropology*. New York: Free Press.

Stoczkowski, W. ([1994] 2002). *Explaining Human Origins: Myth, Imagination and Conjecture*. Cambridge: Cambridge University Press.

Strauss, C., and Quinn, N. (1997). *A Cognitive Theory of Cultural Meaning*. Cambridge: Cambridge University Press.

Stroop, J. R. (1935). Studies of Interference in Serial-Verbal Reaction. *Journal of Experimental Psychology*, *18*(6), 643–62.

Sulloway, F. J. (1996). *Born to Rebel: Birth Order, Family Dynamics, and Creative Lives*. New York: Pantheon.

Sun, R., and Hélie, S. (2013). Psychologically Realistic Cognitive Agents: Taking Human Cognition Seriously. *Journal of Experimental and Theoretical Artificial Intelligence*, *25*(1): 65–92.

Swann, W. B., Jr, Gómez Á., Seyle, D. C., Morales, J. F., Huici, C. (2009). Identity Fusion: The Interplay of Personal and Social Identities in Extreme Group Behavior. *Journal of Personality and Social Psychology, 96*(5), 995–1011.

Swann, W. B., Jr, Jensen, J., Gómez, Á., Whitehouse, H., and Bastian, B. (2012). When Group Membership Gets Personal: A Theory of Identity Fusion. *Psychological Review, 119*(3): 441–56.

Swann, W. B., Jr, Buhrmester, M., Gómez, Á., Jetten, J., Bastian, B., Vázquez, A. Ariyanto, A., Besta, T., Christ, O., Cui, L., Finchilescu, G., González, R., Goto, N., Hornsey, M., Sharma, S., Susianto, H., and Zhang, A. (2014). What Makes a Group Worth Dying For? Identity Fusion Fosters Perception of Familial Ties, Promoting Self-Sacrifice. *Journal of Personality and Social Psychology, 106*(9): 912–26.

Switek, B. (2010). Review of *The Paleobiological Revolution: Essays on the Growth of Modern Paleontology*, D. Sepkoski and M. Ruse (Eds.). Chicago: University of Chicago Press.

Symons, D. (1979). *The Evolution of Human Sexuality*. New York: Oxford University Press.

Szechtman, H., and Woody, E. Z. (2004). Obsessive-Compulsive Disorder as a Disturbance of Security Motivation. *Psychological Review, 111*(1): 111–27.

Talarico, J. M., and Rubin, D. C. (2003). Confidence, Not Consistency, Characterizes Flashbulb Memories. *Psychological Science, 14*(5): 455–61.

Taleb, N. N. (2007). *The Black Swan: The Impact of the Highly Improbable*. New York: Random House.

Talmot-Kaminski, K. (2013). *Religion as Magical Ideology*. Durham, UK: Acumen Publishing Limited.

Taves, A. (2009). *Religious Experience Reconsidered: A Building-Block Approach to the Study of Religion and Other Special Things*. Princeton: Princeton University Press.

Taves, A. (2015). Reverse Engineering Complex Cultural Concepts: Identifying Building Blocks of "Religion." *Journal of Cognition and Culture, 15*(1–2): 191–216.

Tetlock, P. E. (2003). Thinking the Unthinkable: Sacred Values and Taboo Cognitions. *Trends in Cognitive Sciences, 7*(7): 320–24.

Thagard, P. (1992). *Conceptual Revolutions*. Princeton: Princeton University Press.

Thagard, P. (1996). *Mind: Introduction to Cognitive Science*. Cambridge: MIT Press.

Thagard, P. (2006). *Hot Thought: Mechanisms and Applications of Emotional Cognition*. Cambridge: MIT Press.

Thomas, K. (1997). *Religion and the Decline of Magic: Studies in Popular Beliefs in Sixteenth and Seventeenth Century England*. New York: Oxford University Press.

Tomasello, M. (1999). *The Cultural Origins of Human Cognition*. Cambridge, MA: Harvard University Press.

Tremlin, T. (2006). *Minds and Gods: The Cognitive Foundations of Religion*. New York: Oxford University Press.

Tribble, E. B., and Keene, N. (2011). *Cognitive Ecologies and the History of Remembering: Religion, Education and Memory in Early Modern England*. London: Palgrave Macmillan.

Turchin, P. (2008). Arise "Cliodynamics". *Nature, 454*(7200): 34–5.

Turchin, P., Whitehouse, H., François, P., Slingerland, E., and Collard, M. (2012). A Historical Database of Sociocultural Evolution. *Cliodynamics: Journal of Theoretical and Mathematical History, 3*(2): 271–93.

Turchin, P., Brennan, R., Currie, T. E., Feeney, K. C., François, P. Hoyer, D., Manning, J. G., Marciniak, A., Mullins, D., Palmisano, A., Peregrine, P., Turner, E. A. L., and

Whitehouse, H. (2015). Seshat: The Global History Databank. *Cliodynamics: Journal of Quantitative History and Cultural Evolution, 6*(1): 77–107.
Turing, A. M. (1950). Computing Machinery and Intelligence. *Mind*, 50(236): 433–60.
Tylor, E. B. (1871) *Primitive Culture: Researches into the Development of Mythology, Philosophy, Religion, Art, and Custom*. London: J. Murray.
Upal, M. A., Gonce, L. O., Tweney, R. D., and Slone, D. J. (2007). Contextualizing Counterintuitiveness: How Context Affects Comprehension and Memorability of Counterintuitive Concepts. *Cognitive Science, 31*(3): 415–39.
Valéry, P. (1919) *La Crise de l'èsprit. Prèmiere lettre*. Extrait de "Europes de l'antiquité au XXe siècle," collection Bouquins, éditions Robert Laffont, 2000, 405–14 Available on-line: http://classiques.uqac.ca/classiques/Valery_paul/crise_de_lesprit/valery_esprit.pdf. (accessed July 9, 2015).
van Elk, M. (2013). Paranormal Believers Are More Prone to Illusory Agency Detection than Skeptics. *Consciousness and Cognition, 22*(3): 1041–6.
van Elk, M. (2015). Perceptual Biases in Relation to Paranormal and Conspiracy Beliefs. *PLoS ONE, 10*(6). doi: 10.1371/journal.pone.0130422.
van Elk, M., Matzke, D., Gronau, Q. F., Guan, M., Vandekerckhove, J., and Wagenmakers, E.-J. (2015). Meta-Analyses Are No Substitute for Registered Replications: A Skeptical Perspective on Religious Priming. *Frontiers in Psychology, 6*(1365). doi: 10.3389/fpsyg.2015.01365.
van Gulik, R. H. (2003). *Sexual Life in Ancient China: A Preliminary Survey of Chinese Sex and Society from ca. 1500 BC till 1644 AD*. Leiden: Brill Academic Publishers.
Versnel, H. S. (Ed.) (1981). *Faith, Hope and Worship: Aspects of Religious Mentality in the Ancient World*. Leiden: Brill.
Versnel, H. S. (2011). *Coping with the Gods: Wayward Readings in Ancient Greek Theology*. Leiden: Brill.
Veyne, P. (1992). *A History of Private Life: From Pagan Rome to Byzantium*. Cambridge, MA: Belknap.
Von Restorff, H. (1933). Über die Wirkung von Bereichsbildungen im Spurenfeld (The Effects of Field Formation in the Trace Field). *Psychological Research, 18*(1): 299–342.
Waddington, C. H. (1975). Mindless Societies. *New York Times Review of Books* (August 7). Available on-line: http://www.nybooks.com/articles/1975/08/07/mindless-societies/ (accessed April 20, 2016).
Wagner, R. (1991). The Fractal Person. In M. Godelier and M. Strathern (Eds.), *Big-Men and Great Men: The Personification of Power* (159–73). Cambridge: Cambridge University Press.
Warren, R. (1970). Perceptual Restoration of Missing Speech Sounds. *Science, 167*(3917), 392–93.
Wason, P. C. (1966). Reasoning. In B. M. Foss (Ed.), *New Horizons in Psychology* (135–51). Harmondsworth, UK: Penguin.
Wason, P. C. (1968). Reasoning about a Rule. *Quarterly Journal of Experimental Psychology, 20*(3): 273–81.
Watson-Jones, R., Legare, C. H., Whitehouse, H., and Clegg, J. (2014). Task-specific Effects of Ostracism on Imitative Fidelity in Early Childhood. *Evolution and Human Behaviour, 35*(3): 204–10.
Watson-Jones, R., Whitehouse, H., and Legare, C. H. (2016). In-Group Ostracism Increases High Fidelity Imitation in Early Childhood. *Psychological Science, 27*(1): 34–42.

Watts, J., Greenhill, S. J., Atkinson, Q. D., Currie, T. E., Bulbulia, J., and Gray, R. D. (2015). Broad Supernatural Punishment but Not Moralizing High Gods Precede the Evolution of Political Complexity in Austronesia. *Proceedings of the Royal Society B*, *282*(1804): 20142556. doi: 10.1098/rspb.2014.2556.

Watts, J., Sheehan, O., Greenhill, S. J., Gomes-Ng, S., Atkinson, Q. D., Bulbulia, J., and Gray, R. D. (2015). *Pulotu*: Database of Austronesian Supernatural Beliefs and Practices. *PLoS ONE*, *10*(9): e0136783. doi: 10.1371/journal.pone.0136783.

Watts, J., Sheehan, O., Atkinson, Q. D., Bulbulia, J., and Gray, R. D. (2016). Ritual Human Sacrifice Promoted and Sustained the Evolution of Stratified Societies. *Nature*, *532*(7598): 228–31.

Weeden, J., and Kurzban, R. (2013). What Predicts Religiosity? A Multinational Analysis of Reproductive and Cooperative Morals. *Evolution and Human Behavior*, *34*(6): 440–5.

Wegner, D. (2005). Who Is the Controller of Controlled Processes? In R. R. Hassin, J. S. Uleman, and J. A. Bargh (Eds.), *The New Unconscious* (19–36). New York: Oxford University Press.

Weinstein, F. (1995). Psychohistory and the Crisis of the Social Sciences. *History and Theory*, *34*(4): 299–319.

Wellman, H. M., and Johnson, C. N. (2008). Developing Dualism: From Intuitive Understanding to Transcendental Ideas. In A. Antonietti, A Corradini, and J. Lowe (Eds.), *Psycho-Physical Dualism Today: An Interdisciplinary Approach* (3–35). New York: Rowman & Littlefield.

West, S. A., Griffin, A. S., and Gardner, A. (2007). Social Semantics: How Useful Has Group Selection Been? *Journal of Evolutionary Biology*, *21*(3): 374–85.

Whitehouse, H. ([1992] 2010). Memorable Religions: Transmission, Codification, and Change in Divergent Melanesian Contexts. *Man* (*N. S.*), *27*(3): 777–97. Reprint in Pachis, P., and Wiebe, D. (Eds.), *In the Sights of History and the Cognitive Sciences* (413–39). Sheffield, UK: Equinox.

Whitehouse, H. (1995). *Inside the Cult: Religious Innovation and Transmission in Papua New Guinea*. Oxford; Oxford University Press.

Whitehouse, H. ([1996] 2014). Rites of Terror: Emotion, Metaphor, and Memory in Melanesian Initiation Cults. *Journal of the Royal Anthropological Institute (N. S.)*, *2*(4): 703–15. Reprint in J. Corrigan (Ed.), *Religion and Emotion: Approaches and Interpretations* (133–48). Oxford: Oxford University Press.

Whitehouse, H. (2000). *Arguments and Icons. Divergent Modes of Religiosity*. Oxford: Oxford University Press.

Whitehouse, H. (2002). Modes of Religiosity: Towards a Cognitive Explanation of the Sociopolitical Dynamics of Religion. *Method & Theory in the Study of Religion*, *14* (3–4): 293–315.

Whitehouse, H. (2004). *Modes of Religiosity*. Walnut Creek, CA: Altamira Press.

Whitehouse, H. (2005a). Cognitive Historiography: When Science Meets Arts. *Historical Reflections*, *31*(2): 307–18.

Whitehouse, H. (2005b). Emotion, Memory, and Religious Rituals: An Assessment of Two Theories. In K. Milton and M. Svasek (Eds.), *Mixed Emotions* (91–108). Oxford: Berg.

Whitehouse, H. (2008). Cognitive Evolution and Religion: Cognition and Religious Evolution. In J. Bulbulia, R. Sosis, C. Genet, R. Genet, E. Harris, and K. Wyman (Eds.), *The Evolution of Religion: Studies, Theories, and Critiques* (31–42). Santa Margarita, CA: Collins Foundation Press.

Whitehouse, H. (2011). The Coexistence Problem in Psychology, Anthropology, and Evolutionary Theory. *Human Development*, *54*(3): 191–9.

Whitehouse, H. (2012a). Ritual, Cognition, and Evolution. In R. Sun (Ed.), *Grounding the Social Sciences in the Cognitive Sciences* (265–84). Cambridge, MA: MIT Press.

Whitehouse, H. (2012b) Human Rites: Rituals Bind Us, in Modern Societies and Prehistoric Tribes Alike. But Can Our Loyalties Stretch to All of Humankind? *AEON* (December 12). Available on-line https://aeon.co/essays/rituals-define-us-in-fathoming-them-we-might-shape-ourselves (accessed May 15, 2017).

Whitehouse, H. (2013a). Rethinking Proximate Causation and Development in Religious Evolution. In *Cultural Evolution: Society, Technology, Language, and Religion* (349–64). Cambridge, MA: MIT Press.

Whitehouse, H. (2013b). Three Wishes for the World. *Cliodynamics: Journal of Theoretical and Mathematical History 4*(2): 282–90. Available on-line: https://escholarship.org/uc/item/2wv6r7v3#page-9 (accessed May 15, 2017).

Whitehouse, H., and Hodder, I. (2010). Modes of Religiosity at Çatalhöyük. In I. Hodder (Ed.) *Religion in the Emergence of Civilization: Çatalhöyük as a Case Study* (122–45). Cambridge: Cambridge University Press.

Whitehouse, H., and Laidlaw, J. (Eds.). (2004). *Ritual and Memory: Towards a Comparative Anthropology of Religion*. Walnut Creek, CA: AltaMira Press.

Whitehouse, H., and Lanman, J. A. (2014). The Ties that Bind Us: Ritual, Fusion, and Identification. *Current Anthropology*, 55(6): 674–95.

Whitehouse, H., and Martin, L. H. (Eds.). (2004a). *Theorizing Religions Past: Archaeology, History, and Cognition*. Lanham, MD: Rowman and Littlefield.

Whitehouse, H., and Martin, L. H. (Eds.). (2004b). Implications of Cognitive Science for the Study of Religion. Special Issue of *Method & Theory in the Study of Religion, 16*(3).

Whitehouse, H., and Martin, L. H. (Eds.). (2005). History, Memory, and Cognition. Special Issue of *Historical Reflections/Réflexions Historiques, 31*(2).

Whitehouse, H and McCauley, R. N. (Eds.). (2005a). *Mind and Religion: Psychological and Cognitive Foundations of Religiosity*. Walnut Creek, CA: AltaMira Press.

Whitehouse, H., and McCauley, R. N. (Eds.). (2005b). The Psychological and Cognitive Foundations of Religiosity. Special Issue of *Journal of Cognition and Culture, 5*(1–2): 1567–7095.

Whitehouse, H., and McQuinn, B. (2013). Divergent Modes of Religiosity and Armed Struggle. In M. Juergensmeyer, M. Kitts, and M. Jerryson (Eds.), *Oxford Handbook of Religion and Violence* (597–619). Oxford: Oxford University Press.

Whitehouse, H., Richert, R. A., and Stewart, E. (2005). Memory and Analogical Thinking in High-Arousal Rituals. In H. Whitehouse and R. N. McCauley (Eds.), *Mind and Religion: Psychological and Cognitive Foundations of Religiosity* (127–45). Walnut Creek, CA: AltaMira Press.

Whitehouse, H., Kahn, K., Hochberg, M. E., and Bryson, J. J. (2012). The Role for Simulations in Theory Construction for the Social Sciences: Case Studies Concerning Divergent Modes of Religiosity. *Religion, Brain & Behavior, 2*(3): 182–201.

Whitehouse, H., Mazzucato, C., Hodder, I., and Atkinson, Q. D. (2013). Modes of Religiosity and the Evolution of Social Complexity at Çatalhöyük. In I. Hodder (Ed.), *Religion at Work in a Neolithic Society: Vital Matters* (134–55). Cambridge: Cambridge University Press.

Whitehouse, H., McQuinn, B., Buhrmester, M., and Swann, W. B. (2014). Brothers in Arms: Warriors Bond Like Family. *Proceedings of the National Academy of Sciences, 111*(50): 17783–5.

Wiebe, D. (1975). Explanation and the Scientific Study of Religion. *Religion, 5*(1): 33–52.

Wiebe, D. (1989). History or Mythistory in the Study of Religion? The Problem of Demarcation. In M. Pye (Ed.), *Marburg Revisited: Institutions and Strategies in the Study of Religion* (31–48). Marburg: diagonal Verlag.
Wiebe, D. (1999). *The Politics of Religious Studies*. New York: St. Martin's Press.
Wiebe, D. (2004). Critical Reflections on the Modes of Religiosity Argument. In H. Whitehouse and L. H. Martin (Eds.), *Theorizing Religions Past: Archaeology, History, Cognition* (197–213). Walnut Creek, CA: AltaMira.
Wiebe, D. (2005). The Politics of Wishful Thinking? Disentangling the Role of the Scholar-Scientist from that of the Public Intellectual in the Modern Academic Study of Religion. *Temenos. Nordic Journal of Comparative Religion, 41*(1): 7–38.
Wiebe, D. (2009). Religious Biases in Funding Religious Studies Research? *Religio: revue pro religionistiku, 17*(2): 125–40.
Wiebe, D. (2011). Beneath the Surface of History? In L. H. Martin and J. Sørensen (Eds.), *Past Minds: Studies in Cognitive Historiography* (167–77). London: Equinox.
Wiebe, D. (2013). The Significance of the Natural Experience of a "Non-Natural" World to the Question of the Origin of Religion. In A. W. Geertz (Ed.), *Origins of Religion, Cognition and Culture* (140–59). Durham, UK: Acumen.
Wiebe, D. (2014). Milestone or Millstone? Does Norenzayan's Book Live Up to the Hype? *Religion, 44*(4): 674–83.
Wigger, J. B., Paxson, K., and Ryan, L. (2013). What Do Invisible Friends Know? Imaginary Companions, God, and Theory of Mind. *International Journal for the Psychology of Religion, 23*(1): 2–14.
Willard, A. K., Henrich, J., and Norenzayan, A. (n.d.). The Role of Memory, Belief, and Familiarity in the Transmission of Counterintuitive Content. Available on-line: http://coevolution.psych.ubc.ca/pdfs/AKWillard_CogSci_Total.pdf (accessed May 15, 2017).
Wilson, D. S. (2002). *Darwin's Cathedral: Evolution, Religion, and the Nature of Society*. Chicago: University of Chicago Press.
Wilson, D. S., and Green, W. S. (2012). Evolutionary Religious Studies: A Beginner's Guide. In E. Slingerland and M. Collard (Eds.), *Creating Consilience: Integrating the Sciences and the Humanities* (225–42). Oxford and New York: Oxford University Press.
Wilson, E. O. (1975a). *Sociobiology: The New Synthesis*. Cambridge, MA: Belknap Press.
Wilson, E. O. (1975b). For Sociobiology. *New York Times Review of Books* (December 11). Available on-line: http://www.nybooks.com/articles/1975/12/11/for-sociobiology/. (accessed April 20, 2016).
Wilson, M. (2010). The *Re-Tooled Mind*: How Culture Re-Engineers Cognition. *Social, Cognitive, and Affective Neuroscience, 5*(2–3): 180–7.
Winterhalder, B., and E. A. Smith. (1992). Evolutionary Ecology and the Social Sciences. In E. A. Smith, B. Winterhalder (Eds.), *Evolutionary Ecology and Human Behavior* (3–23). New York: Aldine de Gruyter.
Winterhalder, B., and Smith, E. A. (2000). Analyzing Adaptive Strategies: Human Behavioral Ecology at Twenty-Five. *Evolutionary Anthropology, 9*(2): 51–72.
Wittgenstein, L. ([1953] 2001). *Philosophical Investigations*. Oxford: Blackwell.
Wolfram, S. (2010). *A New Kind of Science*. Champaign, IL: Wolfram Media.
Woolf, G. (2012). Only Connect? Networks and Religious Change in the Ancient Mediterranean. Unpublished CRASIS lecture (Groningen). Available on-line: http://www.rug.nl/research/centre-for-religious-studies/crasis/activities/annual-meeting/greg_woolf_crasis_lecture.pdf (accessed June 25, 2016).
Wright, R. (1994). *The Moral Animal*. New York: Random House.
Wright, R. (2000). *Nonzero: The Logic of Human Destiny*. New York: Pantheon.

Wright, R. (2009). *The Evolution of God*. New York: Little Brown.
Xygalatas, D. (2012). *The Burning Saints: Cognition and Culture in the Fire-Walking Rituals of the Anastenaria*. Bristol, CT: Equinox.
Xygalatas, D. (2013). Special Issue on the Experimental Research of Religion. *Journal for the Cognitive Science of Religion, 1*(2): 137–9.
Xygalatas, D. (2014a). Cognitive Science of Religion. In D. A. Leeming (Ed.), *Encyclopedia of Psychology and Religion* (2nd ed.) (343–7). New York: Springer.
Xygalatas, D. (2014b). On the Way towards a Cognitive Historiography: Are We There Yet? *Journal of Cognitive Historiography, 1*(2): 193–200.
Xygalatas, D., Konvalinka, I., Roepstorff, A., and Bulbulia, J. (2011). Quantifying Collective Effervescence: Heart-Rate Dynamics at a Fire Walking Ritual. *Communicative and Integrative Biology, 4*(6): 735–8.
Xygalatas, D., Mitkidis, P., Fischer, R., Reddish, P., Skewes, J., Geertz, A. W., Roepstorff, A., and Bulbulia, J. (2013a). Extreme Rituals Promote Prosociality. *Psychological Science, 24*(8): 1602–5.
Xygalatas, D., Schjødt, U., Bulbulia, J., Konvalinka, I., Jegindø, E. M., Reddish, P., and Roepstoff, A. (2013b). Autobiographical Memory in a Fire-Walking Ritual. *Journal of Cognition and Culture, 13*(1–2): 1–16.
Znamenski, A. A. (2007). *The Beauty of the Primitive: Shamanism and the Western Imagination*. New York: Oxford University Press.

Index

Abelson, R. 123
Abrams, D. 178
action representation 21, 60, 206n. 11
afterlife 9–10, 43–4, 73, 197–8
agency detection 8, 11–12, 59–60, 80, 199–201
  agency detection device (ADD) 200–1
Aiden, E. 179
Alcock, J. 185
Alexander, R. D. 113
Ali, S. S. 64
Allen, E. 186
Alvarez, W. 117
Ambasciano, L. 3, 107–8, 112, 113, 115, 121, 197
American Academy of Religion (AAR) 1, 3, 62, 159, 170, 203n. 5, 207n. 1
American Anthropological Association 159, 186
American Society for Cybernetics 171, 208n. 6
Andersen, M. 64
Anderson, A. K. 125, 129
Annales School 90
anthropology 1–2, 8, 18, 20–1, 34, 44–5, 48, 58–9, 61, 65, 76, 82–3, 98–9, 101, 108–9, 123–4, 146, 156–7, 172, 182–6, 188, 190, 197
  cognitive 8, 124, 129–30, 169, 181–3, 187
anthropomorphism 21, 27, 59–60, 64, 73, 77–80, 195, 198–9
antiscientific claims 20, 207n. 1
Anttonen, V. 76
archaeology 2, 47, 51–2, 159, 167, 170, 176
Armitage, D. 116
Armitage, R. 116
Arnheim, R. 79
Asad, T. 72
Astuti, R. 31, 197
Atkinson, Q. D. 21, 51–2, 55
Atran, S. 74, 75, 76, 110, 127, 145, 156, 164
Axial Age doctrines 84–6, 119, 121, 149
Bacon, F. 74, 79, 80
Bainbridge, W. S. 9, 38, 172, 178
Bains, S. 160
Banerjee, K. 203n. 5
Banning, E. B. 115
Bar, M. 59
Barash, D. 184
Bargh, J. A. 61
Barkow, J. 33, 161, 163, 185, 187, 208n. 7
Barrett, H. C. 206n. 10
Barrett, J. L. 3, 12, 14, 24, 27–9, 32, 33, 44, 47, 59–61, 63, 72, 74, 76, 81–4, 102, 110, 112, 119, 124, 127, 129, 133, 157, 159, 170, 193–200, 206nn. 10, 11
Barsalou, L. W. 59
Bartholow, B. D. 125
Bartlett, F. C. 123, 130, 171, 175
Bateson, G. 45, 171, 179
Baumard, N. 55, 81–3, 85, 114–15, 188
Baumeister, R. F. 61
Becher, T. 190
Bechtel, W. 36, 109, 121
Beck, L. W. 136–8
Beck, R. 80, 91
Beit-Hallahmi, B. 75, 76, 113
Bellah, R. N. 28, 103, 112, 113, 163, 164, 166, 205n. 6
Bentley, A. 109
Berger, P. L. 206n. 16
Bering, J. M. 9–10, 25, 29–31, 33, 44, 73, 76, 159, 197–8, 200, 201
Bettencourt, B. A. 124
Bickle, J. 18
big gods theory 113–15, 134, 138–41, 203n. 2
Blake, C. 93
blank slate view of mind 8, 58, 111, 157, 171–2
Blanke, O. 64
Bloch, M. 72, 77, 78, 82, 113, 188
Bloom, P. 76, 196, 198
Blume, M. 11

Boas. F. 184–5
Bolhuis, J. J. 120
Book, L. 190
Bordogna, F. 191
Boudreau, K. 188
Boudry, M. 112
Bourdieu, P. 188
Bower, G. H. 123
Boyd, R. 33, 176
Boyer, P. 1, 3, 10, 13, 21, 24, 26, 31–3, 35, 38, 44, 47–8, 58–61, 63, 72, 74, 76–9, 81–4, 92, 102, 110, 114, 121, 124, 133, 145, 151, 156–7, 159, 164, 170, 175, 178, 182–3, 187–8, 191, 193–5, 197–200, 205n. 4, 206n. 6, 9, 10, 11], 209n. 2
brain 13, 33, 36–8, 40–1, 59, 79, 81, 84, 91, 94, 102, 109, 143–4, 157, 167, 171
  imaging 13, 33, 36–9, 40–1
Braudel, F. 116
Braxton, D. M. 174, 178
Bremmer, J. 90
Brindell, S. 116, 119–20
Broesch, J. 125
Brown, G. R. 185, 186, 187, 209n. 6
Brown, P. 85
Brown, R. 129
Bshary, R. 86
Buddhism 73, 78, 87
Buhrmester, M. D. 50, 53, 55
Bulbulia, J. 33, 34, 67, 71, 72, 109–13, 116, 120, 164, 166–7
Buller, D. J. 161
Burke, P. 90
Burkert, W. 90
Buss, D. M. 33, 34, 161, 163
by-product theory 9, 11, 14, 15, 21, 24, 33, 110, 144, 194

Caldwell, T. L. 62
Campbell, D. 189–90
Carey, S. 75, 77
Carley, K. M. 177
Carroll, J. 184
Carruthers, M. 188
Casler, K. 28
Chambers, C. D. 61
Chang, L. J. 124
Cherry, J. 76, 77
Chevallier, C. 85
Chiao, J. Y. 89, 91
Chomsky, N. 12, 57–8, 60, 170, 172–3, 180
Christian, D. 116
Christianity 22, 27, 37–8, 66, 78, 85, 87, 113, 150
  healing testimonies 37–8, 66
Christiansen, M. H. 179
Christianson, S. A. 125
Claidière, N. 84
Clark, A. 207
Clark, B. R. 190
Claus, P. 92
Clavin, W. 117
Cleland, C. E. 116, 119, 120
cliodynamics 3, 115
cognition, defined 58–9
cognitive historiography 1–2, 91, 121, 149, 170, 176, 188, 207n. 2
cognitive neuroscience 18, 33, 36, 39, 76
cognitive psychology 8, 90, 123, 169, 175, 188, 193
application of cognitive theories to historical material 89–95, 103–4
cognitive psychology of religion 8, 175–6, 193, 197
cognitive science of religion (CSR)
  application of cognitive theories to historical material 89–95, 103–4
  and cognitive psychology of religion 175–6, 197
  and evolutionary psychology 160–3
  focus on cognition 11, 197–8
  founding and development of 3, 24–6, 155–6, 169–70, 182–4
  fractionating religion 43–4, 54, 62, 73, 144, 163–4, 166, 197, 209n. 3
  functionalist approach 72–3, 147, 206nn. 8, 12
  fundamental principles of 157–8
  historiography, *see as main heading*
  hostility towards 143
  and human behavioral ecology (HBE) 34, 85–6, 155, 158–61, 163, 187
  hypothesis-driven research 112–15
  information processing approach 7, 127, 171–3, 175, 177–80
  interdisciplinarity of 189–92, 196–7
  prehistory of 171–3

recommendations 163–6
and religious studies 157–8
second wave of, *see as main heading* cognitive science of religion (CSR), second wave of third modeling option 177–9
cognitive science of religion (CSR), second wave of 108–21
belittling of historiography 110–11, 116–21
interdisciplinary collaboration 121, 196–7
Cohen, E. 28, 39, 55, 76, 134–6, 141, 196–7
Cohn, N. 90
Colborne, N. 110
Collar, A. 109
Collins, A. M. 172
Cosmides, L. 33, 208n. 7
costly signaling theory 35–6, 82
Crary, S. T. 191
credibility-enhancing displays (CREDs) 35–6, 141
CREDs, *see* credibility-enhancing displays (CREDs)
Crick, F. 73
Csibra, G. 80
Cullen, H. 76
cultural epidemiology 10, 54, 74, 179
Cummings, J. N. 188
Cummins, D. D. 125
Cummins, R. 26
Cummins, R. C. 125
Currie, T. E. 52
cybernetics 171, 173, 183, 208n. 6
Czachesz, I. 19, 109

D'Andrade, R. 123–4, 156, 183, 185, 209n. 5
Daly, M. 162
Damasio, A. 92
dance 13, 200–1
Danziger, K. 123
Darwin, C. 3, 73, 77, 79–80, 160, 184, 185, 208n. 4, 209n. 7
Dawes, G. W. 71
Dawkins, R. 10, 11, 163, 183
Day, M. 111
de Certeau, M. 99
De Smedt, J. 112
de Spinoza, B. 74, 76, 78, 80
dead agents' minds, patterns of reasoning about 25, 29–31
Deeley, Q. 64
Dehaene, S. 26
deMause, L. 90
Dennett, D. C. 10, 76, 119, 185, 208n. 4, 209n. 7
Diamond, J. 116, 119, 150
Dietrich, O. 115
Dodds, E. R. 90
Dolan, R. J. 125
Donald, M. 98, 59, 103, 113
Donald, S. 161
Donald, W. 1, 139, 207n. 1
Douglas, M. 13, 73, 171
Doyen, S. 61
Driesenga, A. 29214
Durkheim, E. 7, 10, 11, 43, 72, 77, 78, 103, 175, 178

Ebbinghaus, H. 123
Edelman, G. M. 59
Eidinow, E. 94, 121
Eisenstadt, S. 84
Eldredge, N. 120
Eliade, M. 107, 121, 170
Ellis, B. J. 85
Emmons, N. A. 75
emotion
conceptual incongruities 124–5
experimental study 125–9
influence on memory, 31–2, 124–32, 145, 204n. 10
and rituals 26, 38–40, 46, 50
Engeström, Y. 209n. 10
EP, *see* evolutionary psychology (EP)
Epstein, J. M. 178–9
Evans, E. M. 29, 44
Evans-Pritchard, E. E. 44, 184
evolution 9, 11–12, 93
evolution, culture 45, 52, 63, 65, 90, 98, 103, 110–1, 113, 140, 148–9, 150, 157, 161, 176–7, 184–7
evolution, human 59–60, 93, 110, 146–8
evolutionary psychology (EP) 8, 33, 145, 155, 157, 158, 160–3
evolutionary science of religion 2–3, 11, 8–15, 33–6, 44, 51–2, 63, 71, 77, 82, 85, 86, 102, 110, 182, 202

evolutionary theorizing 2, 3, 9, 10, 12, 33–6, 47, 71, 77, 79, 86, 117, 121, 129, 140, 144–6, 160–2, 170, 175, 184–7, 206n. 8
mass extinctions 117, 121
expectation violation 31–2, 124–32, 145
experimentation, importance of 11, 24–41, 44, 62–7, 90–1, 94, 108–10, 119, 130, 166
example of 123–32
explanatory pluralism 17–20, 23, 34, 187, 204n. 11

Farah, M. J. 76
Feeny, C. 170
Festinger, L. 172
fire walking 39–40, 52, 196
Fischer, D. H. 111
Fish, S. 190
Fisher, H. 186
Fitzgerald, T. 21, 72, 188
fMRI, *see* functional magnetic resonance imaging (fMRI)
Foster, K. 77, 80
Foucault, M. 90
fractionating religion 43–4, 48, 54, 62, 73, 144, 163–4, 166, 197, 209n. 3
Frank, R. H. 14, 27, 60, 125, 193
Freud, S. 11, 43, 78, 134, 143–4, 207n. 20
Frith, C. D. 59
Frith, U. 66
Frodeman, R. 191
functional magnetic resonance imaging (fMRI) 33–7, 64, 90, 109

Gaalen, L. 112–13
Gardner, H. 9, 57
Gaulin, S. J. 160, 161
Geertz, A. W. 3, 57, 59, 62, 91–3, 109, 110, 112, 113, 156, 157, 166, 167, 205n. 2
Geertz, C. 22, 190
Gell, A. 47, 76
Gergely, G. 49
Gervais, W. M. 130
Gibson, J. 79
Gigerenzer, G. 76
Gilles, M. 150
Gil-White, F. J. 36
Gimenéz, M. 31
Gleick, J. 171, 173
Gobet, F. 75, 76
GöblekiTepe 113–15
Golden Rule 85
Goldin, P. 85
Gombrich, E. 79, 90
Gomes, C. M. 62
Gonce, L. O. 32
Gopnik, A. 12
Gottlieb, G. 147
Gould, S. J. 116, 117, 119–20, 186–7
Graham, J. 20, 36
Granet, M. 90
Gray, C. E. 182, 208n. 3
Greco, J. 79
Green, C. D. 61
Green, W. S. 110
Greenfield, P. M. 185
Greenway, T. S. 199
Gregory, J. P. 199
Gregory, R. L. 79
Greig, J. Y. T. 79
Griesemer, J. 206nn. 14, 18
Griffith, A. B. 120
Grobman, A. 121
Gross, M. B. 150
group selection 9–10, 33, 175–6, 201
Guldi, J. 116
Guthrie, S. E. 1, 3, 12, 21, 24, 44, 58, 59–60, 71, 75, 76, 77, 79–80, 82, 102, 110, 121, 145, 156, 182, 188, 195, 197, 198, 199, 201, 206n. 9, 209n. 4

Haddon expedition 171
Haggett, P. 121
Hallowell, I. 72, 75, 78
Hamann, S. B. 125, 129
Han, S. 89, 91
Harmon-Vukić, M. E. 32, 206n. 11
Harris, P. L. 31, 197, 203n. 5, 209n. 2
Harrison, J. 62
Harvey, G. 75
Harvey, V. 78
Hatano, G. 75, 77
HBE, *see* human behavioral ecology (HBE)
Heberlein, A. 76, 79
Heider, F. 80
Heintz, C. 94
Henrich, J. 33, 35–6, 61, 109, 130, 204n. 11

Herrmann, P. A. 50, 55
Heuer, F. 125, 129
Hinduism 27, 52, 85
Hirschfeld, L. A. 84, 209n. 2
Hirshman, E. 124
historiography 1–2, 90–1, 93, 116–17, 140, 149, 170, 176, 188, 206n. 15, 207n. 2
belittling of 110–11, 116–21
as a cognitive enterprise 91–5, 151
and experimentation 66, 133
presentism 3, 99, 110–12, 179
history of mentalities 90
Hodder, I. 51, 52
Hodge, K. M. 31, 196
Hogan, J. A. 120
Hogg, M. 178
Holland, D. 206n. 19
Holtz, T. R., Jr. 117, 118
Hornbeck, R. G. 197
Horton, R. 72, 78
Hrotic, S. 3, 181, 185, 189, 191, 209n. 5
Huang, J. 197
Hughes, A. W. 113
Huizinga, J. 90
human behavioral ecology (HBE) 34, 85–6, 155, 158–61, 163, 187
human evolution 59–60, 93, 110, 146–8
Hume, D. 1, 76, 77, 79, 80, 209n. 4
Hutchinson, J. R. 117
Hyperactive Agency Detection Device (HADD) 12, 59–60, 76–7, 102, 157

IACRS, *see* International Association for the Cognitive Science of Religion (IACSR)
IAHR, *see* International Association for the History of Religions (IAHR)
imagistic practices 46–7, 50–4, 61
fusion via 51, 52, 53–4
Inagaki, K. 75, 77
information processing 7, 127, 171–3, 175, 177–9
International Association for the Cognitive Science of Religion (IACSR) 1, 3, 62, 155, 160, 162–4, 166, 170, 203nn. 1, 5, 207nn. 1, 5
International Association for the History of Religions (IAHR) 3, 62, 204n. 5
intuitionist approaches 11, 12–15, 28–32, 43–4, 60–1, 74–6, 78–87, 92, 100, 102, 133, 145, 196, 198–201, 204n. 6
Inzlicht, M. 63
Irons, W. 35

Jablonski, D. 121
Jack, A. I. 200
James, W. 13, 143, 191
Jaschik, S. 209n. 5
Jaspers, K. 84
Jensen, J. S. 59, 145
John Templeton Foundation (JTF) 3, 160
Johnson, C. N. 2013
Johnson, D. D. P. 76, 202
Johnson, M. 76, 135, 150
Jolly, A. 80
Jong, J. 53, 55, 73, 112
Jordan, F. M. 111
Josephson, J. A. 73
Josiah Macy Jr. Foundation (JMF) 171–2, 173
Journal of Cognitive Historiography 2, 188, 203–4n. 5, 207n. 2
Journal for the Cognitive Science of Religion 159, 203
Judaism 78, 85

Kaiserfeld, T. 191
Kaptchuk, T. J. 66
Karmiloff - Smith, A. 206n. 10
Kaufer, D. S. 177
Keene, N. 187–8
Keightley, T. 184
Keil, F. C. 14, 27, 60, 82, 84, 193, 199
Kelemen, D. 28–9, 30, 44, 78, 80, 195, 198, 199
Kensinger, E. A. 129
kibbutzim, religious versus secular 34–5
Kiesler, S. 188
Kiper, J. 131, 164, 165, 167
Kitzmiller et al. v. Dover Area School District 186
Klass, M. 72
Klein, J. T. 190, 209n. 8
Koch, G. 77
Kokko, H. 77, 80
Konvalinka, I. 39–40, 64, 196
Kramer, K. L. 159

Krebs, J. R. 158
Kroeber, A. L. 182, 208n. 3
Kuhn, T. 19, 26, 71, 90
Kundt, R. 110, 176, 208n. 7
Kurzban, R. 87, 158
Kvadsheim, R. 184

LaBar, K. S. 129
Laboratory for the Experimental Research of Religion (LEVYNA) 155, 167, 203n. 1
Laidlaw, J. 2, 24, 48, 196
Lakatos, I. 7, 81
Lakoff, G. 76, 135, 150
Laland, K. N. 147, 185, 186, 187, 209n. 6
Lane, J. E. 3, 85, 109, 110, 119, 169, 174, 177–80, 183, 197
language 7, 11–12, 45, 49, 135, 141, 151, 170–3, 180, 200
Lanman, J. A. 35, 51, 53, 55, 157, 163
Lawson, E. T. 1–2, 3, 7, 14, 17–18, 20, 21, 24, 25, 34, 38, 47, 58, 59, 60, 81, 82, 83, 93, 119, 121, 123, 146, 156, 169–70, 172, 173, 174–8, 182, 183, 187–9, 190, 191–2, 193–4, 195, 197, 200, 205n. 1, 207n. 1, 209n. 9
Lazarus, R. S. 125
Leder, D. 76
LeDoux, J. E. 12, 129
Lefebvre, G. 90
Legare, C. H. 49, 197, 204n. 5, 206n. 11
Leslie, M. 188
Levine, M. P. 205n. 1
Lévi-Strauss, C. 205n. 3, 206n. 6
Levins, R. 120
Levy-Bruhl, L. 156
Lewontin, R. C. 120, 185–6
Liénard, P. 26, 48, 63, 83, 197
Lienhardt, G. 72
life history theory 84–5, 159
Lillard, A. S. 76, 80
Lindeman, M. 204n. 6
Lohmann, R. 76
Lorenz, K. 209n. 6
Lotman, Y. M. 151
Luckmann, T. 72
Luomanen, P. 162
Lyons, D. E. 49

Macaulay, D. 107–8, 114, 115
Maclaurin, J. 71
Madigan, S. A. 26
magic 44, 54, 83, 98–102, 104, 144, 197–8, 206n. 5
Malafouris, L. 93
Malinowski, B. 44
Malley, B. 47, 170, 194, 197
Mandler, J. M 123
Markússon, G. I. 110
Marlowe, F. 159
Marriott, J. 92
Martin, L. H. 1–4, 17, 19, 24, 47–8, 61, 72, 86, 89, 91, 92–4, 107–8, 109, 110, 111, 112, 113, 116, 121, 131, 133, 140, 160, 174, 183, 188, 194, 196, 203n. 1, 204n. 12, 207n. 1, 209nn. 4, 9
Mayr, E. 119
McBurney, D. H. 160–1
McCauley, R. N. 1–2, 3, 14, 17–38, 47–8, 58–60, 81–3, 93, 109, 111, 113, 123, 146, 156, 156, 169–78, 182–3, 187, 189–202, 204n. 7, 205n. 1, 209n. 1
McCorkle, W. W, 94, 174, 178
McCullough, M. E. 62
McCutcheon, R. T. 72, 113, 188
McGuire, M. B. 66
MCI, *see* minimally counterintuitive representations (MCI)
McKay, R. 44, 53, 55, 77
McMaster, J. 186
McMillan, D. 124
McNeill, W. H 121
McQuinn, B. 53, 55
Mead, M. 171, 173, 179, 183, 186, 189, 208n. 3
Medawar, J. 116
Medawar, P. B. 116, 119
Mehler, J. 26
Melanesian cultures 44–7, 196
Meltzoff, A. N. 77
memory 19, 92, 98, 171, 187
  autobiographical 21, 50–1, 91–2, 182
  episodic 21, 61
  flashbulb 26
  influence of emotion on 31–2, 124–32, 146, 204n. 10
  experimental study of 125–9
  semantic 51, 61, 126, 206n. 11

Menary, R A. 34
Merton, R. K. 189
Mesoudi, A. 111, 119
Methys, C. D. 79
Michel, J.-B. 179
Miller, G. 183
mind
  blank slate view of mind 8, 58, 111, 157, 171–2
  and brain 109
  childrens' minds 29–30
  dead agents' minds 25, 29–31
  group mind 175, 199–201
  mind-body dualism 44, 76
  dead agents' minds 25, 29–31
  theory of 21, 28, 60, 200, 204n. 6
  workings of 8, 19–20, 21, 24, 26, 93–4, 170, 207n. 20
minimally counterintuitive representations (MCI) 10, 25, 31–2, 60–1, 74–7, 123–32, 133, 157, 187, 199, 206n. 11
  criticism of 75–7, 124–5
  violations of 31–2, 124–32, 145
  mnemonic advantage of 25, 31–2, 124–5, 127, 129
Minsky, M. 172, 173, 207n. 3
Mithen, S. 72, 78, 79
Mithraism 1–2, 91
modes of religiosity 2, 46, 50–1, 59, 60, 83, 178, 182, 196, 207n. 3
Modi, J. J. 183
Moore, M. K. 77
Moretti, F. 207n. 22
Morin, O. 84
Morinis, A. 26
Morris, D. 186
Müller, F. M. 207n. 5
Mullins, D. 55
Murray, M. J. 74
Musick, M. A. 61

Näreaho, L. 75
NAASR, *see* North American Association for the Study of Religion (NAASR)
Nagel, T. 22
Nambisan, S. 192
natural science ideal 8, 97, 112, 136–7
Nelson, C. 189
Nettle, D. 85
Newberg, A. B. 112
Newson, L. 55, 110
Nichols, S. 84, 125
Nielbo, K. L. 26, 64, 107, 178, 206nn. 11, 13
Nietzsche, F. 76–9
Nongbri, B. 72, 73, 78, 99
Norenzayan, A. 32, 63, 72, 79, 111, 113–14, 127, 129, 134, 138–41, 200, 203n. 2, 204n. 9
North American Association for the Study of Religion (NAASR) 1, 3, 170, 207 (n. 1, 5)
Northoff, G. 89, 91
Notroff, J. 115
Nyhof, M. A. 32, 124, 129, 199, 206n. 11

O'Toole, G. 185
Öhman, A. 125
Occam's Razor 138
Odling-Smee, F. J. 147
Orians, G. H. 184
Ormerod, P. 109
Oster, E. 86
Otto, B. C. 99, 206n. 5

Pachis, P. 3, 48, 61, 89
Paden, W. E. 170
Padian, K. 117
Paldam, E. 66
Papua New Guinea (PNG) 44–7
Parker, R. 90
parsimony, principle of 76, 117, 137–8, 175, 179
Parsons, K. M. 117
Pearsall, J. 111
penis envy 134–5
Penner, H. 73, 107
Perry, E. 73
Pessoa, L. 125
Petersen, A. K. 97, 98, 197, 205n. 3
Peterson, G. R. 27
Phanerozoic 117–18
Phelps, E. A. 125, 129
physics envy 134–6, 141
Pigliucci, M. 110, 112, 116, 119, 120
Pinker, S. 113, 161, 163
Plotkin, H. 161

Popper, K. 7, 120
Porubanova, M. 3, 123, 124, 127, 129, 131, 199
possession 28, 64, 73, 196
Powell, J. W. 184
prayer 32, 37, 46, 51, 66
Preus, S. J. 121
Price, D. 209n. 5
Proudfoot, W. 157
psychology
 cognitive psychology of religion 8, 57–8, 171–2, 175–6, 195–7, 207n. 5, 209n. 2
 evolutionary psychology, 8, 33, 145, 155, 157–8, 161, 187
 experimental work 36–7, 47, 53, 61–3, 65, 90–1
 folk 12, 28, 31, 75
 psychologism 143
Purzycki, B. G. 75, 110, 123, 124, 125, 129, 130, 131, 164, 167
Pyysiäinen, I. 19, 61, 71, 72, 74, 76, 81, 83, 109, 113, 121, 197, 200

Quillian, M. R. 172
Quinlan, M. B. 86
Quinlan, R. J. 86
Quinn, N. 188, 206n. 19

Raihani, N. J. 86
Ramble, C. 24, 32, 124, 199, 206n. 11
Rappaport, R. A. 147
Raup, D. M. 117, 121
Reisberg, D. 125, 129, 198
relevance theory 11, 171, 207n. 4
Religion Explained (2001) 1, 164, 191
religion
 application of cognitive theories to 89–95, 103–4, 172, 176
 attempts to define 71–4, 174, 188
 cultural expression of 83–4, 92, 157, 194–7, 200–1
 evolutionary science of, *see as main heading*
 fractionation of 43–4, 48, 54, 62, 73, 144, 163–4, 166, 197, 209n. 3
 history of 19, 58, 65–6
 interpretation versus explanation of 176–7
 minimal counterintuitivism theory, *see as main heading*
 reductionist approach to 163–4, 175, 177, 179
 religious experience 13, 33, 38–40, 63, 65, 164
 role in human deep history 144–5
 theory of anthropomorphism, *see* anthropomorphism
Religion, Brain & Behavior 62, 203n. 5
Renfrew, C. 110
Rethinking Religion: Connecting Cognition and Culture 1, 58, 60, 81–2, 156, 169–70, 182–3
Rhode, D. L. 189
Riale, F. N. 183
Richerson, P. J. 33, 110, 176, 179
Richert, R. A. 29, 50, 52
Ridley, M. 163
Ristau, C. A. 77
rituals 8, 13–14, 34, 64, 83, 195, 205n. 4, 206n, 11
 causal opacity of 26, 48–9, 51–3, 64, 196
 competence theory 2, 60, 82
 costs of 10–12, 35–6, 82
 doctrinal practices 27, 46–7, 50–4, 61, 83
 efficacy of 14, 98–104
 field experiments 52–3
 firewalking 39–40, 52, 196
 imagistic practices 46–7, 50–4, 61
 as social glue 25, 50–1
Rizzolatti, G. 40
Roazzi, M. 76, 197
Robinson, J. 116, 119
Robson, C. 185
Roepstorff, A. 89, 91, 93
Roman period of history 1–2, 91, 103
Rosa, H. 206n. 17
Rosen, R. 135–6
Rossano, M. 158
Rosset, E. 28
Rowland, J. 191
Rozin, P. 13
Rubin, D. C. 125
Rudwick, M. J. S. 117
Ruffle, B. J. 34–5, 202
Rumelhart, D. E. 123

Ruse, M. 121
Russell, Y. I. 50, 55, 75, 76
Rydving, H. 71, 73–4, 75, 205n. 1

Sørensen, J. 3, 19, 26, 61, 64, 100–1, 102, 107, 116, 121, 143, 147, 151, 178, 197, 205n. 1, 206nn. 6, 7, 11, 13
Sahlins, M. 185, 190
Saler, B. 22, 72, 77, 78, 82, 133, 198
Sanfey, A. G. 124
Schachter, D. L. 123
Schank, R. C. 123
schema theory 90, 123
Schilbach, L. 76, 79
Schilbrack, K. 71
Schimmack, U. 125
Schjødt, U. 25–6, 37, 8, 57, 59, 64, 66, 109, 188
Schmidt, S. R. 115, 121, 124
Scholl, B. 80
Schultz, W. T. 90
Sepkoski, D. 210
Sepkoski, J. J., Jr. 117
SESHAT: Global History Database 52, 115, 119
sexuality 20, 36, 85–6, 109, 134, 148, 151, 161, 201, 206n. 17, 207n. 20
Shannon, C. E. 171, 173, 179, 207n. 4
Shea, J. J. 110
Shermer, M. 116, 121
Shinto 73
Shryock, A. 116
Simmel, G. 90
Simmel, M. 80
Sjöblom, T. 107
Skibbe, L. 76, 80
Skinner, B. F. 172
Skjoldli, J. 115
Slaughter, V. 75
Slingerland, E. 66, 110, 111, 119, 120, 159, 162, 185, 196
Slone, D. J. 14, 27, 32, 82, 109, 121, 206n. 11
Smail, D. L. 116
Smith, B. H. 75
Smith, E. A. 158
Smith, J. Z. 72, 111, 117, 174, 188
Snow, C. P. 97
Society for the Scientific Study of Religion (SSSR) 2
sociobiology 33, 162–3, 185–7, 189
Sosis, R. 3, 9–10, 11, 34–5, 155, 164, 165, 166, 183, 202, 209n. 3
Souza, A. L. 197, 204n. 11, 206n. 11
Sperber, D.10, 11, 38, 47, 58, 59, 74, 84, 124, 145, 151, 156, 171, 173, 183, 188, 198, 207n. 4
Spinney, L. 116, 78, 156, 174, 183
Spiro, M. 78, 156, 174, 183
SSSR, *see* Society for the Scientific Study of Religion (SSSR)
Staines, J. 208n. 1
Stangor, C. 124
Stark, R. W. 9, 38, 172, 178
Starr, C. G. 90
Stausberg, M. 112, 139, 140, 203n. 2
Stocking, G. 185
Stoczkowski, W. 115
Strauss, C. 78, 188, 205n. 3, 206n. 6
Stroop, J. R. 26, 63
stroop effect 26, 63
Sulloway, F. J. 116
Sun, R. 178
Swann, W. B., Jr. 50–1, 53
Symons, D. 161
Szechtman, H. 13

Talarico, J. M. 125
Taleb, N. N. 116
Talmot-Kaminski, K. 176
Taoism 85
Taves, A. 39, 125, 157, 209n. 3
teleology 79–80
  promiscuous 25, 28–9, 204n. 6
Tetlock, P. E. 165
Thagard, P. 19, 59, 204n. 10
theological correctness 14, 82–3
theological incorrectness 25, 27–8
theory of mind (ToM) *see under* mind
Thomas, K. 19, 86, 184
Tomasello, M. 148, 182, 184
Tremlin, T. 110
Tremoulet, P. 80
Tribble, E. B. 187–8
Trigg, R. 112
Trowler, P. R. 190
Tullett, A. M. 63

Turchin, P. 52, 66, 110, 111, 112, 115, 119
Turing, A. M. 57, 172, 179
Tylor, E. B. 43, 78

ultrasociality 111, 113, 115, 139
see big gods theory
Uro, R. 19

Valéry, P. 98, 204n. 1
van Elk, M. 63, 112
van Gulik, R. H. 85
Van Orman, B. 199
Versnel, H. S. 90
Veyne, P. 85
Von Restorff, H. 26

Wagner, R. 45
Walker, S. 24
Warren, R. 26
Wason, P. C. 33
Watson-Jones, R. 49–50, 55
Watts, J. 66, 113, 115, 116, 119, 121
Weeden, J. 87, 158
Wegner, D. 76, 79
Weinstein, F. 111
Wellman, H. M. 76
West, S. A. 9
Whitehouse, H. 2, 3, 19, 21, 24, 35, 38, 44, 45, 47, 48, 50–4, 59–61, 71–3, 83, 86, 92, 119, 121, 156, 163, 164, 179, 174–5, 177, 178, 182, 194–7, 207n. 3
Wiebe, D. 1–3, 17, 61, 107, 111–13, 121, 139, 160, 188, 203n. 1, 203n. 3, 204n. 12, 207n. 1, 208n. 4, 209n. 9
Wigger, J. B. 197
Willard, A. K. 36, 75, 123, 124, 125
Wilson, D. 11, 84, 171, 207n. 4
Wilson, D. S. 9–10, 110, 162, 163, 175
Wilson, E. O. 33, 185–6
Wilson, M. 93, 94, 162
Winterhalder, B. 158
Wittgenstein, L. 79, 205n. 5
Wolfram, S. 178
Woody, E. Z. 13
Woolf, G. 116
Wright, R. 113, 161, 163

Xygalatas, D. 26, 39–41, 55, 64, 94, 109–10, 119, 157–8, 174

Yonan, E. A. 107

Znamenski, A. A. 113